SUBSURFACE
MICROBIOLOGY
AND BIOGEOCHEMISTRY

WILEY SERIES IN
ECOLOGICAL AND APPLIED MICROBIOLOGY

SUBSURFACE MICROBIOLOGY AND BIOGEOCHEMISTRY

Edited by

James K. Fredrickson
Senior Staff Scientist
Environmental Microbiology
Pacific Northwest National Laboratory
Richland, WA

Madilyn Fletcher
Division of Science and Math
Belle Baruch Institute for Marine Biology
University of South Carolina
Columbia, SC

A JOHN WILEY & SONS, INC., PUBLICATION

New York • Chichester • Weinheim • Brisbane • Singapore • Toronto

Cover Image. A bacterial biofilm mixed with calcite precipitates on a surface exposed to groundwater in the Äspö HRL tunnel. See Chapter 4 for details.

For ordering and customer service, call 1-800-CALL-WILEY.

Library of Congress Cataloging-in-Publication Data:
Subsurface microbiology and biogeochemistry / edited by James K. Fredrickson, Madilyn Fletcher.
   p. cm.
   Includes index.
   ISBN 0-471-31577-X
   1. Geomicrobiology. 2. Biogeochemistry. I. Fredrickson, James K. II. Fletcher, Madilyn.

QR103.S83 2001
579'.16 - - dc 21   00-043582

Printed in the United States of America.

10 9 8 7 6 5 4 3 2 1

CONTENTS

v

PREFACE

During the past decade, the subsurface environment has represented a true frontier for microbiological research. Until recently, our understanding of microorganisms in the subsurface was largely a matter of speculation and based on sparse and sometimes anecdotal data. Much of the region below the terrestrial surface was believed to be hostile to microorganisms and essentially devoid of living organisms. However, enormous strides have been made in investigating the presence and characteristics of subsurface microorganisms, and we have come to appreciate these environments for their microbiological and chemical complexity and potential for harboring novel bacteria of environmental and, possibly, industrial importance.

Scientific investigations into the microbiology of deep subsurface environments early in this past century were stimulated by demands for petroleum and subsequent exploration for and study of oil fields and oil-bearing rocks. One of the first indications that microorganisms inhabited deep subsurface environments emerged in the 1920s when a geologist at the University of Chicago, Edson Bastin, examined the source of hydrogen sulfide and bicarbonate in water from deeply buried oil fields in Illinois. In an experiment reported in *Science* in 1926, Bastin and several colleagues at the University of Chicago submitted samples to bacteriological analysis and were successful in culturing sulfate-reducing bacteria (SRB) from groundwater samples collected from oil fields at depths of 150–600 m. These results suggested that microorganisms were responsible for the *in situ* reduction of sulfate to sulfide. Years later, investigations by Russian scientists also indicated that diverse microbial populations were associated with hydrocarbon-bearing rocks and waters. Many scientists at the time viewed the existence of microorganisms in deep terrestrial environments with skepticism due to the considerable uncertainty as to the origins of the microorganisms cultured from groundwaters collected from developed wells. The process of drilling and well development unavoidably introduced organisms into the deep strata, and the practice of flooding with water to enhance oil recovery led to further contamination and stimulation of indigenous and nonindigenous microorganisms. Although such early studies suggested the presence of microorganisms in the deep subsurface, these findings would not be verified until later in the century.

In the late 1970s and early 1980s, emerging groundwater quality issues in the United States stimulated scientists to further investigate the possibility that microorganisms inhabited shallow water-yielding formations as well as relatively deep aquifers. Moreover, technological and methodological advances were made that allowed researchers to collect deep groundwater, sediment, and rock samples while minimizing microbial contamination and chemical changes. Many researchers also employed various types of tracers during sampling that allowed measurement of the degree of contamination. Initial studies, focused primarily on microbiological characterization, revealed that active and diverse communities of microorganisms were present in shallow and deep (> 50 m) groundwaters and sediments. Most of these studies employed traditional microbiological methods involving culturing of microorganisms, visualization using direct microscopic techniques, or, in some cases, isolation and physiological and/or phylogenetic characterization. Despite the limitations of these techniques, important information was obtained on the presence and distribution of microorganisms, including aerobic heterotrophic bacteria, fungi, protozoa, and on total numbers of microbial cells in subsurface environments. However, the *in situ* activities of subsurface microorganisms and the biogeochemical processes they catalyzed remained poorly understood. The results of many of these early studies were biased by the limitations of techniques available at the time for studying microbial ecology and by the lack of robust methods for probing *in situ* microbial activities and community structure. Researchers were heavily dependent on laboratory cultures, introducing a potentially severe bias into results. It is now widely recognized that only a fraction, often less than 1% of the total microbial population, is typically cultured from environmental samples. Moreover, microorganisms cultured and studied under laboratory conditions may exhibit phenotypes quite different from those expressed in the environment. Regardless of these limitations, many novel aerobic and anaerobic microorganisms with interesting biochemical and genetic traits have been isolated from subsurface environments.

The results of these deep subsurface microbiology studies greatly spurred the interest of scientists from other disciplines, including geology, hydrology, geochemistry, and environmental engineering. Multidisciplinary teams and approaches and more robust analytical methods were increasingly applied to the study of deep subsurface microbiology. These approaches greatly extended the range of subsurface environments that microorganisms were shown to inhabit and allowed exploration of relationships between microbial abundance, physiology, taxonomy, and activity, and the subsurface environment, including geochemical, geological, and hydrological properties. Now it is clear that the subsurface environment provides numerous opportunities for microbial growth and survival, and, through this book, we have endeavored to capture this new understanding of the broad range and diversity of the previously "hidden" subterranean organisms.

We extend our sincere thanks to the contributors to this volume for the time, effort, and creative thought that made this work possible. We are also indebted to Ralph Mitchell for his sustained encouragement and support, and to Jill Walters and Luna Han for their editorial assistance and advice.

JIM FREDRICKSON
MADILYN FLETCHER

CONTRIBUTORS

Fred J. Brockman, Pacific Northwest National Laboratory, Environmental Microbiology Group, Richland, WA

Darrell P. Chandler, Pacific Northwest Northwest National Laboratory, Environmental Microbiology Group, Richland, WA

Frederick S. Colwell, Idaho National Engineering and Environmental Laboratory, Biotechnology Department, Idaho Falls, ID

Steven Desrocher, Golder Associates, Atlanta, GA

Madilyn Fletcher, University of South Carolina, Belle W. Baruch Institute for Marine Biology and Coastal Research, Columbia, SC

James K. Fredrickson, Pacific Northwest National Laboratory, Richland, WA

Ethan L. Grossman, Texas A&M University, Department of Geology and Geophysics, College Station, TX

Thomas L. Kieft, New Mexico Institute of Mining and Technology, Department of Biology, Socorro, NM

Derek R. Lovley, University of Massachusetts, Department of Microbiology, Amherst, MA

Eugene L. Madsen, Cornell University, Department of Microbiology, Ithaca, NY

Christopher P. McKay, NASA Ames Research Center, Space Science Division, Moffett Field, CA

James P. McKinley, Pacific Northwest National Laboratory, Interfacial Geochemistry Group, Richland, WA

Ellyn Murphy, Pacific Northwest National Laboratory, Richland, WA

Tullis C. Onstott, Princeton University, Department of Geosciences, Princeton, NJ

K. Pedersen, Göteborg University, Department of Cell and Molecular Biology, Microbiology, Göteborg, Sweden

INTRODUCTION

1

BIOGEOCHEMICAL AND GEOLOGICAL SIGNIFICANCE OF SUBSURFACE MICROBIOLOGY

JAMES K. FREDRICKSON

Pacific Northwest National Laboratory, Richland, Washington

TULLIS C. ONSTOTT

Department of Geosciences, Princeton University, Princeton, New Jersey

1 INTRODUCTION

Techniques have progressed from culturing microorganisms in groundwater samples to the use of robust molecular approaches for directly probing the characteristics of microorganisms in subsurface sediments, rocks. Analyses of directly extracted nucleic acids and lipids, in combination with sensitive assays for biological activity using radiolabeled compounds, have revealed detailed information on the phylogeny, physiology, and function of subsurface microorganisms. Rigorous geochemical analyses of dissolved inorganic ions, gases, and isotopes in groundwater, sediments, and rocks have demonstrated that microbial processes impact the geochemistry of groundwater, the diagenesis of sediments, and the overall cycling of elements in the subsurface. Scientific investigations of the deep subsurface are no longer solely focused on whether microorganisms exist there. Rather, mechanistic investigations

Subsurface Microbiology and Biogeochemistry, Edited by James K. Fredrickson and Madilyn Fletcher.
ISBN 0-471-31577-X. Copyright 2001 by Wiley-Liss, Inc.

now probe where and how they exist in the subsurface, their contributions to planetary biodiversity, and their role in subsurface biogeochemical processes. The purpose of this chapter is to review the relationships between subsurface microbiology, geology, and the diversity of subsurface environments that microorganisms are currently known to inhabit, or are suspected of inhabiting. In addition, the impacts of microbial activities on geochemistry and geology of the deep subsurface are discussed. This information will be presented in conjunction with the means, known or hypothesized, by which the organisms gain energy in these environments and therefore establish the geochemical and geological significance of microorganisms, and their associated activities, in the subsurface.

No discussion of subsurface microbiology is complete without some attention to the term "deep" as a modifier. The word "deep" as it has been applied to studies of subsurface microbiology has a relative connotation rather than a rigorous absolute meaning. Ten m may be considered deep by soil scientists and microbiologists; hydrologists and geologists would consider it relatively shallow. Perhaps a more meaningful definition was offered by Lovley and Chapelle (1995), who argued that the term deep should be restricted to hydrologic systems where flow is considered regional or even intermediate. Regional and intermediate flow systems, in contrast to local systems, have a relatively poor degree of connectivity with the surface environment.

Our intent here is to focus discussion on deep terrestrial environments where resident microbial populations must rely on endogenous energy sources rather than those replenished by transport from the surface. A combination of geochemical, hydrological, and microbiological parameters is required therefore to categorize a subsurface site as "deep" in this respect. Within an aquifer that is highly connected to the surface, for example, an unconfined aquifer, the concentration and isotopic composition of the dissolved inorganic carbon (DIC) and O_2 are in equilibrium with soil gas. By our definition, therefore, deep vadose zones, such as a > 100-m-deep one in Washington studied by Fredrickson et al. (1993), would not qualify as a deep subsurface environment, even though the hydrological characteristics may reflect regional topography. The presence of cosmogenic ^{14}C in the DIC or dissolved organic carbon (DOC) within the aquifer indicates some connectivity to surface-derived metabolic products. ^{14}C analyses of DOM in deep-sea hydrothermal vents has been utilized to ascertain the origin of the DOM and to determine whether the vent ecosystems were bioenergetically isolated from the surface photosphere (Karl, 1995). The presence of both tritium and modern ^{14}C in the DIC in the V2 borehole at the Stripa Mine in Sweden indicated that recent mining water and its associated DOC had contaminated the groundwater to depths of 1200 m (Fontes, 1994). Ekendahl and co-workers (1994) compared the CO_2-assimilation rates to lactate respiration rates in the same borehole and found that heterotrophic metabolism dominated this deep granite system. By our definition, some regions in the > 200-m-deep Middendorf Aquifer of the Atlantic coastal plain (ACP) (Fredrickson et al., 1991; Murphy et al., 1992) or Cubero Sandstone in northwestern New Mexico (Fredrickson et al., 1997) would not necessarily qualify as deep, although both are regional aquifers. If, however, the rates of microbial activity exceed that which can be supported solely by the surface-derived DIC or DOC, then these sites would still qualify as "deep" subsurface environments. This was shown to be the case for the Black Creek Aquifer by McMahon and Chapelle (1991b).

Microbial populations in such environments are insensitive to 10–100 kyr changes in the surface climate and vegetation. Their activity relies on their ability to access energy from "less altered" organic substrates or reduced inorganic mineral surfaces. These surfaces may quickly become depleted in bioavailable nutrients and coated with the byproducts of microbial and geochemical reactions. As a result, the overall activity of "deep" subsurface microbial communities will reflect long-term changes in their environment induced by volcanic or tectonic events (Tseng et al., 1998) or changes in sea level. Thus, a definition for a "deep" subsurface microbial environment requires evaluation of the extent of hydrological and geochemical isolation relative to the type and level of microbial activity.

2 DISTRIBUTIONS OF MICROORGANISMS IN THE SUBSURFACE

2.1 Microbially Catalyzed Oxidation–Reduction Reactions

Deep subsurface environments inhabited by microorganisms can essentially be considered "aphotic" systems. Photosynthesis has mainly an indirect role in sub-surface microbial metabolism, providing reduced organic compounds that can be metabolized by heterotrophs. Microbial generation of energy in deep subsurface environments is derived from biochemical reactions involving the oxidation of reduced compounds and the subsequent transfer of electrons to an adjacent oxidized compound. It is these metabolic processes that have great impact on groundwater geochemistry and geological processes.

Microbial processes involving energy generation are constrained mainly to electron transfer reactions involving the following elements: H, C, N, O, S, Mn, and Fe. The scale of redox couples involving these elements and the relative energy yield of each reaction are shown in Fig. 1.1. The electron activity, shown as $p\epsilon^0$, or the negative log of the electron activity in the diagram, is defined as

$$p\epsilon = -\log\{\epsilon\} = p\epsilon + 1/n \log\{\text{oxidized}\}/\{\text{reduced}\} \qquad (1)$$

where {oxidized} and {reduced} are the activities of the oxidants and reductants, respectively, and n is the number of electrons. The redox couples are shown from the strongest oxidants at the top to the strongest reductants at the bottom in Fig. 1.1. From this diagram, it can be seen that, from a thermodynamic standpoint, it is feasible to couple the oxidation of organic carbon to CO_2 using O_2, NO_3^-, MnO_2, FeOOH, or SO_4^{2-}. Figure 1.1 provides the ΔG values for the oxidation of organic carbon coupled to various microbial electron acceptors; these values indicate that the oxidation of organic carbon should occur sequentially, beginning with the reduction of O_2, the most thermodynamically favored reaction, in descending order through methane fermentation. From microbial and geochemical standpoints, there are a

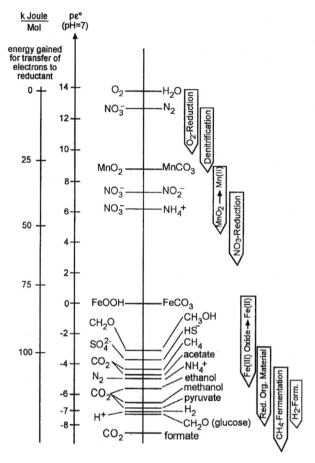

FIGURE 1.1   In a closed aqueous system, the oxidation of organic matter is achieved predominantly via microbial metabolism. Oxidation of CH_2O is coupled to the reduction of electron acceptors in order of decreasing $p\epsilon$, or increasing thermodynamic favorability (from Stumm and Morgan, 1996), with O_2 being consumed first, followed by nitrate, Mn(IV), etc.

number of factors that can influence the sequence, such as the bioavailability of substrates and reaction kinetics. However, at macroscopic scales in systems where microorganisms are active, such as a sediment–water interface or along an aquifer flowpath, the sequential utilization of the oxidants as electron acceptors can be readily observed. The change in relative concentrations of redox-active aqueous species with increasing distance along an idealized groundwater flowpath is shown in Fig. 1.2. For more comprehensive discussions of biogeochemical redox reactions, see Zehnder and Stumm (1988) and Stumm and Morgan (1996).

One of the principal reductants in sedimentary depositional environments is organic matter; this is also true for many subsurface environments. Dissolved or colloidal organic matter is likely to be a significant source of reductant only in

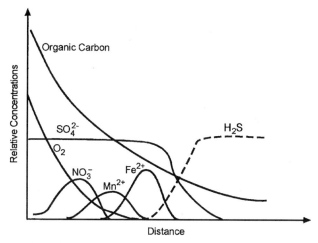

FIGURE 1.2 Schematic representation of changes in groundwater concentrations of ions over distance, reflecting the influence of microbial metabolism along an aquifer flowpath.

shallow (<50 m) aquifers; whereas sediment-associated (detrital) organic matter of terrestrial or marine origin and soluble hydrocarbons are major sources of reductant in deeply buried sediments.

In addition to organic carbon, reduced inorganic species, for example, Fe^{2+}, Mn^{2+}, S^{2-}, or H_2, can provide reducing power for microbial metabolism. These reactions are particularly favorable when coupled to a strong oxidant such as O_2 (Fig. 1.1), but some microorganisms can gain energy from oxidation of some reduced inorganic species coupled to weaker oxidants (such as nitrate). Microorganisms that can derive energy from the oxidation of inorganic compounds for metabolism and growth are termed chemolithotrophs or chemolithoautotrophs. Reduced inorganic substrates may be the dominant source of energy for microorganisms in some deep terrestrial subsurface environments, particularly those dominated by hard rocks such as basalt (Stevens and McKinley, 1995) or granite (Pedersen, 1998) where organic carbon is scarce or absent.

2.2 Physical Considerations

It is clear that microorganisms have evolved or adapted to occupy essentially any niche where energy can be obtained from oxidation–reduction reactions involving organic and inorganic species, as long as those reactions have a favorable ΔG. The ability of microorganisms to survive or even thrive in environments where energy is theoretically available is influenced by a number of biological and abiotic factors. In the deep subsurface, as in other environments, the distributions, community compositions, and activities of microorganisms are strongly a function of the chemical and physical properties of the environment.

Assuming permissive temperatures and available energy, available pore space and water are two primary requirements for microorganisms in the subsurface. Space in rocks or sediments can consist of pores of various sizes, fractures or fissures, and, less often, macroscopic cavities. Pore size and interconnectivity are important factors controlling microbial presence and activity, particularly in sediments and sedimentary rocks. Pore size can have a direct impact on microorganisms. For example, the size of most bacteria, the dominant microbial inhabitants of deep subsurface environments, ranges from several tenths of a micron to approximately 10 μm. Therefore, in rock or sediments where the pores average less than several tenths of a micron in diameter, bacteria are generally absent or inactive. For example, in deep core samples collected from a shale–sandstone sequence in northwestern New Mexico, no metabolic activities were detected in samples where pore throats were 0.2 μm or less (Fredrickson et al., 1997). In contrast, significant activities were detected in cores with most of the pores in the size range of 0.2–15 μm. Interestingly, in some of the less permeable shales with small pore throats, viable sulfate-reducing bacteria were recovered and reduction of $^{35}SO_4^{2-}$ was detected (Krumholz et al., 1997). It has been postulated that bacteria in low-permeability subsurface sediments and sedimentary rocks may be remnants of microbial communities associated with sediments during deposition (Chapelle and Lovley, 1990; Fredrickson et al., 1995), but such hypotheses have proven difficult to test.

Water availability (potential) and water content are also important factors impacting the ability of microorganisms to survive and function in the subsurface. They are important to microorganisms residing in the vadose zone; this subject is discussed in greater detail in Chapter 5.

2.3 Microbial Distributions in the Subsurface: Everything Is *Not* Everywhere

Although geology per se does not usually directly impact the distribution and activities of microorganisms in the subsurface, it does dictate habitat conditions. For example, the geological history of a rock unit influences the rate and age of water flowing through the system, the porosity and permeability of the rock or sediment, and the chemical composition of the rock or sediment that ultimately influences aqueous geochemistry and nutrient availability. Although not a complete listing, Table 1.1 provides a range of geological environments representing regional or intermediate flow systems (Lovley and Chapelle, 1995) or environments otherwise isolated from surface processes and influences that have been sampled for microbiological analyses. Table 1.1 is arranged in sections beginning with sediments, followed by consolidated, indurated rock, and, finally, groundwaters from deep subsurface environments.

Examination of such a limited data set, with broad level measurements such as microbial biomass and viable bacteria, does not allow one to draw specific conclusions regarding relationships between microbial properties and geological environments. However, there are several striking and important aspects of this

TABLE 1.1 Microbial Biomass in Deep Terrestrial Subsurface

Site Description	Geology/Sample Type	Microbial Biomass	Biomass Method	Notes	References
Atlantic coastal plain; Black Creek formation	Lower delta plain sediments; lignite, Fe sulfides	10^5–10^6 CFU g^{-1} 4×10^6 to 4×10^7 cells g^{-1}	Viable plate counts Acridine orange direct counts	Aerobic heterotrophs Total bacteria	Balkwill (1989), Balkwill et al. (1989) Hazen et al. (1991)
Atlantic coastal plain; Middendorf formation	Upper delta plain sediments; fine to coarse sands	10^2 to 4×10^7 CFU g^{-1} 8×10^5 to 7×10^7 cells g^{-1}	Viable plate counts Acridine orange direct counts	Aerobic heterotrophs Total bacteria	Balkwill (1989); Balkwill et al. (1989) Sinclair and Ghiorse (1989)
Atlantic coastal plain; Magothy and Patapsco formations	Fluvial sands, gravels, clays	10^4–10^5 cells g^{-1} 10^2–10^4 cells g^{-1}	Acridine orange direct counts Direct viable counts[a]		Chapelle et al. (1987)
Southcentral Washington; Ringold formation	Lacustrine; compact silty clay	<1–10 MPN g^{-1} <4×10^4 to 10^5 cells mL^{-1} <0.5–45 ρmol PLFA g^{-1}	Most probable number Acridine orange direct counts Total phospholipid fatty acids[b]	Anaerobic (fermenters) Total bacteria Total bacterial biomass	Kieft et al. (1995) Kieft et al. (1995) Fredrickson et al. (1995)
Southcentral Washington; Ringold formation	Paleosol; clay, silt, sand, carbonate nodules	<1–10 MPN g^{-1} <4×10^4 to 2×10^5 cells g^{-1} <0.5–1 ρmol PLFA g^{-1}	Most probable number Acridine orange direct counts Total phospholipid fatty acids	Anaerobic (fermenters) Total bacteria Total bacterial biomass	Kieft et al. (1995) Kieft et al. (1995) Fredrickson et al. (1995)

(continued)

TABLE 1.1 (continued)

Site Description	Geology/Sample Type	Microbial Biomass	Biomass Method	Notes	References
Nevada test site; Rainier Mesa	Zeolitized tuff with perched water	7×10^4 CFU g^{-1}	Viable plate counts	Aerobic heterotrophs	Haldeman et al. (1994)
		6×10^7 cells g^{-1}	Acridine orange direct counts	Total bacteria	Haldeman et al. (1994)
Western Colorado, Piceance Basin Wasatch formation	Cemented sandstone; cross-bedded siltstones and shales	Occasional cell growth	Enrichments	Iron-reducers, fermenters	Colwell et al. (1997)
		<0.4–10.1 ρmol PLFA g^{-1}	Total phospholipid fatty acids	Total bacteria	Colwell et al. (1997)
Northwestern New Mexico,	Shale; low permeability, high total organic carbon and total sulfur	Cell growth in 1 of 4 samples	Enrichments	Sulfate-reducing bacteria	Fredrickson et al. (1997)
Cerro Negro site, Clay Mesa formation		<0.5–4.2 ρmol PLFA g^{-1}	Total phospholipid fatty acids	Total bacteria	Ringelberg et al. (1997)
Northwestern New Mexico,	Sandstone; high permeability		Enrichments, No growth in 5 samples	SRB, fermenters	Fredrickson et al. (1997)
Cerro Negro site, Cubero formation	low total organic carbon and total sulfur	<0.4–1.5 ρmol PLFA g^{-1}	Total phospholipid fatty acids	Total bacteria	Ringelberg et al. (1997)
Northern Virginia, Taylorsville Basin	Shale, siltstone, sandstone	Below detect to 10^5 MPN g^{-1}	Most probable number	Dissimilatory iron reducers	Onstott et al. (1998)

Location	Description	Count	Method	Type	Reference
South Africa, Witwatersrand Basin	Mineralized carbon leader	3×10^7 to 7×10^7 cells g^{-1}	Acridine orange direct counts	Total bacteria	Onstott et al. (1997)
		<0.4–$62\ \rho$mol PLFA g^{-1}	Total phospholipid fatty acids	Total bacterial biomass	
Central Sweden, Stripa mine	Groundwater from quartz-banded hematite/granite	1.6×10^3 to 2.3×10^5 cells mL^{-1}	Acridine orange direct counts	Total bacteria	Pedersen and Ekendahl (1990)
	Glass slides contacted with groundwater for 56–117 days	2.2–5.5×10^6 cells cm^{-2}	Acridine orange direct counts	Total bacteria	Pedersen and Ekendahl (1990)
Southeastern Sweden, Äspö rock laboratory	Groundwater associated with granites	7.5×10^4 to 1.8×10^6 cells mL^{-1}	Acridine orange direct counts	Total bacteria	Pedersen and Ekendahl (1990)
		4.1×10^2 to 3.9×10^4 CFU mL^{-1}	Viable plate counts	Anaerobic heterotrophs	Pedersen and Ekendahl (1990)
South Africa, Witwatersrand Basin	Groundwater	5×10^4 cells mL^{-1}	Acridine orange direct counts	Total bacteria	Onstott et al. (1997)
		$<0.4\ \rho$mol PLFA g^{-1}	Total phospholipid fatty acids	Total bacterial biomass	
Gabon, Africa Oklo natural reactors	Groundwater from wells penetrating Oklo uranium deposit	4.5×10^4 to 5.8×10^5 cells mL^{-1}	Acridine orange direct counts	Total bacteria	Pedersen et al. (1996b)

(continued)

TABLE 1.1 (*continued*)

Site Description	Geology/Sample Type	Microbial Biomass	Biomass Method	Notes	References
Southcentral Washington, Columbia River basalts	Groundwater (artesian) from Priest Rapids Aquifer	3.6×10^3 cells mL^{-1}	Acridine orange direct counts	Total bacteria	Stevens et al. (1993)
		$\geq 10^4$ organisms mL^{-1}	Enrichments	H$_2$ acetogens[c]	Stevens and McKinley (1995)
Southcentral Washington, Columbia River basalts	Groundwater (artesian) from Grande Ronde Aquifer	7.6×10^5 cells mL^{-1}	Acridine orange direct counts	Total bacteria	Stevens et al. (1993)
		$\geq 10^4$ organisms mL^{-1}	Enrichments	H$_2$ acetogens, H$_2$ SRB	Stevens and McKinley (1995)
Northcentral Montana	Groundwater from Madison formation	0.8 to > 240 MPN mL^{-1}	Most probable number	Thermophilic SRB	Olson et al. (1981)
		1.3×10^3 mL^{-1} to 1.3×10^3 mL^{-1}	Acridine orange direct counts	Total bacteria	Olson et al. (1981)
Russia, Volgograd Province	Stratal waters (groundwater from artesian well)	60–250 cells mL^{-1}	Dilution plate counts	Methanotrophic bacteria	Ivanov et al. (1979)

[a] Direct viable counts as determined using the method developed by Kogure et al. (1979).

[b] A conversion factor of 5×10^5 cells ρmol^{-1} for bacterial biomass in subsurface sediments has been reported (Tunlid and White, 1992), although this value is likely to vary by as much as one order of magnitude due to variations in cells size, degree of starvation, etc. 0.4 ρmol PLFA g^{-1} is considered background.

[c] Acteogens that can grow with H$_2$ and CO$_2$.

summary. With the exception of the ACP sediments sampled on DOE's Savannah River Site (SRS) (Balkwill, 1989; Balkwill et al., 1989), the estimates of microbial biomass associated with subsurface sediments and rock are several, and in some cases many, orders of magnitude below the biomass in near surface sediments and soils. Although this result comes as no surprise, it is an important feature of deep subsurface environments and influences how scientists view and study the inter-actions of subsurface microorganisms with their environments.

Another important implication of the results in Table 1.1 is the relative paucity of microorganisms in consolidated, indurated rock; many of the samples investigated had microbial biomass levels below limits of detection, and even when detected, biomass was sparse. Although evidence of microorganisms existing in these environments is strong, there were as many or more samples where there was no evidence of microorganisms. All these studies used extensive precautions to reduce and measure contamination during sampling. Nevertheless, it is extremely difficult to eliminate all contamination (see Section 2.4). In fact, it is a testament to rigorous procedures that microorganisms were below limits of detection in many samples. The paucity of microorganisms in these environments is expected, because the lack of nutrients, water, and pore space, etc., would not support microbial populations other than an occasional few cells. Not surprisingly, scientists investigating these environments have had a tendency to focus on samples where cells are observed directly or indirectly via growth in enrichments. However, it is also important to note environments where there is an apparent or real absence of microorganisms. This negative information provides important constraints on the types of environments that microorganisms can be expected to inhabit and their patchy distribution in the subsurface. The Dutch microbiologist L. M. G. Baas-Becking expressed the concept that bacteria are cosmopolitan in their distribution in his 1934 statement that "everything is everywhere, the environment selects" (Staley, 1999); this concept does not necessarily apply to the subsurface.

In some contrast to rock cores, groundwaters sampled from aquifers in these same materials, as well as in unconsolidated sediments, invariably harbor bacteria at densities between 10^3 and 10^6 mL^{-1}. Valid concerns have been raised as to whether organisms associated with groundwaters sampled from wells represent indigenous subsurface microorganisms, allochthonous microorganisms that colonized the well following construction, or both. It has been demonstrated that sterile material suspended in groundwater wells is rapidly colonized by bacteria (Hirsch and Rades-Rohkohl, 1990), although the source of these organisms was believed to be mainly from indigenous aquifer bacteria. Also, recently established wells contain higher populations of bacteria than do older, established wells (Hirsch and Rades-Rohkohl, 1988), indicative of a disturbance-enhanced population. In spite of these concerns, bacteria are invariably observed in waters flowing from fissures in mines and in groundwater from established wells that have been purged to eliminate any organisms associated with stagnant water in the well bore, indicating they are indigenous groundwater microorganisms.

The common occurrence of microorganisms in groundwaters, almost regardless of environment, should not be surprising since bacteria are dependent on diffusion, through water, of substrates to and products from cells for metabolism and growth. Also, since groundwater from wells or fissures originates from the most transmissive zones of the aquifer, these are also the zones that typically have sufficient pore spaces and higher fluxes of nutrients to support microbial populations. Hence, deep terrestrial environments through which water is readily transmitted are more favorable habitats for microbial populations than are regions where pore spaces or fractures are small or nonconductive (Phelps et al., 1989; Fredrickson et al., 1997).

2.4   Uncertainties: Sampling and Microbial Enumeration

Microbiological investigations of deep subsurface environments present an exceptional challenge to scientists for several reasons. The most difficult obstacle is also the most obvious one: accessibility. Other than access via mines (Onstott et al., 1997; Russell, 1997), caves (Sarbu et al., 1994), or tunnels (Haldeman and Amy, 1993), the main approach for obtaining subsurface rock and sediment samples for microbiological and geochemical analysis is via drilling and coring. Considerable effort has been directed toward adapting and modifying drilling and coring approaches for obtaining subsurface samples suitable for analyses, and for applying tracers, appropriate controls, etc., to measure and account for contamination (Phelps et al., 1989; Colwell et al., 1992; Russell et al., 1992; Griffin et al., 1997). In spite of these efforts, it is extremely difficult to eliminate and measure low levels of microbiological or chemical contamination, especially when sampling environments that are deep enough ($>300\,\mathrm{m}$) to require the use of rotary drilling and coring equipment. This is particularly important considering that microbial populations in deep subsurface environments are typically sparse (Table 1.1). It is often difficult to state unequivocally that a microorganism or population of organisms came from a specific sample, particularly one that was obtained using these methods. The careful use of innovative drilling and coring techniques to minimize penetration of fluids or other contaminating materials into cores, combined with aseptic sampling methods and the use of sensitive tracer techniques, can greatly improve the quality of such samples and the confidence that they are representative. The recovery and analysis of microorganisms whose physiology is consistent with the *in situ* geochemical and physical properties (Boone et al., 1995; Liu et al., 1997a,b; Kieft et al., 1999) also provide evidence that the organisms derived from a given sample are representative of that environment.

Mines, tunnels, and caves can provide excellent opportunities for studying deep subsurface microorganisms and the processes they catalyze. However, there are a number of factors that must be considered when accessing mines for sampling; these factors have been discussed in detail by Russell (1997). One of the principal considerations is that mines typically develop microbial communities that may not be representative of indigenous endolithic microorganisms. The flow of air and increased availability of O_2 associated with ventilation systems can promote the development of lithoautotrophic biofilms on rock surfaces. Although such microorganisms can be of considerable scientific interest in their own right, they may not be indigenous. Ventilation systems and routine mining operations can disperse allochthonous microorganisms throughout the mine. Therefore, it is important to access fresh rock surfaces, well away from any ventilation and wall surface colonization, when sampling for indigenous subsurface microorganisms. Tunnels and mines are particularly well suited for investigations of spatial heterogeneity of microorganisms (Haldeman and Amy, 1993; Haldeman et al., 1993).

Sampling of groundwater can also provide important insights into the microbiology and geochemistry of deep subsurface environments and, in some cases, provides the only means for studying such environments. For example, in low-porosity, highly indurated rock, microbial communities will be concentrated along mineralized fractures. These fractures are extremely susceptible to microbial contamination during drilling and coring. Sampling of groundwater may be the only way of obtaining representative microbial communities in this case. When collecting groundwater samples for analysis, a major question is whether the microorganisms associated with the water are representative of the *in situ* populations. Although it is assumed that most microorganisms in the subsurface are attached to solids, there is little information comparing groundwater-associated, or planktonic, microorganisms with those attached to solids (Hazen et al., 1991). Also, groundwater, depending on the well configuration and screened interval distance, can represent mixed water from several regions within the aquifer (Grenthe et al., 1992), whereas solids are usually considered "point" samples. Hence, mixed waters from different regions may not be characteristic of any particular section, but are a composite sample (Grenthe et al., 1992). It is also recognized that vertical integration of groundwater over a screened interval can result in considerable dilution of solutes and colloids. Because of these issues, collection of samples from discreet zones within the subsurface is usually preferred. There are several approaches for conducting this type of sampling, including the use of a passive multilevel sampler (Ronen et al., 1991), which have been described in detail elsewhere (Fredrickson and Phelps, 1996).

2.5 The Viable–Nonviable Continuum

A frequently stated axiom in microbial ecology is that >99% of the microorganisms existing in nature are not cultivated using standard microbiology techniques (Amann et al., 1995). Such estimates are based on measures of total microbial biomass made using direct microscopic counts of cells or analyses of biochemical components of microbial cells such as lipid phosphate or ATP (Fredrickson and Balkwill, 1998). Although these methods have been extensively employed by microbial ecologists, they are of limited application for studying microbial populations in deep subsurface environments. Because of the requirements for sample dispersion to dislodge cells from particle surfaces and dilution, the practical detection limit for direct microscopic counting of cells in sediment or rock samples is from 10^4–10^5 cells g^{-1}. The measurement of total microbial phospholipid fatty acids (PLFA) is one of the most sensitive techniques for estimating microbial biomass, and it has been extensively applied to subsurface samples (Fredrickson et al., 1995, 1997; Colwell et al., 1997; Kieft et al., 1997). However, even with this sensitive technique, the detection limits are in the low picomolar range (Tunlid et al., 1989) (per gram of sediment or rock), a value that corresponds to between 6×10^4, and 5×10^5 cells g^{-1}, based on reported conversion factors (Tunlid and White, 1992; Kieft et al., 1997). It should be noted that microbial lipid measurements have the added advantage in that, in addition to total microbial biomass, they can provide information on community composition and physiological state (McKinley et al., 1988; White and Ringelberg, 1998). One advantage of sampling microbiota associated with groundwater is that the biomass can be concentrated by filtration (Fry et al., 1997) facilitating the analysis of small populations.

One of the challenges to interpreting biomass measurements is differentiating between viable and dead or moribund cells. A number of staining techniques have been developed to distinguish microscopically viable from nonviable cells. These are based on esterase activity, detected by the ability to cleave fluorescein diacetate (Chrzanowski et al., 1994), or membrane integrity and use commercially available staining kits (Virta et al., 1998). The methods, while having some applications in subsurface microbiology, also have limitations in that they are not universal and do not work for all types of microorganisms. Because they are microscopy-based, they are also subject to the detection limit problems described above. Measurements of diglyceride fatty acids (DGFAs) have been used to indicate dead cells in deep subsurface samples (Kieft et al., 1997) and in microcosm studies investigating bacterial survival (Kieft et al., 1994). Where the ratio of DGFA to PLFA has been used as a measure of dead to viable biomass in deep subsurface environments, the ratio often exceeds 1 (Fredrickson et al., 1997; Kieft et al., 1997; Ringelberg et al., 1997).

Although phospholipids are rapidly turned over in sediments where bacteria are active (White et al., 1979), the turnover rate in low biomass, deep subsurface sediments and rocks is unknown (Fredrickson et al., 1997). Typically, the PLFA concentrations are higher in the clay-rich layers, relative to sandy or coarse-textured sediments. Hedges and Oades (1997) have reported that for sediments that have been size-fractionated, the microbial hydrocarbons and fatty acids are enriched in the clay fractions, whereas vascular plant debris is concentrated in the silt and sand fractions. Also, the C/N ratio decreases and the P concentration increases as grain size decreases. The oxidation of organic matter is inhibited by its adsorption to mineral surfaces. In fine-grained clays, more adsorption occurs due to greater surface area and filtration effects. If the pore diameters are small enough to exclude bacteria and extracellular enzymes, then the turnover rate of PLFA may be substantially less than in unconsolidated sediments.

The rates of *in situ* microbial metabolism in the deep subsurface are among the slowest in environments where microorganisms are known to be active (Chapelle and Lovley, 1990). Phelps et al. (1994) estimated that the average CO_2 respired on a per cell basis in deep aquifer sands of the ACP of the southeastern United States was 2×10^{-5} picomoles $year^{-1}$, which translates into an estimated cell doubling time of several centuries. Numerous other studies, using either direct measurements of microbial activities in core or groundwater samples or indirect measures of *in situ* metabolism based on changes in concentrations of electron acceptors or donors along a groundwater flowpath, generally indicate very slow rates of microbial metabolism compared to other environments. It should be emphasized that direct measurements of microbial activity, such as those that utilize isotopically labeled substrates, can overestimate *in situ* activities based on groundwater analyses by as much as 10^6-fold (Phelps et al., 1994).

In studies with chemoheterotrophic bacteria isolated from deep subsurface environments, Kieft et al. (1997) and Amy et al. (1993) independently found that the cells underwent typical starvation–survival responses observed in other bacteria, including reduction in cell volume and decreasing culturability relative to the total number of cells. Microbial populations in deep subsurface environments are generally characterized by a high proportion of miniaturized, nonculturable cells (Kieft, 1999). Although it is hazardous to generalize, the physiological state of microorganisms that is termed starvation–survival (Morita, 1982; Amy and Morita, 1983) is common to almost all deep subsurface environments examined to date. In effect, deep terrestrial subsurface environments likely harbor a continuum of microbial cells ranging from recently divided, culturable cells to nonculturable cells in a state of starvation–survival to dead or moribund cells that are slowly decaying.

3 GROUNDWATER BIOGEOCHEMISTRY

3.1 Groundwater Geochemical Evolution

Hydrogeologists and geochemists have commonly observed systematic decreases in the E_h of groundwater as it migrates from upland recharge areas to lowland discharge areas under conditions of confined flow, and have attributed the changes to microbial metabolic processes (Champ et al., 1979). Examples of declines in E_h along groundwater flow paths are presented in Fig. 1.3 for the Middendorf formation of the ACP, southeastern United States (Murphy et al., 1992), the Chalk River deltaic sand aquifer located northwest of Ottawa, Ontario (Jackson et al., 1977), and a Late Cretaceous sandstone in the southeastern San Juan Basin in New Mexico (Walvoord et al., 1999). The observed redox zonations that occur are characterized by successive processes that occur over discrete intervals of the flowpaths and are delineated by changes in the concentrations of chemical species involved in oxidation–reduction reactions. Not surprisingly, these species are essentially the same as those identified in Fig. 1.1 as being the principal terminal electron acceptors in microbial metabolism. For example, sulfate and ferric iron reduction were linked to the microbially catalyzed oxidation of organic carbon to CO_2 in the extensive Madison Aquifer of Montana, Wyoming, and South Dakota via geochemical modeling and analysis of C and S isotopes in groundwater (Plummer et al., 1990). Observations reporting similar zonation of redox processes have been reported for other regional groundwater systems, including the Fox Hills-Basal Hell Creek Aquifer of North Dakota and South Dakota (Thorstenson et al., 1979)

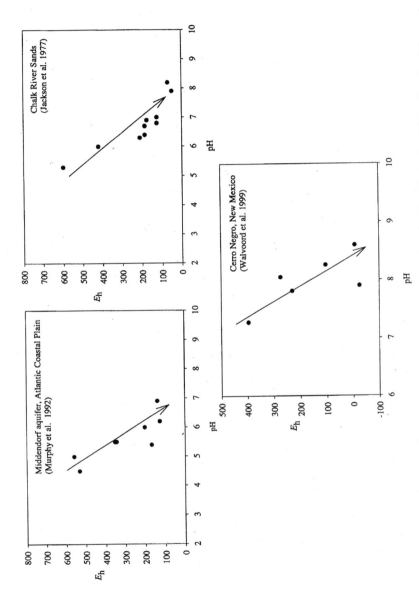

FIGURE 1.3   Changes in groundwater E_h observed along the flowpath of the Middendorf Aquifer in the ACP sediments of the southeastern United States (Murphy et al., 1992), the Chalk River sands near Ottawa, Ontario (Jackson et al., 1977), and a Late Cretaceous sandstone aquifer in the southeastern San Juan Basin (northwestern New Mexico) (Walvoord et al., 1999). Arrows denote groundwater flowpath.

and the Chalk Aquifers of the United Kingdom including the Lincolnshire Limestone (Edmunds et al., 1984).

One of the first studies that closely linked microbial analyses of sediment core samples to CO_2 increases in groundwater of deeply buried coastal plain sediments of the eastern United States was conducted by Chapelle et al. (1987). In this study, the presence of bacterial cells, including methanogens and sulfate-reducing bacteria, was unequivocally identified in deep sediment cores, indicating that the source of CO_2 was microbial oxidation of organic matter (possibly sedimentary lignite). Later work by Murphy et al. (1992) identified geochemical changes along a groundwater flowpath in the Middendorf Aquifer of South Carolina that were consistent with the physiologies of microorganisms cultured from Middendorf core samples, including aerobic chemoheterotrophs and Fe-reducing and sulfate-reducing bacteria. They suggested that lignite inclusions in the aquifer sediments were the principal source of organic carbon for microbial fermentation and respiratory processes. Redox processes within the Middendorf were heterogeneous and likely linked to lignite distribution in the sediments. Subsequent research confirmed the presence of active bacteria in core samples by measuring the oxidation of ^{14}C-labeled glucose or acetate (Chapelle and Lovley, 1990) and incorporation of ^{14}C-labeled acetate into lipids and of ^{3}H-thymidine into DNA of bacterial cells (Phelps et al., 1989). Jones et al. (1989) found that some subsurface sediment slurries accumulated acetate in the absence of organic amendments, indicating the presence of endogenous fermentable C compounds. Later, McMahon and Chapelle (1991b) identified that microbial fermentation of organic matter was occurring in fine-grained aquitard sediments and was likely responsible for driving bacterial respiration in adjacent course-grained aquifers. McMahon (1991a) also demonstrated the biological generation of CO_2 from endogenous carbon in ACP sediments incubated with O_2, NO_3^-, or SO_4^{2-} as the electron acceptor, providing additional evidence for sedimentary organic carbon as the principal electron donor driving microbial metabolism in deeply buried ACP sediments.

It is now clear that sediment-associated organic carbon is an important source of electron donors for *in situ* bacterial metabolism in many deep subsurface sediments and rocks. In low-temperature ACP sediments, microbial fermentation of organic-rich clay produces acetate, formate, and dissolved H_2 (McMahon and Chapelle, 1991). Because the enzymatic electron acceptor processes have varying affinities for H_2, the concentration of H_2 can reflect the dominant electron acceptor reaction (Lovley and Goodwin, 1988). In fact, in subsurface environments that are electron-donor-limited, the H_2-concentration appears to be thermodynamically controlled by the principal terminal electron acceptor reaction so as to maintain a constant ΔG (Hoehler et al., 1998). As a result, the H_2-concentration also varies as a function of temperature and pH. The segregation of the dominant electron acceptor processes along the flowpaths of many aquifers also reflects this competition for acetate and H_2 (Lovley and Chapelle, 1995). Other examples include lacustrine and paleosol sediments beneath the U.S. Department of Energy's Hanford Site in southcentral Washington state (Fredrickson et al., 1995; Kieft et al., 1995) and in Cretaceous rocks of the southwestern United States (Krumholz et al., 1997) that include the extensive Mancos shale. In addition to pristine environments, microbial processes have been linked to changes in redox chemistry in subsurface environments that have been contaminated with organic carbon from landfill leachate (Ludvigsen et al., 1998) or petroleum hydrocarbons (Vroblesky and Chapelle, 1994). In deeper, hydrocarbon-bearing, sedimentary formations at temperatures of 100°–150°C, H_2-concentrations may primarily reflect abiotic, irreversible, hydrolytic disproportionation, hydrolysis of light paraffins, and metastable equilibrium between organic acids (Shock, 1994). Organic acid concentrations are buffered by metastable equilibrium with DIC (Shock, 1988; Helgeson et al., 1993).

Although there is no solid phase-associated organic carbon, active and diverse microbial populations are present in deep igneous rock environments, including the Columbia River Basalt Aquifer (CRB) (Stevens et al., 1993; Fry et al., 1997) and Precambrian granite of the Svenoscandian Shield (Pedersen and Ekendahl, 1990; Pedersen et al., 1996). The groundwater movement in these environments is predominantly through fractures, and the waters are typically low E_h ($< -200\,mV$) with relatively high concentrations of Fe(II) and HS^- (Pedersen and Ekendahl, 1992a,b; Stevens et al., 1993). Interestingly, the H_2-concentrations in the groundwaters from these fractured rock environments can be relatively high, that is, up to 60 µM in the Columbia River basalt aquifers of southcentral Washington and 70 µM in the granite aquifer of southeastern Sweden (Stevens and McKinley, 1995; Pedersen, 1998). H_2 has even been reported to comprise up to 30% of the natural gas discharging freely from deep subsurface fractures (Sherwood et al., 1988). These values are several orders of magnitude greater than in sediments where microbial fermentation is occurring (Lovley and Goodwin, 1988). Many of the bacteria that reside in these environments can grow autotrophically with H_2 as the sole electron donor and CO_2 as the carbon source, including relatively large populations of acetogenic bacteria and autotrophic methanogens (Stevens and McKinley, 1995). Although the source of H_2 in these environments is unknown, Stevens and McKinley (1995) provided evidence that H_2 may be generated *in situ* in basalt aquifers as a result of water disproportionation coupled to oxidation of Fe(II)-bearing silicates. Controversy exists, however, over whether quantities of H_2 required to support chemoautotrophic bacteria are produced by this mechanism at alkaline pH values typical of groundwater in the CRB aquifer (Anderson et al., 1998). It is clear that additional investigations are required to resolve the significance of H_2-based autotrophy in deep igneous rock environments. There has been considerable speculation that reduced gases from the Earth's mantle, including H_2, CH_4, and H_2S, may provide energy sources for lithoautotrophic microbial communities in the Earth's crust (Gold, 1992). Unlike shallow aquifers, many deep subsurface environments are electron donor-rich and electron acceptor-limited, resulting in the close juxtaposition of microorganisms utilizing distinct electron acceptors (Onstott et al., 1998b). The abundance of H_2 in the CH_4-occurrences in the Canadian and Fennoscandian Precambrian Shields indicates that the residence time for H_2 as a separate gas phase in this deep subsurface environment is many orders of magnitude greater than the residence time measured for dissolved H_2 in the pore waters of marine sediments (seconds) (Hoehler et al., 1998) or in sludge and lake sediments (minutes) (Conrad et al., 1985).

3.2 Geochemical Approaches for Determining *In Situ* Microbial Activities

Geochemical modeling of deep subsurface environments is essential for determining to what extent the extant microbial community relies on indigenous energy sources by (1) estimating the rates of various microbial respiration processes and (2) comparing that rate to the flux of electron donors and acceptors from the surface. Geochemical estimates for microbial respiration rates within aquifers are on the order of nmoles L^{-1} year^{-1}, which are a factor of a 10^3 less than those derived from modern lake or marine sediments (Phelps et al., 1994; Murphy and Schramke, 1998). Microbial activity measurements derived from laboratory incubations with radiolabeled substrates using homogenized sediment from the same aquifer (Phelps et al., 1994) yielded rates that are 10^6 times more than those derived from geochemical mass balance and are unrealistically high. This situation contrasts with modern depositional settings where microbial activity measurements based on geochemical and radiolabeled substrate incubations are comparable.

Many of the constitutive equations utilized for modeling redox gradients and quantifying microbial activity in modern depositional environments (Van Cappellen and Wang, 1995) are applicable to aquifers, with some qualifications. For subsurface bacteria to exploit the above exothermic reactions for maintenance and growth, using enzymes to break the bonds of the electron donors and to transfer electrons to appropriate acceptors, the abiotic reaction rates must be exceedingly slow. For example, the minimum time required for abiotic chemical and isotopic equilibration of dissolved S^{2-} with SO_4^{2-} is on the order of 10^8 years for neutral pH and temperatures $< 100°C$ (Ohmoto and Lasaga, 1982). The spatial scale for the transition from one redox zone to the other in aquifers may extend over kilometers in contrast to the centimeters typical for modern sediments. The redox status of the groundwater at any specific geographic position will represent the integrated effect of a heterogeneous mixture of redox couples varying over scales from centimeters to millimeters. Finally, accurate knowledge of groundwater flowpath and detailed delineation of the redox gradients are far more difficult to obtain in the deep subsurface than in most surface sediments, due to the spatial heterogeneity of redox properties in combination with aquifer inaccessibility and limitations associated with sampling water from wells.

The concentration C of an electron acceptor or donor at a specific position in an aquifer can be described by the following:

$$C/t = D^2C/x + VC/x - R(x) \qquad (2)$$

where D is the dispersivity of the substrate (the sum of the mechanical dispersion, diffusivity), V the groundwater velocity, x the distance along the flowpath, t time, and R the net production or consumption rate of a microbial metabolite. The abiotic rates may, in part, be a function of the calculated free energy available from the source and sink reactions, the surface area of a reactant, and/or the activation energy for the reaction. The rate of microbially driven reactions will also decrease as the reactants are depleted and the products increase; the process will eventually cease when the effective free energy is too small to exploit. Microbially catalyzed reaction rates are typically more sensitive to factors such as temperature, salinity, and pH than are analogous abiotic reactions. Microbial reactions can also be influenced by the physical accessibility of the substrate or lack of a trace chemical species critical to the metabolic process, such as an enzyme cofactor. Identifying the subsurface microorganism(s) responsible for a specific biogeochemical reaction and determining its abundance are crucial for developing a quantitative understanding of a specific reaction rate.

Groundwater reaction rates are typically derived by assuming a steady-state solution ($\partial C / \partial t = 0$), by measuring changes in the groundwater chemistry ($\partial C / \partial x$) as a function of age along the flowpath V, and by ignoring the dispersion term ($D = 0$) of the above expression (Plummer, 1977; Plummer et al., 1990; Murphy et al., 1992). Of the electron acceptors and donors, O_2, NO_3^-, CO_2, SO_4^{2-}, and carboxylic anions are potentially surface-derived, whereas Fe(II), Mn(II), Fe(III), Mn(IV), HS^-, CH_4, and H_2 represent indigenous subsurface electron donors and acceptors.

For confined aquifers where the DIC is not equilibrating with soil CO_2 (referred to as a "closed" system for carbon cycling), the DIC represents a balance between subsurface microbial and inorganic processes. These processes include the flux of soil CO_2 from the recharge zone, the depletion of CO_2 during dissolution of calcite and dolomite, the production of CO_2 during calcite precipitation, the microbial oxidation or fermentation of either dissolved organic matter (DOM) or sedimentary organic matter (SOM), the disproportionation of acetate during methanogenesis (Grossman et al., 1989), the depletion of CO_2 during methanogenesis via CO_2-reduction, and oxidation of CH_4 (i.e., methanotrophy, Alperin et al., 1988). For deep subsurface environments at temperatures $> 100°C$, the abundance of DIC reflects metastable equilibrium reactions between DIC, carbonate, feldspar, and clay and is inversely proportional to the Ca^{2+}-concentration (Smith and Ehrenberg, 1989; Ben Baccar et al., 1993; Helgeson et al., 1993). In this case, as the temperature and depth increase, the pH will decrease, pCO_2 will increase, and these two geochemical properties will be decreasingly impacted by microbial activity.

Groundwater SO_4^{2-}-concentrations typically reflect a balance between SO_4^{2-}-production via gypsum dissolution, pyrite oxidation versus bacterial SO_4^{2-}-reduction, pyrite or barite precipitation, and possibly, sulfur or thiosulfate disproportionation reactions. Variations in the dissolved O_2 and NO_3^- have been used to assess the microbial respiration rates with these electron acceptors in shallow aquifers. Dissolved Fe(II) reflects the production of Fe(II) by microbial reduction of Fe(III)-bearing mineral phases, although the sorption of Fe(II) to mineral surfaces or precipitation of Fe(II)-bearing minerals such as FeS, $FeCO_3$, and Fe_3O_4 may be sinks for Fe(II). The groundwater H_2-concentrations reflect the production of H_2 water disproportionation reactions (Seewald, 1994; Stevens and McKinley, 1995), pore-water radiolysis (Hoffman, 1992), microbial fermentation of organic matter, and the microbial consumption of H_2 linked to various terminal electron-accepting processes (Lovley and Goodwin, 1988). Similarly, acetate concentrations represent the balance between production by microbial fermentation, consumption by microbial respiration in aquifers (McMahon and Chapelle, 1991b), and metastable equilibrium with DIC (Helgeson et al., 1993). Finally, in natural gas reservoirs, the dissolved CO_2-, H_2-, and CH_4-concentrations will be controlled by partition into the gas phase and by the total pressure (Colwell et al., 1997).

Elements commonly involved in biological redox reactions contain more than one isotope, and the reaction rate varies slightly with the atomic weight. Reactions that break bonds preferentially break those between lighter isotopes of an element, for example, ^{12}C versus ^{13}C. This is true regardless of whether the reaction is biologically driven or in thermodynamic equilibrium, but the degree of preference for the lighter isotope is different for the enzymatically driven microbial reactions and thermodynamic equilibrium between products and reactants. To distinguish among the multiple sources and sinks for carbon and SO_4^{2-}, geochemical estimates of the microbial CO_2-production and SO_4^{2-}-reduction rates in aquifers have relied on stable isotope analyses of DIC, SO_4^{2-}, and S^{2-} (Plummer, 1977; Plummer et al., 1990; Murphy et al., 1992). Isotopic analyses of other electron acceptors, such as O_2, and NO_3^-, have been used to delineate biogeochemical processes in marine environments, but their application to subsurface environments has been limited to contaminated shallow aquifers (e.g., Bottcher et al., 1990).

4   MICROBIAL ROCK–SEDIMENT INTERACTIONS

With increasing depth or distance along flowpaths, groundwater tends to become supersaturated with respect to carbonate, sulfides or sulfates, silica, and clay (Plummer et al., 1990; McMahon and Chapelle, 1991; Walvoord et al., 1999). Although the formation of these phases is thermodynamically favored, abiotic precipitation may be kinetically inhibited. In this case, subsurface microbial participation in these solid–liquid reactions could play a significant role in diagenesis. Microbially mediated changes in the isotopic composition of the DIC and HS^- are usually reflected in the isotopic composition of precipitated carbonates and Fe sulfides, but whether the mineral type or form is also specific to biotic precipitation processes, as is the case for aragonite precipitation in marine systems, has not been determined.

In vitro experiments indicate that a variety of extracellular Fe sulfides (see the review by Bazylinski and Moskowitz, 1997) and dolomite (Vasconcelos et al., 1995) are formed during microbial SO_4^{2-}-reduction. Bacterial Fe(III)-reduction experiments have yielded extracellular formation of siderite (Mortimer et al., 1997), magnetite (Lovley et al., 1987; Liu et al., 1997a), vivianite, and green rust (Fredrickson et al., 1998). Slobodkin and Wiegel (1997) demonstrated that extracellular magnetite and siderite could be produced by Fe(III)-reducing and H_2-oxidizing bacteria growing at temperatures as high as $87°C$. Bacteria have also been implicated in the direct or indirect nucleation and precipitation of silica and various carbonate and sulfate minerals in the laboratory (Morita, 1980; Novitsky, 1981) and in natural surface and aqueous environments (Konhauser et al., 1992; Coleman and

Raisewell, 1993; Ferris et al., 1994; Schultze-Lam and Beveridge, 1994) (also see the review by Barker et al., 1998). These studies indicate that the types of minerals, their crystal symmetry, size, and morphologies are a complex function of biotic and abiotic factors, including solution and cell surface chemical properties such as pH, Eh, ion composition, temperature, and type of microbial metabolism.

Because many of these experiments are performed under artificial conditions in the laboratory, it is difficult to determine the extent to which microbial mineral precipitation reactions occur in the subsurface relative to abiotic processes. It is also unclear whether microbial biominerals are distinguishable from those produced as a result of abiotic reactions. Experiments utilizing subsurface bacteria, solutions of well-defined composition (as opposed to complex microbiological media), and mineral or rock substrates are becoming more common (Zachara et al., 1998). Dong et al. (2000) demonstrated that a subsurface bacterial strain, *Shewanella putrefaciens* CN32, could reduce the Fe(III) in fine-grained magnetite. Kostka et al. (1996) reported bacterial reduction of Fe(III) within smectite and later determined that this process had a considerable effect on the chemical properties of the clay (Kostka et al., 1999). Both phases are more recalcitrant and representative of deep subsurface environments than amorphous or microcrystalline forms of Fe(III). Detailed investigation of bacterial dissolution and precipitation of minerals under subsurface environmental conditions, however, is still in its infancy.

An alternative approach to examining subsurface bacterial mineralization is to study recently contaminated aquifers where microbial activities are elevated above background levels by the pollutant. In a hydrocarbon-contaminated sandy aquifer, Bennett et al. (1996) found, by using submersed *in situ* mineral microcosms, that groundwater-inhabiting bacteria enhanced the weathering of silicates and the precipitation of clay minerals. Because contaminants impose geochemical conditions within aquifers that are distinct from those that prevailed throughout the aquifer's geological history, diagenetic phases formed during the contamination history are often distinguishable from those occurring over geologic time. Tobin et al. (1999) used this approach to detect traces of pore-filling ferran calcite in a trichloro-ethylene-contaminated basaltic aquifer, where a normally aerobic environment had been replaced by anaerobic conditions, resulting in Fe(III)-reduction.

Pore or fracture-filling diagenetic minerals, therefore, can put limits on the type and magnitude of subsurface microbial activity. Diagenetic carbonate and sulfide assemblages are particularly informative, because the relative contributions of microbial Mn(IV)-, Fe(III)-, and SO_4^{2-}-reduction and methanogenesis can be delineated by determining the Fe and Mn content, the $\delta^{13}C$ of the carbonate, the abundance and $\delta^{34}S$ of the sulfide, and by calculating the mass balance as constrained by the redox equations (Coleman et al., 1993). High positive values of $\delta^{13}C$ (ca. 10–20‰ PDB) for the carbonate reflect methanogenesis. Intermediate negative values of $\delta^{13}C$ (ca. -15 to -30‰ PDB) for carbonate reflect the combined effects of fermentation and microbial oxidation of formate or acetate coupled to microbial Mn(IV)-, Fe(III)-, and SO_4^{2-}-reduction. Finally, very low $\delta^{13}C$ values in carbonate (ca. < -35‰ PDB) are indicative of methanotrophy (Ritger et al., 1987; Pedersen et al., 1997).

Measurement of ^{13}C values in carbonate has been successfully used to model the organic oxidation and inorganic mineralization in low-temperature, near-surface marine sediments (Curtis, 1977; Irwin et al., 1977; Coleman et al., 1981), as well as the formation of carbonate concretions in shallow nonmarine settings (Curtis et al., 1986; Coleman et al., 1993) where microbial processes dominate most redox reactions. Mineralogical evidence of low-level, deep subsurface microbial activity can be inferred by determining whether diagenetic cement was formed at depth or near a surface or shallow subsurface sediment–water interface in the remote past. Several approaches exist for deriving the age of diagenetic carbonate/sulfide formation, relative to a rock unit's geological history. The active precipitation of diagenetic carbonate in an aquifer can be verified by comparing the δ^{18}O and trace metal concentration, ^{87}Sr/^{86}Sr of the carbonate to that of the groundwater for the present-day formation temperature (Budd et al., 1993). The timing of pore-filling carbonate formation can also be related to a basin's burial history by measuring the detrital porosity, because differential burial compaction of partly cemented sediments produces a heterogeneous porosity distribution. If the burial and thermal histories of a rock unit are known, then determining the temperature of diagenetic carbonate formation can indicate its age. The δ^{18}O can be used to estimate the temperature of carbonate precipitation, if the δ^{18}O of the paleogroundwater can be assessed. Analyses of O isotopes in primary aqueous fluid inclusions in the carbonate can establish the temperature at the time of their formation; isotope analyses of secondary fluid inclusions can assist in establishing whether the temperature of the rock has exceeded the upper limit for microbial life since the time of formation (ca. 120°C) (Tseng and Onstott, 1997). Finally, in carbonate that contains hydrocarbon inclusions and/or is associated with sulfide, U(IV) can substitute for Ca(II) (Sturchio et al., 1998). Consequently, the secondary carbonate formed in anoxic environments is amenable to radiometric dating by either the U series disequilibrium method, as in the case of the 500 kyr Devil's Hole vein calcite (Winograd et al., 1992), or the U-Pb approach, as in the case of secondary calcite in Paleozoic limestone (DeWolf and Halliday, 1991; Smith et al., 1991).

McMahon (1991a) related the diagenetic mineral assemblages present in a 50–80-m-deep confined aquifer to microbial activity. The volumetric abundance, isotopic composition of calcite and pyrite cement, and the distribution of secondary porosity of 70 million-year-old marine ACP sediments were used to derive the rates of microbial fermentation and SO_4^{2-}-reduction. The sparry calcite tended to have light ^{18}O and ^{13}C signatures, characteristic of precipitation in present-day meteoric water, whereas the micritic calcite preserved a marine origin. When averaged over 70 myr, the volumetric abundances yielded rates for CO_2-production and SO_4^{2-}-reduction of 10^{-6} and 10^{-8} moles L^{-1} year^{-1}, respectively. These estimates are similar to those derived from present-day pore water fluxes of acetate and SO_4^{2-} from the clays into the sandy aquifer (McMahon and Chapelle, 1991). Furthermore, the cement and secondary porosity was concentrated near the interfaces between the clay aquitards and sandy aquifers, consistent with observations that present-day fermentation in the clay layers supplies organic acids to the sandy layers, where they are oxidized by SO_4^{2-}-respiring bacteria. Because the small pore sizes in the clays restrict access of bacteria to organic matter and because the organic matter that is accessible is consumed and/or becomes more recalcitrant with continued bio-degradation, the subsurface microbial respiration rate should decrease with time. Consequently, much of the cement may have formed early in the ACP burial history, and respiration rates derived from the diagenetic cements only provide maximum estimates of the present-day subsurface microbial activity.

Analyses of many carbonate concretions in sandstone indicate they grew at depths sufficient to prevent the diffusive transport of solute from the sediment–water interface and precipitated from meteoric groundwater (Wilkinson and Dampier, 1990; Coleman et al., 1993). Textural features preserved in some of these carbonates suggest bacterial involvement in the precipitation process (McBride et al., 1994) and may reflect subsurface microbial activity. Reduction halos are centimeter-sized, bleached, carbonate-cemented spheroids, with heavy metal and/or organic-rich cores that typically occur in terrestrial red beds, but are also reported in igneous rocks (Hoffman, 1990). Hoffman (1990) has hypothesized that many reduction halos form as a result of microbial reduction of Fe(III) minerals in the host rock and precipitation of reduced metals in the core of the reduction halo. This supposition is based in part on Th-Pb ages of 100 million years for reduction halos in Permian red beds, based on fluid inclusion analyses and on sulfur isotope composition of sulfide in the core, which reveal significant fractionation. This would indicate a formation depth of at least 1000 m (Hoffman and Frei, 1996) and formation temperatures of 40–70°C.

The maximum depth and temperature at which microbially mediated redox reactions significantly contribute to diagenetic mineral precipitation are uncertain. In sandstone aquifers of the Gulf Coast, Carothers and Kharaka (1980) reported evidence for aceticlastic methanogenesis at temperatures as high as 80°C, based on the $\delta^{13}C$ of DIC. Plummer et al. (1990) suggested that microbial SO_4^{2-}-reduction is occurring in the Madison Aquifer at depths of 2.2 km and temperatures of 87°C, based on the S isotopic differences between SO_4^{2-} and HS^-. Microbial degradation of crude oil manifests itself by the progressive depletion of n-alkanes, followed by branched alkanes, and alkyl aromatics, leaving a residue enriched in napthenes and polynuclear aromatic hydrocarbons, resins, and asphaltenes (Connan, 1984). These features have been reported for subsurface petroleum reservoirs with maximum formation temperatures of 88°C and are normally associated with aerobic (O_2-respiring) conditions (Connan, 1984). More recently, however, Aeckersberg et al. (1991) and Reuter et al. (1994) demonstrated anaerobic oxidation of n-alkanes and alkyl-benzenes to CO_2 by mesophilic and thermophilic SO_4^{2-}-reducing bacteria. Hyperthermophilic SO_4^{2-}-reducing archaea and thermophilic SO_4^{2-}-reducing bacteria have been reported from oil fields worldwide (Stetter et al., 1993; L'Haridon et al., 1995; Leu et al., 1996, 1998) to depths as great as 3 km beneath the sea floor. Such cultures have been reported to grow at temperatures as high as 102°C (Stetter et al., 1993). Even though these studies did not establish whether the microorganisms were indigenous to the oil-bearing strata, the weight of experimental data suggest that sulfate-reducing bacteria and archaea exist to great depths and at temperatures exceeding 100°C (Stetter et al., 1993; L'Haridon et al., 1995; Leu et al., 1996, 1998).

Evidence for the involvement of subsurface and hyperthermophilic SO_2^{2-}-reducing archaea in the formation of diagenetic carbonate has been reported for a North Sea chalk oil reservoir (Jensius and Burruss, 1990). Biodegraded hydrocarbon inclusions were found within fracture-filling calcite, composed of light $\delta^{13}C$ values and were associated with pyrite depleted in ^{34}S. The aqueous inclusion data and $\delta^{18}O$ of the carbonate indicated that the biodegraded oil was trapped at temperatures between 95° and 130°C sometime during the late Miocene and up to the present. These results suggest that hyperthermophilic SO_4^{2-}-reducing bacteria have been active in the subsurface for an extended period.

In the deep subsurface, the rates of microbial activity can be as low as 10^6-fold less than in relatively shallow unconfined aquifers (Kieft and Phelps, 1997). At elevated temperatures favoring hyperthermophilic microorganisms, the rates for abiotic redox reactions begin to approach those for bacterial catalyst. For example, in sandstone/shale systems, at temperatures of $>100°C$, the abiotic oxidation rate of organic matter by Fe(III) in smectites during conversion to illite (Crossey et al., 1986) may be comparable to that for thermophilic microbial Fe(III)-reduction coupled to organic matter oxidation and production of ferroan calcite or ankerite. Metastable equilibrium between carboxylate and carbonate species in oil field waters at $>100°C$ may also compete with microbial heterotrophy and promote the precipitation of calcite with intermediate negative values of $\delta^{13}C$ (Helgeson et al., 1993). The $\delta^{13}C$ and δD of CH_4-occurrences in the subsurface of the Canadian and Fennoscandian Precambrian shields indicate that only a minor proportion of the CH_4 is microbially produced. The majority of the CH_4 probably represents abiogenic synthesis (Sherwood-Lollar et al., 1993a, 1993b). Consequently, estimates of microbial activities in deep subsurface environments derived from analyses of diagenetic cements should be considered maximum values.

5 SUMMARY

Microorganisms are present in a wide range of deep terrestrial subsurface environments, but are limited primarily by the availability of energy sources, pore space, and water. They obtain energy for growth and maintenance via oxidation of sediment- or rock-associated organic matter (heterotrophy), or from reduced inorganic substrates such as H_2, CH_4, or S^{2-} (lithoautotrophy). These metabolic reactions include, for example, oxidation of organic matter to CO_2 and reduction of Fe(III) or Mn(IV) oxides to Fe(II) or Mn(II), and of SO_4^{2-} to S^{2-}. Thus, such processes also impart major changes in the geochemistry of the aqueous and solid phases of the subsurface. These changes are most easily recognized via progressive decreases in the electron activity $p\varepsilon$ of groundwater as it moves from the point of recharge to distal points in a confined aquifer. Microbial processes are also evident in groundwater from the isotopic composition of elements such as C and S, which are fractionated during enzymatic catalysis.

Microbial populations in the deep terrestrial subsurface are generally characterized as being dispersed, or present at relatively low population densities, heterogeneously distributed (everything is *not* everywhere), and having very low metabolic activities, even in comparison to extremely oligotrophic environments such as deep pelagic sediments. There are, of course, exceptions, such as natural deposits of petroleum hydrocarbons or high concentrations of H_2 or CH_4 outgasing from the mantle or produced *in situ* via abiotic reactions in the Earth's crust.

In spite of recent advances in our scientific understanding of the microbiology of the deep terrestrial subsurface and of the biogeochemical processes microorganisms catalyze in this environment, there is still much we do not know. Although there is increasing evidence that lithoautotrophic microbial communities may be common in the deep subsurface, we understand relatively little about their distribution and ecology or their primary energy source(s). There has been considerable speculation, but little actual scientific evidence, that such microbial communities may make significant contributions to carbonaceous deposits in the Earth's crust. While it is clear that microorganisms in the deep subsurface have extremely slow metabolic rates, there has been little evidence to date indicating that they are more highly adapted than microbes in other oligotrophic environments to life under conditions of extreme starvation, although there have been relatively few investigations of this type on which to base strong conclusions. By focusing future scientific investigations on addressing such questions, applying developing technologies for deep drilling and sampling, and using robust analyses for characterizing microbial communities and the processes they catalyze, it should be possible to achieve a more thorough scientific understanding of the distribution and activities of microorganisms deep within the Earth's crust.

Acknowledgments   Some of the research described in this chapter was supported by the Office of Biological and Environmental Research, U.S. Department of Energy (DOE). Pacific Northwest National Laboratory is operated for the DOE by Battelle Memorial Institute under Contract DE-AC06-76RLO 1830.

REFERENCES

Aeckersberg F, Bak F, and Widdel F (1991) Anaerobic oxidation of saturated hydrocarbons to CO_2 by a new type of sulfate-reducing bacterium. *Arch Microbiol 165*:5–14.

Alperin MJ, Reeburgh WS, and Whiticar MJ (1988) Carbon and hydrogen isotope fractionation resulting from anaerobic methane oxidation. *Global Biogeochem Cycles 2*:279–288.

Amann RI, Ludwig W, and Schleifer K-H (1995) Phylogenetic identification and *in situ* detection of individual microbial cells without cultivation. *Appl Environ Microbiol 59*:143–169.

Amy P and Morita RY (1983) Starvation-survival patterns of sixteen freshly isolated open-ocean bacteria. *Appl Environ Microbiol 45*:1109–1115.

Amy PS, Durham C, Hall D, and Haldeman DL (1993) Starvation-survival of deep subsurface isolates. *Curr Microbiol 26*:345–352.

Anderson RT, Chapelle FH, and Lovley DR (1998) Evidence against hydrogen-based microbial ecosystems in basalt aquifers. *Science 281*:976–977.

Balkwill DL and Ghiorse WC (1985) Characterization of subsurface bacteria associated with two shallow aquifers in Oklahoma. *Appl Environ Microbiol 50*:580–588.

Balkwill DL (1989) Numbers, diversity, and morphological characteristics of aerobic, chemoheterotrophic bacteria in deep subsurface sediments from a site in South Carolina. *Geomicrobiol J 7*:33–51.

Balkwill DL, Fredrickson JK, and Thomas JM (1989) Vertical and horizontal variations in the physiological diversity of the aerobic chemoheterotrophic bacterial microflora in deep Southeast Coastal Plain subsurface sediments. *Appl Environ Microbiol 55*:1058–1065.

Barker WW, Welch SA, and Banfield JF (1998) Biogeochemical weathering of silicate minerals. In Banfield JF and Nealson KH (eds): *Interactions between Microbes and Minerals*. Washington, DC: The Mineralogical Society of America, pp 391–428.

Bastin E (1926) The presence of sulphate reducing bacteria in oil field waters. *Science 63*:21–24.

Bazylinski DA and Moskowitz M (1997) *Microbial Biomineralization of Magnetic Iron Minerals: Microbiology, Magnetism and Environmental Significance*. Washington, DC: Mineralogical Society of America.

Beloin RM, Sinclair JL, and Ghiorse WC (1988) Distribution and activity of microorganisms in subsurface sediments of a pristine study site in Oklahoma. *Microbial Ecol 16*:85–97.

Ben Baccar M, Fritz B, and Madè B (1993) Diagenetic albitization of K-feldspar and plagioclase in sandstone reservoirs: Thermodynamic and kinetic modeling. *J Sed Petrol 63*:1100–1109.

Bennett PC, Hiebert FK, and Choi WJ (1996) Microbial colonization and weathering of silicates in a petroleum-contaminated groundwater. *Chem Geol 132*:45–53.

Boone DR, Liu Y, Zhao Z, Balkwill DL, Drake GR, Stevens TO, and Aldrich HC (1995) *Bacillus infernus* sp. nov., an Fe(III)- and Mn(IV)-reducing anaerobe from the deep terrestrial subsurface. *Int J Syst Bacteriol 45*:441–448.

Bottcher J, Strebel O, Voerkelius S, and Schmidt H-L (1990) Using isotope fractionation of nitrate-nitrogen and nitrate-oxygen for evaluation of microbial denitrification in a sandy aquifer. *J Hydrol 114*:413–424.

Budd DA, Hammes U, and Vacher HL (1993) Calcite cementation in the upper Floridan aquifer: a modern example for confined-aquifer cementation models? *Geology 21*:33–36.

Carothers WW and Kharak YK (1980) Stable carbon isotopes of HCO_3^- in oil-field waters— implications for the origin of CO_2. *Geochim Cosmochim Acta 44*:323–332.

Champ DR, Gulens J, and Jackson RE (1979) Oxidation–reduction sequences in ground water flow systems. *Can J Earth Sci 16*:12–23.

Chapelle FH and Lovley DR (1990) Rates of microbial metabolism in deep coastal plain aquifers. *Appl Environ Microbiol 56*:1865–1874.

Chapelle FH, Zelibor J, Grimes DJ, and Knobel LL (1987) Bacteria in deep coastal plain sediments of Maryland: A possible source of CO_2 to groundwater. *Water Resources Res 23*:1625–1632.

Chrzanowski TH, Crotty RD, Hubbard JG, and Welch RP (1994) Applicability of fluorescein diacetate method of detecting active bacteria in freshwater. *Microbial Ecol 10*:179–185.

Coleman DD, Risatty JB, and Schoell M (1981) Fractionation of carbon and hydrogen isotopes by methane-oxidizing bacteria. *Geochim Cosmochim Acta 45*:1033–1037.

Coleman ML, Hedrick DB, Lovley DR, White DC, and Pye K (1993) Reduction of Fe(III) in sediments by sulphate-reducing bacteria. *Nature 361*:436–438.

Coleman ML and Raisewell R (1993) Microbial mineralization of organic matter: Mechanisms of self-organization and inferred rates of precipitation of diagenetic minerals. *Phil Trans Roy Soc London 344*:69–87.

Colwell FS, Onstott TC, Delwiche ME, Chandler D, Fredrickson JK, Yao Q-J, McKinley JP, Boone D, Griffiths R, Phelps TJ, Ringelberg D, White DC, LaFreniere L, Balkwill D, Lehman RM, Konisky J, and Long PE (1997) Microorganisms from deep, high temperature sandstones: Constraints on microbial colonization. *FEMS Microbiol Rev 20*:425–435.

Colwell FS, Stormberg GJ, Phelps TJ, Birnbaum SA, McKinley J, Rawson SA, Veverka C, Goodwin S, Long PE, Russell BF, Garland T, Thompson D, Skinner P, and Grover S (1992) Innovative techniques for collection of saturated and unsaturated subsurface basalts and sediments for microbiological characterization. *J Microbiol Meth 15*:279–292.

Connan J (1984) Biodegradation of crude oils in reservoirs. In Brooks J and Welte D (eds): *Advances in Petroleum Geochemistry*, pp 299–335.

Conrad R, Phelps TJ, and Zeikus JG (1985) Gas metabolism evidence in support of the juxtaposition of hydrogen-producing methanogenic bacteria in sewage sludge and lake sediments. *App Environ Microbiol 50*:595–601.

Crossley LJ, Surdam RC, and Lahann R (1986) Application of organic/inorganic diagenesis to porosity prediction. In Gautier DL (ed): *Roles of Organic Matter in Sediment Diagenesis.* SEPM Special Publication 38, Tulsa, OK, pp 147–155.

Curtis C (1977) Sedimentary geochemistry: Environments and processes dominated by involvement of an aqueous phase. *Phil Trans Roy Soc London 286A*:353–372.

Curtis C, Coleman ML, and Love LG (1986) Pore water evolution during sediment burial from isotopic and mineral chemistry of calcite, dolomite and siderite concretions. *Geochim Cosmochim Acta 50*:2321–2334.

DeWolf CP and Halliday AN (1991) U-Pb dating of a remagnetized Paleozoic limestone. *Geophys Res Lett 18*:1445–1448.

Dong H, Fredrickson JK, Kennedy DW, Zachara JM, Kukkadapu RK, and Onstott TC (2000) Mineral transformation associated with the microbial reduction of magnetite. *Chem Geol 169*:299–318.

Edmunds WM, Miles DL, and Cook JM (1984) A comparative study of sequential redox processes in three British aquifers. *IAHS Publ 150*:55–70.

Ekendahl S, Arlinger J, Ståhl F, and Pedersen K (1994) Characterization of attached bacterial populations in deep granitic groundwater from the Stripa research mine by 16SrRNA gene sequencing and scanning electron microscopy. *Microbiology 140*:1575–1583.

Ferris FG, Wiese RG, and Fyfe WS (1994) Precipitation of carbonate minerals by micro-organisms: Implications for silicate weathering and the global carbon dioxide budget. *Geomicrobiol J 12*:1–13.

Fontes J-C (1994) Isotope palaeohydrology and the prediction of long-term repository behaviour. *Terra Nova 6*:20–36.

Fredrickson JK and Balkwill DL (1998) Sampling and enumeration techniques. In Burlage RS, Atlas R, Stahl D, Geesey G, and Sayler G (eds): *Techniques in Microbial Ecology.* New York: Oxford University Press, pp 239–254.

Fredrickson JK, Galkwill DL, Zachara JM, Li SW, Brockman FJ, and Simmons MA (1991a) Physiological diversity and distributions of heterotrophic bacteria in deep Cretaceous sediments of the Atlantic Coastal Plain. *Appl Environ Microbiol 57*:402–411.

Fredrickson JK, Brockman FJ, Bjornstad BN, Long PE, Li SW, McKinley JP, Wright JV, Conca JL, Kieft TL, and Balkwill DL (1993) Microbiological characteristics of pristine and contaminated deep vadose sediments from an arid region. *Geomicrobiol J 11*:95–107.

Fredrickson JK, Brockman FJ, Workman DJ, Li SW, and Stevens TO (1991b) Isolation and characterization of a subsurface bacterium capable of growth on toluene, naphthalene, and other aromatic compounds. *Appl Environ Microbiol 57*:796–803.

Fredrickson JK, McKinley JP, Bjornstad BN, Ringelberg DB, White DC, Krumholz LR, Suflita JM, Colwell FS, Lehman RM, and Phelps TJ (1997) Pore-size constraints on the activity and survival of subsurface bacteria in a late Cretaceous shale-sandstone sequence, northwestern New Mexico. *Geomicrobiol J 14*:183–202.

Fredrickson JK, McKinley JP, Nierzwicki-Bauer SA, White DC, Ringelberg DB, Rawson SA, Li S-W, Brockman FJ, and Bjornstad BN (1995) Microbial community structure and biogeochemistry of Miocene subsurface sediment: Implications for long-term microbial survival. *Molec Ecol 4*:619–626.

Fredrickson JK and Phelps TJ (1996) Subsurface drilling and sampling. In Hurst CJ, Knudsen GR, McInerney MJ, Stetzenbach LD, and Walter MV (eds): *Manual of Environmental Microbiology*. Washington, DC: American Society for Microbiology, pp 526–540.

Fredrickson JK, Zachara JM, Kennedy DW, Dong H, Onstott TC, Hinman NW, and Li SW (1998) Biogenic iron mineralization accompanying the dissimilatory reduction of hydrous ferric oxide by a groundwater bacterium. *Geochim Cosmochim Acta 62*:3239–3257.

Fry NK, Fredrickson JK, Fishbain S, Wagner M, and Stahl DA (1997) Population structure of microbial communities associated with two deep, anaerobic, alkaline aquifers. *Appl Environ Microbiol 63*:1498–1504.

Gold T (1992) The deep, hot biosphere. *Proc Natl Acad Sci USA 89*:6045–6049.

Grenthe I, Stumm W, Laaksuharju M, Nilsson A-C, and Wikberg P (1992) Redox potential and redox reactions in deep groundwater systems. *Chem Geol 98*:131–150.

Griffin WT, Phelps TJ, Colwell FS, and Fredrickson JK (1997) Methods for obtaining deep subsurface microbiological samples by drilling. In Amy PS and Haldeman DL (eds): *The Microbiology of the Terrestrial Deep Subsurface*. Boca Raton, FL: Lewis Publishers, pp 23–44.

Grossman EL, Coffman BK, Fritz SJ, and Wada H (1989) Bacterial production of methane and its influence on ground-water chemistry in east-central Texas aquifers. *Geology 17*:495–499.

Haldeman DL and Amy PS (1993) Bacterial heterogeneity in deep subsurface tunnels at Ranier Mesa, Nevada Test Site. *Microbial Ecol 25*:183–194.

Haldeman DL, Amy PS, Ringelberg D, and White DC (1993) Characterization of the microbiology within a 21 m^3 section of rock from the deep subsurface. *Microbial Ecol 26*:145–159.

Haldeman DL, Pitonzo BJ, Story SP, and Amy PS (1994) Comparison of the microbiota recovered from surface and deep subsurface rock, water, and soil along an elevational gradient. *Geomicrobiol J 12*:99–111.

Hazen TC, Jimeniz L, Lopez de Victoria G, and Fliermans CB (1991) Comparison of bacteria from deep subsurface sediment and adjacent groundwater. *Microbial Ecol 22*:293–304.

Hedges JI and Oades JM (1997) Comparative organic geochemistries of soils and marine sediments. *Org Geochem 27*:319–361.

Helgeson HC, Knox AM, Owens CE, and Shock EL (1993) Petroleum, oil field waters, and authigenic mineral assemblages: Are they in metastable equilibrium in hydrocarbon reservoirs? *Geochim Cosmochim Acta 57*:3295–3339.

Hirsch P and Rades-Rohkohl E (1988) Some special problems in the determination of viable counts of groundwater microorganisms. *Microbial Ecol 16*:99–113.

Hirsch P and Rades-Rohkohl E (1990) Microbial colonization of aquifer sediment exposed in a groundwater well in northern Germany. *Appl Environ Microbiol 56*:2963–2966.

Hoehler TM, Alperin MJ, Alber DB, and Martens CS (1998) Thermodynamic control on hydrogen concentrations in anoxic sediments. *Geochim Cosmochim Acta 62*:1745–1756.

Hoffman BA (1990) Reduction spheroids from northern Switzerland: Mineralogy, geochemistry, and genetic models. *Chem Geol 81*:55–81.

Hoffman BA (1992) Isolated reduction phenomena in red beds: A result of porewater radiolysis? In Haraka and Maest (eds): *Water–Rock Interaction.* Proceedings of 7th International Symposium on Water–Rock Interaction, Balkema/Rotterdam/Brookfield, pp 503–506.

Hoffman BA and Frei R (1996) Age constraints of reduction spot formation from Permian red bed sediments, northern Switzerland, inferred from U-Th-Pb systematics. *Schweiz Mineralogy Petrogr Mitt 76*:235–244.

Irwin H, Curtis C, and Coleman M (1977) Isotopic evidence for source of diagenetic carbonates formed during burial of organic-rich sediments. *Nature 269*:209–213.

Ivanov MV, Belyaev SS, Laurinavichus KS, and Namsaraev BB (1979) Microbiological oxidation of methane in stratal waters of the Lower Povolzh'ye. *Mikrobiologiya 48*:129–132.

Jackson RE, Merritt WF, Champe DR, Gulens J, and Inch KJ (1977) The distribution coefficient as a geochemical measure of the mobility of contaminants in a groundwater flow system. The use of nuclear techniques in water pollution studies. Cracow, Poland: International Atomic Energy Agency, Vienna, Austria.

Jensius J and Burruss RC (1990) Hydrocarbon–water interactions during brine migration: Evidence from hydrocarbon inclusions in calcite cements from Danish North Sea oil fields. *Geochim Cosmochim Acta 54*:705–713.

Jones RE, Beeman RE, and Suflita JM (1989) Anaerobic metabolic processes in the deep terrestrial subsurface. *Geomicrobiol J 7*:117–130.

Karl DM (1995) Ecology of free-living, hydrothermal vent microbial communities. In Karl DM (ed): *The Microbiology of Deep-Sea Hydrothermal Vents.* Boca Raton, FL: CRC Press, pp 35–124.

Kieft TL (2000) Size matters: dwarf cells in soil and subsurface terrestrial environments. In Colwell RR and Grimes DJ (eds): *Noncultural Microorganisms in the Environment.* Washington D.C.: American Society for Microbiology, pp. 19–46.

Kieft TL, Fredrickson JK, McKinley JP, Bjornstad BN, Rawson SA, Phelps TJ, Brockman FJ, and Pfiffner SM (1995) Microbiological comparisons within and across contiguous lacustrine, paleosol, and fluvial subsurface sediments. *Appl Environ Microbiol 61*:749–757.

Kieft TL, Fredrickison JK, Onstott TC, Gorby YA, Kostandarithes HM, Bailey TJ, Kennedy DW, Li SW, Plymale AE, Spadoni CM, and Gray MS (1999) Dissimilatory reduction of Fe(III) and other electron acceptors by a *Thermus* isolate. *Appl Environ Microbiol 65*: 1214–1221.

Kieft TL and Phelps TJ (1997) Life in the slow lane: Activities of microorganisms in the subsurface. In Amy PS and Haldeman DL (eds): *The Microbiology of the Terrestrial Deep Subsurface.* CRC Press, Boca Raton, FL, pp 137–164.

Kieft TL, Ringelberg DB, and White DC (1994) Changes in ester-linked phospholipid fatty acid profiles of subsurface bacteria during starvation and desiccation in a porous medium. *Appl Environ Microbiol 60*:3292–3299.

Kieft TL, Wilch E, O'Connor K, Ringelberg DB, and White DC (1997) Survival and phospholipid fatty acid profiles of surface and subsurface bacteria in natural sediment microcosms. *Appl Environ Microbiol 63*:1531–1542.

Kogure K, Simidu U, and Taga N (1979) A tentative direct microscopic method for counting living bacteria by fluorescence microscopy. *Can J Microbiol 25*:415–420.

Konhauser KO, Mann H, and Fyfe WS (1992) Prolific organic SiO_2 precipitation in a solute-deficient river: Rio Negro, Brazil. *Geology 20*:227–230.

Kostka JE, Nealson KH, Wu J, and Stucki JW (1996) Reduction of the structural Fe(III) in smectite by a pure culture of the Fe-reducing bacterium, *Shewanella putrefaciens* strain MR-1. *Clays Clay Miner 44*:522–529.

Kostka JE, Wu J, Nealson KH, and Stucki JW (1999) The impact of structural Fe(III) by bacteria on the surface chemistry of smectite clay minerals. *Geochim Cosmochim Acta 63*:3705–3713.

Krumholz LR, McKinley JP, Ulrich GA, and Suflita JM (1997) Confined subsurface microbial communities in Cretaceous rock. *Nature 386*:64–66.

Leu JY, McGovern-Traa CP, Porter A Jr, and Hamilton WA (1996) Thermophilic sulfate reducers present in production water from North Sea Oil Field. *Acta Ocean Tai 35*:395–407.

Leu JY, McGovern-Traa CP, Porter A Jr, Harris WJ, and Hamilton WA (1998) Identification and phylogenetic analysis of thermophilic sulfate-reducing bacteria in oil field samples by 16S rDNA gene cloning and sequencing. *Anaerobe 4*:165–174.

L'Haridon S, Reysenbach AL, Glenat P, Prieur D, and Jeanthon C (1995) Hot subterranean biosphere in a continental oil reservoir. *Nature 377*:223–225.

Liu SV, Zhou J, Zhang C, Cole DR, Gajdarziska-Josifovska M, and Phelps TJ (1997a) Thermophilic Fe(III)-reducing bacteria from the deep subsurface: The evolutionary implications. *Science 277*:1106–1109.

Liu Y, Karnauchow TM, Jarrell KF, Balkwill DL, Drake GR, Ringelberg D, Clarno R, and Boone DR (1997b) Description of two new thermophilic *Desulfotomaculum* spp., *Desulfotomaculum putei* sp. nov., from a deep terrestrial subsurface, and *Desulfotomaculum luciae* sp. nov., from a hot spring. *Int J Syst Bacteriol 47*:615–621.

Lovley DR and Chapelle FH (1995) Deep subsurface microbial processes. *Rev Geophys 33*:365–381.

Lovley DR and Goodwin S (1988) Hydrogen concentration as an indicator of the predominant terminal electron-accepting reactions in aquatic environments. *Geochim Cosmochim Acta 51*:2993–3003.

Lovley DR, Stolz JF, Nord GL Jr, and Phillips EJP (1987) Anaerobic production of magnetite by a dissimilatory iron-reducing microorganism. *Nature 330*:252–254.

Ludvigsen L, Albrechtsen H-J, Heron G, Bjerg PL, and Christensen TH (1998) Anaerobic microbial processes in a landfill leachate contaminated aquifer (Grinsted, Denmark). *J Contam Hydrol 33*:273–291.

McBride EF, Picard MD, and Folk RL (1994) Oriented concretions, Ionian coast, Italy: Evidence of groundwater flow direction. *J Sed Res A64*:535–540.

McKinley VL, Costerton JW, and White DC (1988) Microbial biomass, activity, and community structure of water and particulates retrieved by backflow from a waterflood injection well. *Appl Environ Microbiol 54*:1383–1393.

McMahon PB and Chapelle FH (1991a) Geochemistry of dissolved inorganic carbon in a Coastal Plain aquifer. 2. Modeling carbon sources, sinks, and d13C evolution. *J Hydrol 127*:109–135.

McMahon PB and Chapelle FH (1991b) Microbial production of organic acids in aquitard sediments and its role in aquifer geochemistry. *Nature 349*:233–235.

Morita RY (1980) Calcite precipitation by marine bacteria. *Geomicrobiol J 2*:63–82.

Morita RY (1982) Starvation-survival of heterotrophs in the marine environment. *Adv Microbial Ecol 6*:117–198.

Mortimer RJG, Coleman ML, and Rae JE (1997) Effect of bacteria on the elemental composition of early diagenetic siderite: Implications for paleoenvironmental interpretations. *Sedimentology 44*:759–765.

Murphy EM and Schramke JA (1998) Estimation of microbial respiration rates in groundwater by geochemical modeling constrained with stable isotopes. *Geochim Cosmochim Acta 62*:3395–3406.

Murphy EM, Schramke JA, Fredrickson JK, Bledsoe HW, Francis AJ, Sklarew DS, and Linehan JC (1992) The influence of microbial activity and sedimentary organic carbon on the isotope geochemistry of the Middendorf Aquifer. *Water Resources Res 28*:723–740.

Novitsky JA (1981) Calcium carbonate precipitation by marine bacteria. *Geomicrobiol J 2*:375–388.

Ohmoto H and Lasaga AC (1982) Kinetics of reactions between aqueous sulfates and sulfides in hydrothermal systems. *Geochim Cosmochim Acta 46*:1727–1745.

Olson GJ, Dockins WS, McFeters GA, and Iverson WP (1981) Sulfate-reducing and methanogenic bacteria from deep aquifers in Montana. *Geomicrobiol J 2*:327–340.

Onstott TC, Phelps TJ, Colwell FS, Ringelberg D, White DC, Boone DR, McKinley JP, Stevens TO, Long PE, Balkwill DL, Griffin WT, and Kieft T (1998a) Observations pertaining to the origin and ecology of microorganisms recovered from the deep subsurface of Taylorsville Basin, Virginia. *Geomicrobiol J 15*:352–285.

Onstott TC, Phelps TJ, Kieft TJ, Colwell FS, Balkwill DL, Fredrickson JK, and Brockman FJ (1998b) A global perspective on the microbial abundance and activity in the deep subsurface. In Seckbach J (ed): *Microorganisms and Life in Extreme Environments* Norwell, MA: Kluwer Academic Publishers, pp 1–14.

Onstott TC, Tobin K, Dong H, DeFlaun MF, Fredrickson JK, Bailey T, Brockman FJ, Kieft T, Peacock A, White DC, Balkwill D., Phelps TJ, and Boone DR (1997) The deep gold mines of South Africa: Windows into the subsurface biosphere. In Hoover RB (ed): *Instruments, Methods, and Missions for the Investigation of Extraterrestrial Microorganisms*. San Diego, CA: The International Society for Optical Engineering.

Pedersen K (1998) Evidence for a hydrogen-driven, intra-terrestrial biosphere in deep granitic rock aquifers. 8th International Symposium on Microbial Ecology, Halifax, Nova Scotia: Canadian Society for Microbiology.

Pedersen K, Arlinger J, Ekendahl S, and Hallbeck L (1996a) 16S rRNA gene diversity of attached and unattached bacteria in boreholes along the access tunnel to the Äspö hard rock laboratory, Sweden. *FEMS Microbiol Ecol 19*:249–262.

Pedersen K, Arlinger J, Hallbeck L, and Pettersson C (1996b) Diversity and distribution of subterranean bacteria in groundwater at Oklo in Gabon, Africa, as determined by 16S rRNA gene sequencing. *Molec Ecol 5*:427–436.

Pedersen K and Ekendahl S (1990) Distribution and activity of bacteria in deep granitic groundwaters of southeastern Sweden. *Microbial Ecol 20*:37–52.

Pedersen K and Ekendahl S (1992a) Assimilation of CO_2 and introduced organic compounds by bacterial communities in groundwater from Southeastern Sweden deep crystalline bedrock. *Microbial Ecol 23*:1–14.

Pedersen K and Ekendahl S (1992b) Incorporation of CO_2 and introduced organic compounds by bacterial populations in groundwater from the deep crystalline bedrock of the Stripa mine. *J Gen Microbiol 138*:369–376.

Pedersen K, Ekendahl S, Tullborg E-L, Furnes H, Thorseth I, and Tumyr O (1997) Evidence of life at 207 m depth in a granitic aquifer. *Geology 25*:827–830.

Phelps TJ, Fliermans CB, Garland TR, Pfiffner SM, and White DC (1989) Recovery of deep subsurface sediments for microbiological studies. *J Microbiol Meth 9*:267–280.

Phelps TJ, Murphy EM, Pfiffner SM, and White DC (1994) Comparison between geochemical and biological estimates of subsurface microbial activities. *Microbial Ecol 28*:335–349.

Phelps TJ, Raione EG, White DC, and Fliermans CB (1989) Microbial activities in deep subsurface environments. *Geomicrobiol J 7*:79–92.

Plummer LN (1977) Defining reactions and mass transfer in part of the Floridian aquifer. *Water Resources Res 13*:801–812.

Plummer NL, Busby JF, Lee RW, and Hanshaw BB (1990) Geochemical modeling of the Madison aquifer in parts of Montana, Wyoming, and South Dakota. *Water Resources Res 26*:1981–2014.

Reuter P, Rabus R, Wilkes H, Aeckersberg F, Rainey FA, Jannasch HW, and Widdel F (1994) Anaerobic oxidation of hydrocarbons in crude oil by new types of sulphate-reducing bacteria. *Nature 3727*:455–458.

Ringelberg DB, Sutton S, and White DC (1997) Biomass, bioactivity and biodiversity: Microbial ecology of the deep subsurface: Analysis of ester-linked phospholipid fatty acids. *FEMS Microbiol Rev 20*:371–377.

Ritger S, Carson B, and Suess E (1987) Methane-derived authigenic carbonates formed by subduction-induced pore-water expulsion along the Oregon/Washington margin. *Geol Soc Am Bull 98*:147–156.

Ronen DM, Magaritz M, and Molz FJ (1991) Comparison between forced gradient tests to determine the vertical distribution of horizontal transport properties of aquifers. *Water Resources Res 27*:1309–1314.

Russell BF, Phelps TJ, Griffin WT, and Sargent KA (1992) Procedures for sampling deep subsurface microbial communities in unconsolidated sediments. *Ground Water Monitor Rev 12*:96–104.

Russell CE (1997) The collection of subsurface samples by mining. In Amy PS and Haldeman DL (eds): *The Microbiology of the Terrestrial Deep Subsurface*. Boca Raton, FL: CRC Press, pp 45–59.

Sarbu SM, Kinkle BK, Vlasceanu L, Kane TC, and Popa R (1994) Microbiological characterization of a sulfide-rich groundwater ecosystem. *Geomicrobiol J 12*:175–182.

Schultze-Lam S and Beveridge TJ (1994) Nucleation of celestite and strontianite on a cyanobacterial S-layer. *Appl Environ Microbiol 60*:447–453.

Seewald JS (1994) Evidence for metastable equilibrium between hydrocarbons under hydrothermal conditions. *Nature 370*:285–287.

Sherwood B, Fritz P, Frape SK, Macko SA, Weise SM, and Welhan JA (1988) Methane occurrences in the Canadian Shield. *Chem Geol 71*:223–236.

Sherwood-Lollar B, Frape SK, Fritz P, Macko SA, Welhan JA, Blomqvist R, and Lahermo PW (1993a) Evidence for bacterially generated hydrocarbon gas in Canadian shield and Fennoscandian shield rocks. *Geochim Cosmochim Acta 57*:5073–5085.

Sherwood-Lollar B, Frape SK, Weise SM, Fritz P, Macko SA, and Welhan JA (1993b) Abiogenic methanogenesis in crystalline rocks. *Geochim Cosmochim Acta 57*:5087–5097.

Shock EL (1988) Organic acid metastability in sedimentary basins. *Geology 16*886–890.

Shock EL (1994) Application of thermodynamic calculations to geochemical processes involving organic acids. In Pitman ED and Lewn MD (eds): *Organic Acids in Geological Processes*. New York: Springer-Verlag.

Sinclair JL and Ghiorse WC (1989) Distribution of aerobic bacteria, protozoa, algae, and fungi in deep subsurface sediments. *Geomicrobiol J 7*:15–32.

Slobodkin AI and Wiegel J (1997) Fe(III) as an electron acceptor for H_2 oxidation in thermophilic anaerobic enrichment cultures form geothermal areas. *Extremophiles 1*:106–109.

Smith JT and Ehrenberg SN (1989) Correlation of carbon dioxide abundance with temperature in clastic hydrocarbon reservoirs: Relationship to inorganic chemical equilibrium. *Mar Petrol Geol 6*:129–135.

Smith PE, Farquhar RM, and Hancock RG (1991) Direct radiometric age determinations of carbonate diagenesis using U-Pb in secondary calcite. *Earth Planet Sci Lett 105*:474–491.

Staley JT (1999) Bacterial biodiversity: a time for place. *ASM News 65*:681–687.

Stetter KO, Huber R, Blochl E, Kurr M, Eden RD, Fielder M, Cash H, and Vance I (1993) Hyperthermophilic Archaea are thriving in deep North Sea and Alaskan oil reservoirs. *Nature 365*:743–745.

Stevens TO and McKinley JP (1995) Lithoautotrophic microbial ecosystems in deep basalt aquifers. *Science 270*:450–454.

Stevens TO, McKinley JP, and Fredrickson JK (1993) Bacteria associated with deep, alkaline, anaerobic groundwaters in southeast Washington. *Microbial Ecol 25*:35–50.

Stumm W and Morgan JJ (1996) *Aquatic Chemistry.* New York: John Wiley & Sons.

Sturchio NC, Antonio MR, Soderholm L, Sutton SR, and Brannon JC (1998) Tetravalent uranium in calcite. *Science 281*:971–973.

Thorstenson DC, Fisher DW, and Croft MG (1979) The geochemistry of the Fox Hills-Basal Hell Creek Aquifer in southwestern North Dakota and northwestern South Dakota. *Water Resources Res 15*:1479–1498.

Tobin KJ, Colwell FS, Onstott TC, and Smith R (2000) Recent calcite spar in an aquifer waste plume: A possible example of contamination driven calcite precipitation. *Chem Geol 169*:449–460.

Tseng H-Y and Onstott TC (1997) A tectonic origin for the deep subsurface microorganisms of Taylorsville Basin: Thermal and fluid flow model constraints. *FEMS Microbiol Rev 20*:391–397.

Tseng H-Y, Person M, and Onstott TC (1998) Hydrogeologic constraint on the origin of deep subsurface microorganisms with a Triassic basin. *Water Resources Res 34*:937–948.

Tunlid A, Ringelberg D, Phelps TJ, Low C, and White DC (1989) Measurement of phospholipid fatty acids at picomolar concentrations in biofilms and deep subsurface sediments using gas chromatography and chemical ionization mass spectrometry. *J Microbiol Meth 10*:139–153.

Tunlid A and White DC (1992) Biochemical analysis of biomass, community structure, nutritional status, and metabolic activity of microbial communities in soil. In Stotzky G and Bollag J-M (eds): *Soil Biochemistry.* New York: Marcel Dekker, pp 229–262.

Van Cappellen P and Wang Y (1995) Metal cycling in surface sediments: Modeling the interplay of transport and reaction. In Allen HE (ed): *Metal Contaminated Sediments.* Chelsea, MI: Ann Arbor Press, pp 21–64.

Vasconcelos C, Mckenzie JA, Bernasconi S, Grujic D, and Tien AJ (1995) Microbial mediation as a possible mechanism for natural dolomite formation at low temperatures. *Nature 377*:220–222.

Virta M, Lineria S, Kankaanpää P, Karp M, Pelotonen K, Nuutila J, and Lilius E-M (1998) Determination of complement-mediated killing of bacteria by viability staining and bioluminescence. *Appl Environ Microbiol 62*:515–519.

Vroblesky DA and Chapelle FH (1994) Temporal and spatial changes of terminal electron-accepting processes in a petroleum hydrocarbon-contaminated aquifer and the significance for contaminant biodegradation. *Water Resources Res 30*:1561–1570.

Walvoord MA, Pegram P, Phillips FM, Person M, Kieft TL, Fredrickson JK, McKinley JP, and Swenson JB (1999) Groundwater flow and geochemistry in the southeastern San Juan Basin: Implications for microbial transport and activity. *Water Resources Res 35*:1409–1424.

White DC, Davis WM, Nickels JS, King JD, and Bobbie RJ (1979) Determination of the sedimentary microbial biomass by extractable lipid phosphate. *Oecologia 40*:51–62.

White DC and Ringelberg DB (1998) Signature lipid biomarker analysis. In Burlage RS, Atlas R, Stahl D, Geesey G, and Sayler G (eds): *Techniques in Microbial Ecology.* New York: Oxford University Press, pp 255–272.

Wilkinson M and Dampier MD (1990) The rate of growth of sandstone-hosted calcite concretions. *Geochim Cosmochim Acta 54*:3391–3399.

Winograd IJ, Coplen TB, Landwehr JM, Riggs AC, Ludwig KR, Sazbo BJ, Kolesar PT, and Revesz KM (1992) Continuous 500,000-year climate record from vein calcite in Devils Hole, Nevada. *Science 258*:255–260.

Zachara JM, Fredrickson JK, Li SW, Kennedy DW, Smith SC, and Gassman PL (1998) Bacterial reduction of crystalline Fe^{3+} oxides in single phase suspensions and subsurface materials. *Amer Mineral 83*:1426–1443.

Zehnder AJB and Stumm W (1988) Geochemistry and biogeochemistry of anaerobic habitats. In Zehnder AJB (ed): *Biology of Anaerobic Microorganisms.* New York: John Wiley & Sons, pp 1–38.

2

TRANSPORT OF MICROORGANISMS IN THE SUBSURFACE: THE ROLE OF ATTACHMENT AND COLONIZATION OF PARTICLE SURFACES

MADILYN FLETCHER

Belle W. Baruch Institute for Marine Biology and Coastal Research, University of South Carolina, Columbia, South Carolina

ELLYN MURPHY

Pacific Northwest National Laboratory, Richland, Washington

1 INTRODUCTION

It is likely that large numbers of diverse microorganisms are carried into the subsurface along with water percolated through the surface soils and sediments. As these microbes begin their journeys through the subsurface, nutrients become increasingly scarce, desiccation may be a threat in the vadose zone, and encounters with particle surfaces are probably common. The movement of these organisms through the subsurface is likely to be influenced by their tendency to adhere to solid

---

Subsurface Microbiology and Biogeochemistry, Edited by James K. Fredrickson and Madilyn Fletcher.
ISBN 0-471-31577-X. Copyright 2001 by Wiley-Liss, Inc.

surfaces, while their survival will be determined by their physiological capabilities and the stresses they encounter. Observations confirm that some bacteria can move substantial distances through porous media and subsurface sediments, but field studies are still few in number, and there is much to be learned about the factors that control transport in subsurface environments. Much of our understanding of the relationship between transport and bacterial characteristics is based on laboratory studies, and only recently has the science reached the point of relating laboratory-based studies on transport with field-scale events. Our understanding at this point is that bacteria with strong tendencies to adhere to surfaces are retained on particles and their movement is relatively limited. However, we are only beginning to resolve the many factors that come together to determine bacterial retention or movement in the diverse conditions encountered in the subsurface.

2 SUBSURFACE ENVIRONMENTS

To understand factors that control movement and distribution of microorganisms in the subsurface, one must have a clear understanding of the environmental char- acteristics found in the sediments and rocks comprising this complex ecosystem. The subsurface is generally divided into two primary zones that are delineated by the degree of water saturation. Near-surface sediments are usually unsaturated and are referred to as the vadose zone. Deep sediments, which are saturated and therefore below the water table, are referred to as groundwater systems. Designating these subsurface systems by the degree of water saturation is appropriate because water is the principal medium through which bacteria are transported in the vadose zone and in groundwater.

The vadose zone is a three-phase system consisting of solid, liquid, and gas, whereas the groundwater zone consists of only the solid and liquid phases. The physical method of water movement is quite different in vadose versus groundwater systems (see Sections 2.1 and 2.2), and these differences may affect bacterial transport. When water moves through porous media, dissolution, reaction, and precipitation processes take place, which influence the structure of the porous media, the chemistry of the water, and the surface chemistry of the solid. The resulting physical and chemical environment can have a significant effect on the retardation or transport of bacteria (Section 3.1.3).

The subsurface chemical environment can be quite harsh to bacteria. These environments are often limited in key nutrients such as phosphate and nitrogen. Deep vadose zone and groundwater systems are generally referred to as oligotrophic, that is, readily usable carbon substrates are scarce. Below the plant root zone the concentration of organic carbon can become severely limiting in natural environ- ments. Dissolved organic carbon is the most readily usable form of carbon for heterotrophic bacteria. In contrast, inorganic carbon tends to be much higher in groundwaters and, depending on the reaction path of the groundwater, bicarbonate can be a dominant anion (Freeze and Cherry, 1979). In highly reducing environ- ments, particulate organic carbon is often found in the subsurface in the form of kerogen, coal, and lignitic-type compounds. These compounds are recalcitrant, but may form the basis of a microbial food web in some subsurface systems (Murphy et al., 1992). In these oligotrophic systems, energy expenditure is limited and may be largely relegated to maintenance of the organism. As a result, the ability of an organism to adapt and survive in this environment may play a key role in the transport of a microbial population.

2.1 Vadose Zone

Both the degree of water saturation and the velocity of the water influence bacterial transport. The importance of water saturation can be seen in the vadose zone where the matric potential, or retention of water in the pore spaces, controls the transport of water. As the soil water content decreases, water is contained in increasingly smaller capillary spaces between the sediment grains. Accordingly, bacterial movement may cease because the cells are too large to pass through the pores, a mechanism referred to as straining. Likewise, as the thickness of the water film decreases, inhibition of bacterial transport through their adsorption to interfaces becomes significant. Reducing the distance between the bacteria and the sediment surface increases the probability that the bacteria will come in contact with the surface. Studies have shown that the amount of bacterial retention in unsaturated sediments is related to cell properties, such as hydrophobicity (Huysman and Verstraete, 1993) and size (Gannon et al., 1991a). These cell properties result in removal by straining and adhesion mechanisms (Fig. 2.1). Also, in partially saturated systems the surface area of gas–water interfaces increases and the dynamics of the air–water interface can

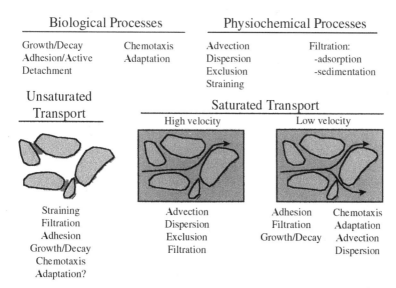

Biological Processes		Physiochemical Processes	
Growth/Decay	Chemotaxis	Advection	Filtration:
Adhesion/Active	Adaptation	Dispersion	-adsorption
Detachment		Exclusion	-sedimentation
		Straining	

Unsaturated Transport

Saturated Transport

High velocity

Low velocity

Straining	Advection	Adhesion	Chemotaxis
Filtration	Dispersion	Filtration	Adaptation
Adhesion	Exclusion	Growth/Decay	Advection
Growth/Decay	Filtration		Dispersion
Chemotaxis			
Adaptation?			

FIGURE 2.1 Biological and physiochemcial processes involved in bacterial transport and the potential importance of these processes under unsaturated and saturated flow conditions.

clearly influence bacterial distribution (see Section 3.1.3). In a series of papers Wan and others (Wan and Wilson, 1994a, 1994b; Wan et al., 1994) showed that bacteria will preferentially sorb to gas–water interfaces over silica surfaces. It has also been shown that the attachment of cells at the gas–water interface increases with cell hydrophobicity (Schafer et al., 1998; Wan et al., 1994).

In temperate climates water films may be sufficient to promote the advection of bacteria in unsaturated systems during natural rainfall events. Because the time scales are usually short for laboratory and field experiments, the focus in both temperate and arid climates has often been on bacterial transport during preferential flow (Breitenbeck et al., 1988; Natsch et al., 1996; Smith et al., 1985). Preferential flow occurs where large pores, old root channels, insect and worm burrows, or fractures in the porous media become saturated with water; usually from an intense rainfall event that causes ponding and/or surface run-on. When saturated, these large pore structures preferentially conduct the water in a vertical soil column. Although one might expect the nutrient flux to be enhanced during preferential flow, the effect of these nutrients on the indigenous microbial population has not been found to produce a metabolically active community. Instead, microorganisms undergoing rapid transport from preferential flow have been observed to be under greater nutrient stress, as measured by their phospholipid fatty acid ratios, than their counterparts that were not subjected to rapid transport (Balkwill et al., 1998). Balkwill et al. (1998) speculated that during preferential flow, bacteria can be transported by advection to depths in the vadose zone where it is difficult for the microorganisms to compete effectively with an established community.

2.2 Groundwater Systems

In groundwater systems bacteria are primarily transported by the bulk motion of the flowing groundwater; a process known as advection. Groundwater velocity varies widely and is dependent on the hydraulic gradient, a function of the overlying pressure head of water and frictional forces along its flowpath. Under natural conditions, groundwater may flow as fast as a few meters per day or as slow as less than a meter per year. In a pumping well, a forced gradient exists and groundwater velocities may be on the order of meters per second.

The velocity of the groundwater will influence the relative importance of physiochemical and biological processes, which in turn can either promote or retard bacterial transport in porous media (Fig. 2.1). Under high-velocity regimes, the transport timeframes may be insufficient for the expression of processes associated with growth or the expenditure of energy (e.g., adhesion or chemotaxis). On the other hand, high velocities may promote physicochemical processes, such as physical filtration, that is the removal of particles from solution by collision with, and deposition on, the grain surfaces. According to filtration theory, mass transfer generally increases with groundwater velocity (McDowell-Boyer et al., 1986). Filtration theory also predicts an increase in attachment rate with velocity; however, this linear relation may not hold under conditions where (1) shear stresses associated with high velocities (e.g., pumped groundwater) prevent deposition, or (2) other coefficients in filtration theory that are inversely dependent on velocity (e.g., diffusion and sedimentation) effectively counter the linear velocity relationship. The linear dependence of the bacterial attachment rate on velocity, predicted by filtration theory, has simply not been experimentally evaluated at relevant ground-water velocities.

"Exclusion" occurs when the transported bacteria move faster than the average water velocity. Groundwater velocity within a pore throat is generally parabolically distributed, with the maximum velocity at the centerline of the pore and velocities of zero at the grain surface (de Marsily, 1986). Large particles, such as bacteria, may preferentially experience the higher velocities near the pore centerlines, yielding an average velocity that is higher than that of a conservative salt tracer. As groundwater velocities increase, the region of high velocity in the pore throat will also increase. On the other hand, such increases may also result in increased deposition on particle surfaces (above). Thus, the combined effect of exclusion and physical filtration on bacterial transport in groundwaters would be earlier breakthrough of bacteria (relative to a conservative tracer) from exclusion, but a lower mass recovery of bacteria (due to the greater physical filtration).

Under natural gradients, groundwater velocities are generally quite low and advection may be a minor transport process. In these cases biological processes such as growth/decay, active attachment/detachment, chemotaxis, and adaptation may play a more important role in the transport and/or retardation of bacteria (Fig. 2.1). This is because at low groundwater velocities, the time scale over which transport occurs is sufficiently long for survival of the organism to become a factor in its transport. Survival or maintenance of a bacterial population requires some minimal expenditure of energy, possibly to adapt the organism's metabolism to the oligo-trophic environment, adhere to a grain surface to avoid predation, or move toward an energy source (chemotaxis). Removal of bacteria may occur through death/decay or predation (Harvey, 1997). Under more favorable nutrient conditions, growth may facilitate bacterial transport through cell division (Jenneman et al., 1985, 1986; Reynolds et al., 1989; Sharma et al., 1993). It has been estimated that the generation time in oligotrophic environments may be on the order of hundreds to thousands of years (Phelps et al., 1994); therefore, transport initiated by growth may be rare in natural groundwater environments.

3 BACTERIAL INTERACTIONS WITH SURFACES

Much of our understanding of bacterial interactions with solid surfaces is derived from laboratory studies of model systems. Many such investigations have attempted to identify the factors that control adhesion by manipulating and measuring variables relating to substratum characteristics, bacterial surface charge or hydrophobicity, or solution chemistry. There has been a strong incentive to devise strategies for controlling bacterial attachment, as bacterial colonization of surfaces is a serious and costly problem in a range of situations, such as on ship bottoms, pipelines, food processing surfaces, clinical instrumentation, or implants and prosthetic devices.

Presumably many of the data from these fundamental studies are relevant to bacterium–substratum interactions in subsurface environments.

3.1 Attachment to Surfaces

3.1.1 *Physicochemical Attraction and Repulsion Interactions* When bacteria closely approach a solid surface, the probability of attachment is strongly influenced by physical and chemical (physicochemical) interactions. These may be attractive or repulsive, depending on the complex interplay of the chemistries of the bacterial and substratum surfaces and the aqueous phase (Fletcher, 1996). Marshall et al. (1971) made an important conceptual leap in our understanding of the forces that determine microbial adhesion by experimentally addressing whether bacterial attachment to surfaces is governed by the same physicochemical interactions that determine deposition of nonliving colloidal particles. This pivotal study was followed by numerous additional applications of physicochemical theory to bacterial attachment phenomena. Two theoretical approaches have been used (Fletcher, 1996): the DLVO model for lyophobic colloid stability (Derjaguin and Landau, 1941; Verwey and Overbeek, 1948) and thermodynamic theory (e.g., Neumann et al., 1974). In the DLVO approach, the bacterial and substratum surfaces each have a net negative surface charge, as well as adsorbed counterions from the adjacent solution, which is termed the electric double layer. Such like-charged surfaces will be prevented from coming together because of electrostatic repulsion unless the solution electrolyte concentration is sufficiently high, which results in a decrease in the thickness of the electric double layer. Thus, the relevance of DLVO theory to bacterial adhesion can be tested by measuring substratum and bacterial surface charge (usually as electrophoretic mobility; see, e.g., van Loosdrecht et al., 1987) and determining adhesion in solutions of known and controlled ionic strength. As predicted, some studies have shown attachment numbers to increase with an increase in ionic strength of the suspending solution (Weerkamp et al., 1988; van Loosdrecht et al., 1989), whereas other studies do not show such relationships (McEldowney and Fletcher, 1986). Such inconsistencies suggest that electrostatic repulsion may be a primary obstacle preventing adhesion only in situations where negatively charged cells and surfaces are unable to approach each other close enough for adhesive interactions to come into play. When electrostatic repulsion does not occur or can be overcome, whether adhesion will occur depends on other attractive and repulsive interactions, such as hydrophobicity or steric effects, not accounted for by surface charge measurements. Nevertheless, we would expect electrostatic interactions to be significant in subsurface environments that are characterized by charged mineral surface and low-ionic-strength pore water. Indeed, electrostatic interactions would be attractive and favor attachment when one surface is positively charged. By coating negatively charged surfaces, for example, quartz sand with Fe^{3+}, numbers of attached cells can be significantly increased (Mills et al., 1994).

The second physicochemical approach that has been used to describe bacterial attachment to surfaces predicts that bacterial attachment is a spontaneous event that can be described by the second law of thermodynamics, and thus is accompanied by a decrease in free energy of the system. This thermodynamic approach is more inclusive than an electrostatic approach and takes into account the various types of attractive and repulsive interactions, such as van der Waals, electrostatic, or dipole, but expresses them collectively in terms of free energy. The approach requires estimation of numerical values of thermodynamic measures, such as the interfacial free energies at the bacterium and substratum surfaces, usually determined by measurement of contact angles of liquids on lawns of bacterial cells and the test substrata (Fletcher and Marshall, 1982; van der Mei et al., 1991). In some reports, measurements of bacterial adhesion in specific laboratory systems were consistent with thermodynamic predictions (Absolom et al., 1983), but others were not (Bellon-Fontaine et al., 1990). One possible cause for these inconsistencies is the difficulty in obtaining accurate values for bacterial surface free energies because of their complex chemistry and hydration *in vivo*. Thus, calculations of free energy changes during adhesion, which are based on these estimates, may not be accurate.

However, there is an inherent and serious difficulty in trying to apply thermodynamic theory to biological systems. Clearly, a living organism or biological community does not exemplify the second law of thermodynamics, which predicts an increase in randomness or disorder. Rather, living systems are characterized by order and were described by Prigogine (Prigogine and Stengers, 1984) as self-ordering systems whose structure is maintained by a constant flow of matter and energy through them. Bacteria are living organisms that convert substrate into energy, produce new biomass and increase populations through replication, and synthesize surface polymers that stabilize (or in some cases destabilize) adhesion. Furthermore, interactions among cells set up complex synergies or antagonisms that affect the stability of adhesive interactions over time. Any theoretical model that adequately describes or predicts bacterial adhesion in the long term must take account of the complex interplay of components, rarely described by linear mathematics and frequently changing in time along with progressive changes in environmental conditions.

Although the application of physicochemical theory has not been entirely successful in explaining or predicting various bacterial attachment behaviors, many of these studies led to an increasingly common observation: In numerous cases, hydrophobicity of the solid surfaces or of the bacterial surfaces tended to result in increased numbers of attached cells (Rosenberg and Kjelleberg, 1986; Fletcher, 1996). Similarly, high degrees of hydration of a surface tended to decrease bacterial attachment (Pringle and Fletcher, 1986). For a hydrated surface to bind to another surface, water must be displaced, which is energetically unfavorable and may not be offset by the forces driving attractive interactions. Thus, the presence of hydrophobic sites on either bacterial or substratum surfaces should facilitate contact and subsequent adhesion of the two surfaces.

3.1.2 *Bacterial Adhesion Mechanisms*

A diversity of polymers and structures is involved in the adhesion of different bacterial types. Extracellular polysaccharides are produced by attached bacteria (Allison and Sutherland, 1987) and apparently strengthen their binding to surfaces and to other organisms. Exopolysaccharides can also initiate attachment to surfaces, but binding may depend on surface composition. For example, the polymer produced by *Deleya marina* facilitated its attachment to hydrophilic surfaces but not to a hydrophobic surface, whereas a mutant deficient in polysaccharide had reduced adhesion to hydrophilic surfaces, but increased attachment to a hydrophobic surface (Shea et al., 1991). Because of the polar nature of these polymers, they also tend to remain hydrated and may provide protection from desiccation should the surface become exposed to air (Roberson and Firestone, 1992), as in the vadose zone. There is a considerable range in the compositions of bacterial extracellular polysaccharides. In aquatic organisms, they are generally heteropolysaccharides, containing largely neutral sugars and various amounts of uronic acids and acetate or pyruvate substitutions (Sutherland, 1983; Boyle and Reade, 1983; Read and Costerton, 1987), which will influence their solubility and adhesive properties.

A number of different types of polymers appear to be involved in the initial attachment event. Studies on the effects of proteases on attachment indicate that protein can be involved in the early stages of attachment (Danielsson et al., 1977; Fletcher and Marshall, 1982; Paul and Jeffrey, 1985). Microscopic observations have indicated that lateral flagella on vibrios (Lawrence et al., 1992) and polar flagella on pseudomonads (Korber et al., 1989; Fletcher, 1996) act as attachment structures. Moreover, analyses of a transposon mutant with altered adhesion properties to soil and a variety of plant seeds demonstrated that alterations in flagellin were responsible for reduced attachment (DeFlaun et al., 1994). The ability of the protein appendages fimbriae, or pili, to act in binding is well documented among clinically important bacteria and some environmental isolates (Vesper, 1987; Weerkamp et al., 1986), but their role in attachment to inanimate surfaces, and particularly in the subsurface, is not known.

Lipopolysaccharide (LPS) structure apparently can influence adhesion of pseudomonads in some cases. Alterations in the length of the polysaccharide chain (or O-antigen) on the LPS can affect adhesion properties, either because it plays a primary role in adhesive interactions or it masks groups near the membrane, such as phosphate groups in the core-lipid A region (Makin and Beveridge, 1996) or outer membrane proteins (Arredondo et al., 1994). Increases in adhesion (Stenström, 1989; Williams and Fletcher, 1996) or hydrophobicity (Hermansson et al., 1982; Stenström, 1989) have been observed with a loss or attenuation of O-antigen. However, attenuation of O-antigen has also correlated with decreased adhesion and increased transport (DeFlaun et al., 1999). Because of the inconsistencies in these results, at this point we can only conclude that LPS composition can influence attachment, but that this effect is very dependent on the specific chemistries of the LPS and other closely associated polymers in the outer membrane.

Because of the extreme complexity in bacterial surface chemistry and the difficulty in identifying specific polymers that act as adhesives, attempts have been made to characterize bacterial surface properties according to parameters that approximate gross characteristics, such as surface charge (James, 1991; Gannon et al., 1991a, 1991b; Baygents et al., 1998; Glynn et al., 1998), hydrophobicity

(Rosenberg and Kjelleberg, 1986; Gannon et al., 1991a), or surface free energy (Busscher et al., 1984). For example, attachment to soil and sediment particles has been shown to be positively related to bacterial surface hydrophobicity (Stenström, 1989; Williams and Fletcher, 1996; DeFlaun et al., 1999). However, such measurement of average surface properties has not necessarily been consistent with attachment and transport through porous media. In a study of a variety of organisms (*Enterobacter*, *Pseudomonas*, *Bacillus*, *Achromobacter*, *Flavobacterium*, and *Arthrobacter* species), no consistent relationship was found between transport and bacterial surface charge, hydrophobicity, or extracellular polymers (Gannon et al., 1991a). This difficulty in finding specific parameters that describe or predict attachment (and hence transport) is probably due to the overwhelming complexity of the system and the fact that multiple, sometimes interacting factors influence the process in a nonlinear manner. For example, the ability to model bacterial attachment was improved by considering cell surface hydrophobicity and electrokinetic potential together, rather than individually, when analyzing the attachment of 23 strains to negatively charged polystryrene (van Loosdrecht et al., 1987). Attachment was greatest with the most hydrophobic cells, regardless of their electrokinetic potential. However, with the more hydrophilic cells, attachment decreased with an increase in negative surface potential. It is possible that these two properties—hydrophobicity and surface charge—are of prime importance because of the role they play in determining whether the cell can get close enough to the surface for additional adhesive interactions to come into play. Then additional surface properties, as well as physiological processes, determine the actual adhesive interaction.

Not only do different bacteria adhere by means of different surface polymers or appendages, but for a given population of an organism, there can be significant variations in attachment even among clones from a single cell (Williams and Fletcher, 1996). Capillary electrophoresis measurements of clonal populations of two subsurface isolates indicated differences in surface charge density among subpopulations of each organism (Baygents et al., 1998). Altered adhesion phenotypes can be selected by serial passaging in liquid medium (Cuperus et al., 1992), by repeated transfer through porous media columns (DeFlaun et al., 1990), or by collection on surfaces in a low-dilution-rate chemostat (Ragout et al., 1996). Thus, there is considerable heterogeneity in attachment properties among bacteria, which promotes their dispersal to both surfaces and the aqueous phase.

Much of the variability in experimental observations of bacterial attachment may be explained by the complexity in bacterial surface polymer composition, as well as the change in polymer composition and synthesis with changing environmental conditions or time. Clearly, the chemical compositions of cell surface polymers can be influenced by environmental conditions, particularly types of nutrients and their concentrations. Often, after bacteria have been attached to a surface for hours or days, hydrated amorphous polymers accumulate along with increasing numbers of attached cells. These polymers form an intercellular matrix in which the cells are embedded and constitute the highly hydrated, "slimey" matrix that constitutes a major portion of the bacterial biofilm. The polymers seem to build up after attachment has occurred, suggesting that attachment to a surface in some way signals the switching on of polymer synthesis that strengthens cell–surface attachment, thus enabling biofilm formation. So far, attachment to surfaces has been shown to induce expression of genes related to lateral flagella formation in *Vibrio* species (Belas et al., 1986), swarming in *Proteus* species (Belas, 1996), and *alg*C expression in *Pseudomonas aeruginosa* (Davies and Geesey, 1995). Researchers have also demonstrated that attachment to surfaces can be accompanied by expression of new suites of genes (Dagostino et al., 1991), as well as morphology changes (Dalton et al., 1994), and synthesis of macromolecules, including proteins (Brözel et al., 1995). It is likely that these reflect physiological adjustments to the altered microenvironment at the surface and the need for different colonization strategies, for example, extracellular polymer synthesis. These limited examples may be just the first observations of what may be a common phenomenon, that is, the switching on of genes that result in the conversion of cells from a single-cell, free-swimming form to a complex multicellular, surface-associated mode of existence.

There is also increasing evidence that the properties of bacteria in cell clusters can be influenced by chemical signaling factors, termed autoinducers (chemicals which above a particular threshold concentration induce specific types of activity). These signal chemicals allow bacteria to initiate certain metabolic activities (e.g., bioluminescence, capsular production, surfactant production, antibiotic synthesis) only when they reach sufficient population densities, and thus the phenomenon has been termed "quorum sensing" (Greenberg, 1997). Since, surface colonization facilitates clustering of bacteria, and thus presumably chemical signaling, quorum sensing may play a role in the development and succession of biofilm communities (Davies et al., 1998), but this is an area that is only beginning to be explored.

In summary, although the intercellular polymers in biofilms are largely polysaccharides, numerous types of polymers appear to be involved in the initial adhesive interaction between the single cell and the attachment substratum (Fletcher, 1996). Experiments with nonadhesive mutants of various *Pseudomonas* species have indicated that flagella or lipopolysaccharides may act as adhesives, and proteinaceous microfilaments, called pili or fimbriae, can also mediate attachment. However, because of their ubiquitous presence on bacterial surfaces, polysaccharides and cell surface proteins are also likely to play some role in the adhesive interaction. So far, it has been impossible to ascribe adhesion to any one type of polymer on the bacterial surface, and it is likely that multiple polymers contribute to the overall adhesive characteristics of a given organism.

3.1.3 Physicochemical Factors

In the deep subsurface environment, the physicochemical factors most likely to influence surface interactions are mineralogy of the surfaces and pore water chemistry. There is abundant evidence that numbers of cells and strength of binding are influenced by surface properties, particularly electrostatic charge (Rijnaarts et al., 1993; Jansen and Kohnen, 1995), surface free energy (Jansen and Kohnen, 1995; Taylor et al., 1997; van Pelt et al., 1985; Wiencek and Fletcher, 1995, 1997), a related parameter, hydrophobicity (Pringle and Fletcher, 1986; Taylor et al., 1997; Wiencek and Fletcher, 1995, 1997), and texture (Baker, 1984). Most of the studies have been done with surfaces composed of organic polymers (e.g., polystyrene, other types of plastic), glass, or metals, and observations with minerals are relatively rare.

Dissolved solutes in the suspending medium are clearly important. Electrolytes are significant for their influence on electrostatic interactions, discussed above (see Section 3.1.1). In bench-top studies of bacterial transport through sand, a ten-fold increase in ionic strength resulted in cell recoveries reduced by approximately 50–100% (Fontes et al., 1991). Surface-active agents, such as Tween (Marshall et al., 1989), or solutes that alter surface tension (Fletcher, 1983) have also been shown to reduce attachment. Dissolved organic substances can influence adhesion through their adsorption on the surface, thereby altering its properties. Such "conditioning films" that result in an increase in hydration of the surface may be particularly effective in reducing attachment of cells (Fletcher, 1976; Fletcher and Marshall, 1982). Dissolved solutes may also indirectly affect bacterial attachment characteristics through their involvement in physiological processes. Dissolved organics could serve as nutrients, or conversely, if they are recalcitrant, they could mask other surface localized substrates (e.g., mineral electron acceptors) through adsorption on the surface (see Chapter 1 and Chapter 6). Nutritional changes can influence adhesion in a number of ways, including polymer synthesis, cell growth, or other cell surface characteristics (McEldowney and Fletcher, 1986).

In the vadose zone, the presence of unsaturated pore regions and a fluctuating air–water interface can have a considerable effect on the movement of cells. Surface-active solutes and particles tend to be adsorbed at the air–water interface because of the accompanying decrease in interfacial free energy. Studies in aquatic systems have found that organic solutes, particles, and microorganisms are concentrated at the water surface in a thin microlayer approximately $100\,\mu m$ thick (Sieburth, 1976; Norkrans, 1980), but relatively little is known about the types of materials concentrated at the air–water interface in subsurface systems. Recent studies by Wan and Wilson (1994b) and Wan et al. (1994) demonstrated that accumulation of microorganisms at the air–water interface could influence their transport characteristics. In glass micromodels of porous systems, the movement of colloidal polystyrene beads, clay particles, and bacteria was observed microscopically (Wan and Wilson, 1994b). Enrichment at the interface tended to increase with particle hydrophobicity, solution ionic strength, and positive electrostatic charge of the particles. They hypothesized that the sorption of bacteria onto the gas–water interface occurred in two energy stages: (1) the DLVO (van der Waals attraction and diffuse double layer repulsion) and hydration forces control the first stage where the interface is ruptured by the bacteria, and (2) a capillary force fixes the bacteria at the interface in essentially an irreversible position (Wan and Wilson, 1994b). In bench-top studies, increases in the volume of pore space occupied by air resulted in a decrease in bacterial breakthrough concentrations (Wan et al., 1994) and in bacteriophage (Powelson et al., 1990). Presumably, as the interface moves across particle surfaces, bacteria are trapped in the receding water phase and deposited at the surface, precluding their further transport possibly until the pores become saturated again. On the other hand, moving air bubbles have also been seen to dislodge bacteria from glass and polymer surfaces (Pitt et al., 1993). Our observations are that loosely attached bacteria are easily removed from hydrophobic polystyrene—but not from a hydrophilic surface—by being swept off the surface as it passes through an air–water interface. Collectively, these limited observations indicate that bacteria are removed from a surface by a receding air–water interface when the bacteria are bound relatively weakly, but they tend to remain attached when firmly bound and/or when the surface remains hydrated due to additional adsorbed organic material or polymers (Fletcher, 1976; Fletcher and Marshall, 1982).

3.2 Physiological Effects of Bacterial Adhesiveness

Probably the greatest single reason for our inability to predict bacterial attachment behavior *in situ* is the physiological plasticity of these organisms in changing environmental conditions or with time. Because of the multiple variables that can influence adhesiveness and our inability to identify, measure, and control them in laboratory systems, it is extremely difficult to devise an experimental system that takes account of physiological and phenotypic changes in bacteria. Attachment properties can be affected by nutrient sources, concentration, and flux (Knox et al., 1985; Molin et al., 1982; McEldowney and Fletcher, 1986; Ellwood et al., 1974). In batch culture, adhesion has been shown to be affected by the growth phase (Powell and Slater, 1982; Rosenberg and Rosenberg, 1985); with some organisms attachment is greater in log phase, whereas with others attachment properties increase into stationary phase. Mutations or phase variations in cell surface components, such as the fibrillar layer on *Streptococcus salivarius* (Weerkamp et al., 1986) or in lipopolysaccharide in *Pseudomonas* species (Williams and Fletcher, 1996; Makin and Beveridge, 1996), have been shown to alter adhesion characteristics. Factors that influence extracellular polymer production can also influence adhesion. Carbon source, carbon:nitrogen ratio, carbon flux, or nutrient concentration may influence polymer production or adhesive ability in various ways, depending on the organism (Marshall et al., 1971; Molin et al., 1982; McEldowney and Fletcher, 1986; Bonet et al., 1993).

Most of the subsurface is deficient in organic nutrients, and such starvation conditions will affect the attachment properties of many types of bacteria (Section 2.1). However, the nature of the effect depends on the specific organism and can lead to an increase, decrease, or even no effect on attachment properties. Adhesiveness increased with substrate availability for the anaerobes *Syntrophomonas wolfei* and *Desulfovibrio* sp. G11 (van Schie and Fletcher, 1999) and *C. amalonaticus* (Meier-Schneiders et al., 1993). The proximate cause for such changes in adhesion is not known for most cases, but alterations in surface polymer synthesis or composition are likely explanations. This was observed in *Pseudomonas* sp. S9, where starvation appeared to result in a shift from a closely bound extracellular polymer to one that was loosely associated with the cells and induced detachment from surfaces (Wrangstadh et al., 1990).

For anaerobes, because of the toxic influence of oxygen on physiological processes, a matter of interest is whether oxygen influences attachment. Again, the answer depends on the organism in question. For *S. wolfei*, aeration severely inhibited attachment, whereas it had no effect on adhesion of *Desulfomonile tiedjei* (van Schie and Fletcher, 1999). This illustrates differences in the extent to which physiological activity is involved in the attachment process and/or the stability of cell surface adhesion properties in the presence of air. The stability of attachment properties in the presence of air has significance for bioremediation strategies, as anaerobic bacteria have the greatest potential for subsurface systems contaminated with chlorinated hydrocarbons, such as tetrachloroethylene (PCE) and trichloroethy-lene (TCE) (de Bruin et al., 1992; DiStefano et al., 1991; El Fantroussi et al., 1998).

Because of the long time periods involved in transport of bacteria along flow fields in subsurface environments, the considerations are quite different from those applying to short-term adhesion studies in laboratory settings. Although bacteria may be relatively "well fed" when first introduced through the surface soil, organic nutrients are soon depleted and conditions favor those organisms with very low maintenance requirements or chemosynthetic autotrophs (Chapter 3). Would there be a selective advantage in attachment to a surface, where the stationary cell could take advantage of sparse nutrients delivered by groundwater flow or could establish mutualistic colonies of interactive functional microorganisms? Are those organisms found in the deep subsurface adapted to that environment, even with respect to their adhesion abilities, or are they found there because of their ability to survive long periods and to remain freely suspended in the aqueous phase? The answers to these questions are still matters for speculation and offer some interesting insights into how organisms can survive seemingly hostile environments.

3.3 Significance of Porosity and Cell Size on Transport

The porosity, the presence of fractures, and the nature of flowpaths are clearly major factors influencing the movement of cells through the subsurface (Fontes et al., 1991; Barton and Ford, 1995). Pores in fine sediments and those with significant clay and silt fractions (DeFlaun et al., 1999) should be more easily clogged than course sediments, as was indicated in comparisons of 0.40–0.33 mm and 1.14–1.00 mm sand fractions (Fontes et al., 1991). In turn, the size and shape of cells can influence their transport in some situations. For example, in bench-top studies, small coccoid cells had considerably higher recovery rates than larger, rod-shaped cells (Fontes et al., 1991). Similarly, in measurements of transport of 19 bacterial isolates, bacterial retention was statistically related to only size, and not to electrostatic charge, cell surface hydrophobicity, or presence of capsules or flagella (Gannon et al., 1991a). Cell shape, expressed as the ratio of cell width to length, may be even more significant than just size itself (Weiss et al., 1995). Cell size has not always been found to relate to transport (Camper et al., 1993), but it may only be expected to be a factor when the bacterial size is on the same scale as the flow channels. However, it is also possible that large cells are more likely to attach to surfaces, as their biomass is sufficient for gravitational forces to facilitate their deposition on particles surfaces (van Schie and Fletcher, 1999).

3.4 Motility

Bacterial motility has been shown to influence attachment of bacteria to surfaces, apparently by increasing the probability of the cells encountering the surface (Fletcher, 1980). However, motility of cells does not appear to be significant in transport through porous media, as shown with *P. fluorescens* moving through columns packed with glass beads (Camper et al., 1993). The presence of a gradient of a chemical substrate also did not affect the measured transport of chemotactic bacteria (Barton and Ford, 1995), indicating that bacterial motility occurs at a scale that is irrelevant in transport systems.

4 EXPERIMENTAL APPROACHES TO MEASURING BACTERIAL TRANSPORT IN POROUS SYSTEMS

The processes controlling bacterial distribution have been conducted at many different scales, ranging from micron-scale interactions of bacteria with a surface, to field-scale investigations of bacterial transport. At the micron scale the basic cell–level interactions with surfaces are generally investigated in batch experiments (Fig. 2.2). The centimeter scale adds the complexity of a porous media matrix and flow. At the meter scale, which is most closely representative of field-scale processes, chemical and physical heterogeneity are added, which lead to spatial and temporal variations in microbes/nutrients and reaction rates. Each higher scale of observation encompasses the processes at the smaller scales and therefore adds complexity to the experimental system.

Most measurements of bacterial transport have been in relatively simple bench-top systems using glass or plastic columns on the order of 20 cm or fewer in length. Such systems can be constructed to determine the influence of specific variables, such as solution chemistry, particle size or composition, or bacterial types or surface properties, but the results from such simple systems are not readily extrapolated to *in situ* conditions. Accordingly, larger-scale intermediate scale experiments (ISEs) have been designed to incorporate assessment of larger flow fields, as well as defined heterogeneities. Finally, some field studies have been conducted to measure movement of microorganisms *in situ*. These are inherently difficult to perform, and the interpretation of results can be obscured by undefined heterogeneities in the flowpath and the presence of resident microorganisms. Nevertheless, we are reaching the stage where synthesis of the information gained from studies at these three levels of scales can provide us with improved understanding of the factors that influence microbial transport in natural systems.

4.1 Bench-top column studies

It is likely that the factors which determine attachment in static bench-top systems are relevant to adhesion to particles in porous media. However, additional factors, particularly increased surface area and convection to surfaces (Rijnaarts et al., 1993),

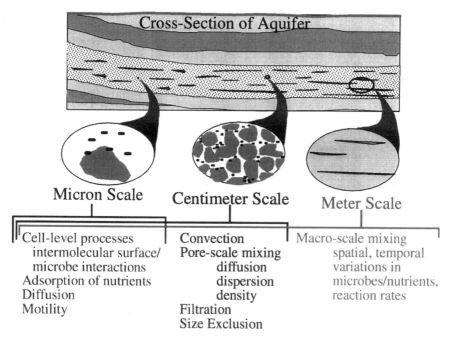

FIGURE 2.2 The different scales of experimentation and the processes that are expressed at each scale of observation.

will come into play during transport through sediments. Thus, experimental systems utilizing particulates and one-dimensional flow have been devised to evaluate transport of numerous organisms in defined conditions. Measurements of bacterial transport in bench-top columns include several variations in design and parameters measured. Columns, usually 1–30 cm long (Gannon et al., 1991b; Rijnaarts et al., 1993; DeFlaun et al., 1999), can be homogeneously packed with sand or sediment (Gannon et al., 1991b; Jenkins and Lion, 1993; DeFlaun et al., 1999), or glass or plastic beads (Camper et al., 1993; Rijnaarts et al., 1993). Systems have been operated by introducing cells and then keeping the system under constant flow (Fontes et al., 1991), or cell suspensions may be allowed to sit in the column and be subsequently run out and nonattached cells collected (Stenström, 1989; DeFlaun et al., 1990). Transport can be measured as the numbers of bacteria that emerge with the effluent, and often the distribution of bacterial breakthrough over time has been measured by determining cell numbers in successive volume fractions (Fontes et al., 1991; Rijnaarts et al., 1993). Column sediment has also been destructively sampled at the end of the experiment and the bacteria retained on sediment fractions determined (DeFlaun et al., 1999). Thus, transport can be expressed as bacterial numbers transported (or conversely retained) in the column or as breakthrough curves over time. Unfortunately, in such experiments, mass balances between input and output plus cells retained in the column are not always obtained, possibly from the inherent difficulty in estimating bacteria retained on particle surfaces. Bacterial transport has also been expressed as "collision efficiency," α, which is the fraction of bacteria retained after passage of a known cell suspension through a column (Baygents et al., 1998).

Although there are exceptions and inconsistencies among the data, the following trends emerge from these bench-top studies. The transport of bacteria through porous media is reduced for relatively adhesive cells, and they do appear to attach and be retained on surfaces of particles. It has not been possible to identify any single type of interaction that is responsible for retention on particle surfaces, although electrostatic repulsion between highly negatively charged bacteria and surfaces may well inhibit deposition, and hydrophobic interactions may facilitate retention. The relationship between porosity and bacterial size and shape is a factor: Clearly, the larger the ratio between bacterial size and flow channel diameter, the more likely the pores are to become clogged. There is considerable variability among the adhesion properties and transport characteristics of different organisms and even subpopulations of the same organism. Pore water chemistry affects transport; higher ionic strength may increase retention, and other solutes may condition particle surfaces through adsorption or may serve as bacterial substrates. Variations in mineralogy, particle size, and heterogeneity of the porous matrix also influence transport in, as yet, undefined ways.

4.2   Intermediate Scale Experiments

Intermediate scale experiments (ISEs) are designed to provide a bridge between the largely homogeneous experiments conducted in columns and highly complex, heterogeneous field experiments. These experiments are conducted in meter(s)-long laboratory flow cells with multiple sampling ports or instrument connections for measuring system parameters. ISEs have the advantage of simulating complex chemical and physical heterogeneities in a very controlled laboratory experimental environment. Heterogeneous physical structures that occur on the sub-facies and facies scales may have a profound effect on bacterial transport (Harvey et al., 1993). However, because of the length scales and boundary effects inherent in column experiments, we often cannot evaluate the impact of these structures on transport. It is also often difficult to separate the individual processes affecting the transport of bacteria in natural field environments.

Intermediate-scale experiments have other advantages. Processes can be evaluated in multiple dimensions at simulated groundwater flowrates. High-density sampling is possible without disturbing the flowfield because of the large cross-sectional area available in the flow cell. Where possible, *in situ* microelectrode sensors are used to monitor specific chemical or pressure changes in the flowfield. This type of sampling yields large data sets for modeling individual processes. The data sets can also be used to develop theories for up-scaling pore-scale processes so that we can understand the manifestation of these processes at larger, field-relevant scales.

Although flow cells have been used for many different studies related to reactive flow and transport (Bruch, 1970; Shamir and Harleman, 1967; Silliman et al., 1987; Silliman and Simpson, 1987; Starr et al., 1985; Sudicky et al., 1985; Szecsody et al., 1994; Wood et al., 1994), only a few experiments have been conducted where bacterial transport was evaluated (Murphy et al., 1997; Shonnard et al., 1994). Shonnard et al. (1994) showed that by raising the ionic strength of a phosphate buffer, the attachment of *Methylosinus trichosporium* OB3b to an aquifer sand increased by fifty-fold. The attachment was not electrolyte-specific, but rather a result of compression of the electric double layer due to solution ionic strength (see Section 3.1.1). The bacterial attachment in these experiments primarily occurred in regions near the injection port and in a flowfield created between the injector and withdrawal ports that simulated screened wells. Pore clogging did not occur, and the authors concluded that the introduction of biomass may actually produce less pore clogging than *in situ* biostimulation of indigenous microbial populations (Shonnard et al., 1994).

In another intermediate-scale experiment, the effects of physical heterogeneity on biodegradation of a substrate and on bacterial distribution were investigated (Murphy et al., 1997). In this experiment, low-permeability, silt-size inclusions were randomly distributed in a high permeability sand matrix, a heterogeneity pattern commonly found in natural systems. Many shallow unconfined groundwaters are in equilibrium with oxygen. As a contaminant plume moves through these ground-waters, the oxygen is rapidly depleted, creating anoxic conditions. Due to slow transport, the low-permeability areas in the aquifer will retain oxygen longer. Therefore, the heterogeneity pattern simulated in the intermediate-scale flow cell created regions of slow transport that prolonged the dual availability of the electron acceptor and donor, which in turn enhanced bacterial growth in these regions. These mixed redox zones may actually enhance the degradation of complex contaminants that require both anaerobic and aerobic metabolic processes (Bradley and Chapelle, 1995; Bruns-Nagel et al., 1998; Essaid et al., 1995; Lyngkilde and Christensen, 1992; Vroblesky and Chapelle, 1994).

In the Murphy et al. (1997) experiment, the sand was uniformly inoculated with *Pseudomonas cepacia* 866A before packing the flow cell. This created an initial condition of uniform biomass in the flow cell. A plume of substrate was then injected into the flow stream, as can be seen in Fig. 2.3*a*. The low-permeability silt inclusions tended to shear the substrate plume, which increased the surface area of the plume. The aqueous biomass increased at the leading edge of the substrate plume, decreased in the middle of the plume, and increased again at the tail of the plume (Fig. 2.3*b*). The decrease in the middle of the plume was attributed to a salt tracer that reduced the electrostatic repulsion, enhancing bacterial attachment. The increase in aqueous biomass at the leading and tailing edges of the plume was likely due to growth, because mixing at the surface of the plume enhanced the dual availability of substrate and oxygen. It appeared, although not verified in this experiment, that the aqueous biomass may have resulted from cell division-mediated transport. Essentially all the predicted growth in biomass could be accounted for in the aqueous stream exiting the flow cell. Finally, the low-permeability silt inclusions

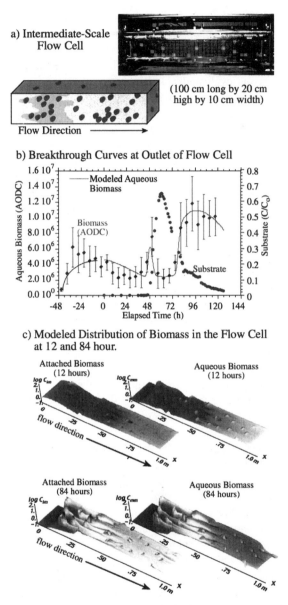

FIGURE 2.3   Intermediate-scale experiments in laboratory flow cells. (*a*) Flow cell and low-permeability silt inclusions randomly distributed in a sandy matrix. (*b*) Breakthrough curves of a substrate pulse and biomass in this flow cell experiment. (*c*) Modeled distribution of aqueous-phase and attached biomass in the flow cell at 12 and 84 hours.

also prolonged the residence time of oxygen and substrate, and higher concentrations of attached and aqueous biomass were associated with the silt sediments (Fig. 2.2c). This experiment illustrated the usefulness of ISEs for identifying coupled biological and physicochemical processes in complex systems.

Another series of intermediate-scale experiments were used to study the transport a nonadhesive *P. fluorescens* strain relative to a conservative tracer during flow in a highly heterogeneous, porous medium (Silliman, Dunlap, Fletcher and Schneegurt, submitted). ISEs (1.6 m × 0.67 m × 10 cm) were packed with a complex mixture of ten sands exhibiting different permeabilities and surface mineralogies. Separate bench-top column tests on homogeneous samples of each sand showed significantly greater transport potential in the coarse (high-permeability) sands compared to the fine sands. The sands were packed to create a correlated random field, and flow was driven by constant head from the inflow (where the tracers were introduced) to an outflow well 1.2 m from the inflow boundary. Results showed that the initial breakthrough of the bacteria coincided with the arrival of the conservative chemical tracer, and this early breakthrough was followed by late-time tailing. Initial modeling of results using a kinetic attachment–detachment model indicated that the fundamental shape and timing of the bacterial breakthrough curves are reproducible. Continuing analysis is focused on whether the bacterial breakthrough curves observed are amenable to analysis with a homogeneous model representing large-scale, averaged behavior.

4.3 Field Studies of Bacterial Transport

Scientists have been aware of the transport of microorganisms and viruses in the environment for over a century. Many of the early observations of bacterial transport were in response to human epidemics and resulted from leaking septic or sewage treatment systems. Groundwater transport of bacterial pathogens from a contaminated public well was responsible for a cholera epidemic in Central London in 1854 (Price, 1996). Davis et al. (1985) reported that some of the first dye tracer experiments were used to establish the water origin of typhoid fever in France in 1882. Cadwell and Parr (1937) used coliform bacteria for examining the potential transport of pathogens from a latrine. These early studies established the ability of bacteria to be transported in the natural environment, but it has been only in the last decade that several field investigations have attempted to identify the actual processes controlling bacterial transport. There have been many excellent reviews of field bacterial transport experiments, which we will not attempt to duplicate here (Gerba, 1984; Harvey, 1997; Keswick et al., 1992).

Selected field studies of colloid transport are listed in Table 2.1. This list is not meant to be inclusive of all field studies of bacterial or virus transport, but rather we focused on studies where an attempt was made to identify a specific transport process. Exclusion and attachment (or attenuation) have been the primary processes identified in field studies. Size exclusion (see Section 2.2) is the phenomenon of transported particles moving faster than the average pore-water velocity as indicated by the breakthrough of an inert molecular-scale tracer. Conventional transport theory

TABLE 2.1 Transport Processes Indicated by Field Studies of Colloid Transport

Transport Process	Distance Traveled	Forced or Natural Gradient	Media	Microorganism	Tracer	Citation
Exclusion	1.5–7.6 m	Forced	Sand and gravel	Yeast	20,000 ppm Br^-	Wood and Ehrlich (1978)
Exclusion	12.7–29.8 m	Both	Fractured rock	E. coli	$^{82}Br^-$, NO_3^-, Na-fluorescein	Champ and Schroeter (1988)
Exclusion	1.7–3.2 m	Forced	Cape Cod interbedded lenses of sand and gravel	DAPI-bacteria and microspheres	1525 ppm Br^-	Harvey et al. (1989)
Attachment	6.9 m	Natural	Cape Cod interbedded lenses of sand and gravel	Microspheres only	1425 ppm Cl^-	Harvey and Garabedian (1991)
Attachment	7 m	Natural	Cape Cod interbedded lenses of sand and gravel	DAPI-bacteria	150 ppm Br^-	
Attachment	100 m	Natural	Sand and gravel[a]	PX174 coliphage	52,000 ppm uranine	Oetzel et al. (1991)
Attachment/ adhesion	25 m	Natural		Serratra marcescens		
Attachment/ detachment	6 m	Natural	Cape Cod interbedded lenses of sand and gravel	DAPI-bacteria and microspheres	154 ppm Br^-	Harvey et al. (1993)
Exclusion	4 m	Forced	Fractured, clay-rich till	Bacteriophages	Br^- concentration not reported	McKay et al. (1993)
Attachment/ detachment (no exclusion)	6–12 m	Natural	Cape Cod; sand with some gravel	Bacteriophage, bacteria, microspheres	208 ppm Br^-	Bales et al. (1995)
Attachment/ detachment	3.6 m	Natural[b]	Cape Cod interbedded lenses of sand and gravel	^{32}P-bacteriophage	150 ppm Br^- or Cl^-	Pieper et al. (1997)

[a] Type of porous media not specified but likely sand and gravel from reported K_{sat} = 1.4 to 2 × 10⁻³ m/sec.

[b] Authors used an injection withdrawal system that created a velocity of 0.4–1.3 m/day, very similar to the natural velocity of 0.3–1.0 m/day.

assumes that molecular-scale solutes represent the full distribution of velocities. Bacteria and large colloids, by virtue of their size, preferentially experience the higher velocities near pore centerlines, yielding an average velocity that is higher than that of a dissolved tracer.

Evaluation of these data (Table 2.1) indicates that the process of exclusion dominates when (1) a forced gradient was applied and/or (2) the test was conducted in a fractured media system (Champ and Schroeter, 1988; Harvey et al., 1989; McKay et al., 1993; Wood and Ehrlich, 1978). It is not clear why the higher velocities under the forced gradient would accentuate an exclusion process. One might expect exclusion to be augmented at low flow velocities where the streamlines maximize the potential for the dissolved tracer to sample the smallest pores. Although flow theory presumes that the streamlines do not change under a forced gradient, it is also possible that the pumping is creating preferential flowpaths in the bulk porous media. Exclusion may occur along these preferential flowpaths. Clearly, careful experimentation is needed to address these issues.

Processes of attachment–detachment dominate in field studies that were conducted under natural (or near-natural) gradient conditions and in porous media groundwater systems. In many of these studies, only a fraction of the starting concentration of bacteria is recovered in the field experiment ($< 0.1\%$; Harvey et al., 1989, 1993; Harvey and Garabedian, 1991). In some field experiments, scientists have taken advantage of contrasting properties of tracers and colloids to investigate relative transport and infer processes. In a study near Freiburg, Germany, Oetzel et al. (1991) injected actively growing cells of *Serratia marcescens* and a metabolically inactive coliophage, ΦX174. The authors concluded that the greater attenuation of *Serratia marcescens* was due to a combination of passive attachment and active adhesion resulting from synthesis of extracellular materials that bind to mineral surfaces.

In most field experiments, the amount of biomass attenuated is calculated relative to the recovery of a conservative tracer. This assumption is reasonable because it is extremely difficult to control all the variables that affect bacterial transport in field experiments. Caution should be applied to interpreting bacterial transport processes relative to a conservative tracer, however, when the tracer is applied in high concentrations that create a density effect. Density effects have been demonstrated at anion concentrations as low as $250\,ppm$ (Istok and Humphrey, 1995; Murphy et al., 1997). Whether or not density affects the flow and transport properties is a function of the relative density of the tracer versus the background groundwater and the velocity and length of the flowpath over which the tracer is transported. Generally, a conservative (e.g., nonreacting) anion is mixed with background groundwater to create a tracer solution for use in the field. Most investigators have recognized the potential problems associated with high tracer concentrations and have significantly reduced the concentrations of anions used in field experiments (Table 2.1).

The purpose of using a conservative tracer in field tests is to obtain a benchmark to assess the behavior of the colloid or bacteria. Therefore, all transport processes such as exclusion and attachment–detachment are determined relative to the idealized behavior of the conservative tracer. When density affects the flow and transport of the conservative tracer, we no longer have the benchmark of idealized flow and transport, and the transport properties of the bacteria derived from the breakthrough curve of the bacteria are erroneous. Continuum models that are based on the convective dispersion equation (CDE) are routinely used to derive the velocity and dispersivity from the conservative tracer; however, these models are no longer valid when density is a factor and will underpredict the density effect. This is because there is no intrinsic representation in a continuum model of the pore-scale instabilities that arise due to density or viscosity variations. Thus, if density effects are significant, deterministic models are taxed beyond their theoretical capability and the velocity and dispersivity derived from the conservative tracer are incorrect.

The problem of density effects from the conservative tracer is accentuated in laboratory column experiments. In some cases, investigators do not report the concentration of the conservative tracer, yet use that tracer in a simple one-dimensional model to establish the flow properties against which the attachment–detachment parameters of the bacteria are calculated. When reported, the concentration of the conservative tracer may range from several molar salt to 0.01 M salt, all of which will create significant density effects. Murphy et al. (1997) described this effect in a flow cell where the direction of water flow was vertically upward. As a dense tracer moves upward through the column, gravity forces cause the tracer to sink. If a pulse of tracer has been added, then the effects of sinking will be most evident in the tailing of the breakthrough curve. If the dense tracer is applied vertically downward in a column experiment, then the tracer may likewise sink downward into the lighter groundwater, accentuating the velocity. Finally, if a dense tracer is applied at a field site into a horizontal flowfield, the mass of the tracer may result in sinking of the plume (Fig. 2.4). This behavior may be erroneously interpreted as heterogeneity or higher-conductivity flowpaths in the lower portion of the flowfield. A general rule for tracer concentration when working with dilute

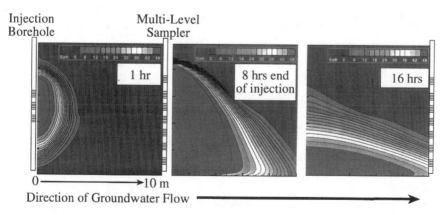

FIGURE 2.4 Cross-section view of a numerical simulation of the density effects on a 0.06-M NaCl ($\sim 4\,g/L$) tracer solution (M. Williams, personal communication).

groundwater is to limit the conservative tracer to no more than 50 mg/L in controlled laboratory experiments and no more than 100 mg/L in field experiments.

2.5 Conclusions

The mechanisms that affect the distribution of microorganisms in subsurface systems are complex and multidisciplinary, spanning the basics of physics, biology, and chemistry. Understanding and coupling these processes with the physical flow and transport of microorganisms in porous media remain a challenge. Bacterial interactions with surfaces are most often experimentally represented as bulk kinetic attachment–detachment rates. These composite rates, however, represent a complex multistep process that may include (1) the approach of a microorganism to the solid surface by advection or motility; (2) intermolecular–surface interactions between the mineral surface and microorganism; (3) bacterial adhesion through the production of extracellular materials; (4) growth or cell division that may result in colonization and/or biofilm formation; and (5) growth or cell division that may result in the detachment of daughter cells or the production of lyases that may facilitate detachment. The complexity of each of these steps and the dynamic nature of bacteria in the subsurface environment require advances in both experimental and theoretical aspects of transport, attachment–detachment to surfaces, and colonization and survival in oligotrophic environments.

Experiments on bacterial interactions with surfaces have been conducted at several scales, from micron-scale interactions of bacteria with a specific surface, to one-dimensional homogeneous column studies, to intermediate-scale experiments with physically and chemically heterogeneous porous media, to, finally, field-scale experiments. Although the field-scale experiments afford the least amount of control of variables, these studies often provide insight into mechanisms and processes that are not evident in laboratory investigations. Our understanding of the specific processes controlling the distribution of microorganisms in the subsurface will lead to advancements in the theoretical representation of these processes in reactive transport models. An iterative, multidisciplinary approach between experimentation and theoretical representation of processes will ultimately provide a greater understanding of the distribution of microorganisms in the subsurface.

ACKNOWLEDGMENT

Our recent work reported here was supported by the U.S. Department of Energy.

REFERENCES

Absolom DR, Lamberti FV, Policova Z, Zingg W, van Oss CJ, and Neumann AW (1983) Surface thermodynamics of bacterial adhesion. *Appl Environ Microbiol 46*:90–97.

Allison DG and Sutherland IW (1987) The role of exopolysaccharides in adhesion of freshwater bacteria. *J Gen Microbiol 133*:1319–1327.

Arredondo R, García A, and Jerez CA (1994) Partial removal of lipopolysaccharide from *Thiobacillus ferrooxidans* affects its adhesion to solids. *Appl Environ Microbiol 60*:2846–2851.

Baker JH (1984) Factors affecting the bacterial colonization of various surfaces in a river. *Can J Microbiol 30*:511–515.

Bales RC, Li S, Maguire KM, Yahya MT, Gerba CP, and Harvey RW (1995) Virus and bacteria transport in a sandy aquifer, Cape Cod, Massachusetts. *Ground Water 33*:653–661.

Balkwill DL, Murphy EM, Fair DM, Ringelberg DB, and White DC (1998) Microbial communities in high and low recharge environments: Implications for microbial transport in the vadose zone. *Microbial Ecol 35*:156–171.

Barton JW and Ford RM (1995) Determination of effective transport coefficients for bacterial migration in sand columns. *Appl Environ Microbiol 61*:3329–3335.

Baygents JC, Glynn JR Jr, Albinger O, Biesemeyer BK, Ogden KL, and Arnold RG (1998) Variation of surface charge density in monoclonal bacterial populations: Implications for transport through porous media. *Environ Sci Technol 32*:1596–1603.

Belas R (1996) Sensing, response, and adaptation to surfaces: Swarmer cell differentiation and behavior. In Fletcher M (ed): *Bacterial Adhesion: Molecular and Ecological Diversity.* New York: John Wiley & Sons, pp 281–331.

Belas R, Simon M, and Silverman M (1986) Regulation of lateral flagella gene transcription in *Vibrio parahaemolyticus. J Bacteriol 167*:210–218.

Bellon-Fontaine M-N, Mozes N, van der Mei HC, Sjollema J, Cerf O, Rouxhet PG, and Busscher HJ (1990) A comparison of thermodynamic approaches to predict the adhesion of dairy microorganisms to solid substrata. *Cell Biophys 17*:93–106.

Bonet R, Simon-Pujol MD, and Congregado F (1993) Effects of nutrients on exopolysaccharide production and surface properties of *Aeromonas salmonidica. Appl Environ Microbiol 59*:2437–2441.

Boyle CD and Reade AE (1983) Characterization of two extracellular polysaccharides from marine bacteria. *Appl Environ Microbiol 46*:392–399.

Bradley PM and Chapelle FH (1995) Factors affecting microbial 2,4,6-trinitrotoluene mineralization in contaminated soil. *Environ Sci Technol 29*:802–806.

Breitenbeck GA, Yang H, and Dunigan EP (1988) Water-facilitated dispersal of inoculant *Bradyrhizobium japonicum* in soils. *Biol Fertil Soils 7*:58–62.

Brözel VS, Strydom GM, and Cloete TE (1995) A method for the study of *de novo* protein synthesis in *Pseudomonas aeruginosa* after attachment. *Biofouling 8*:195–201.

Bruch JC (1970) Two-dimensional dispersion experiments in a porous medium. *Water Resources Res 6*(3):791–800.

Bruns-Nagel D, Drzyzga O, Steinbach K, Schmidt TC, Low EV, Gorontzy T, Blotvogel K-H, and Gemsa D (1998) Anaerobic/aerobic composting of 2, 4, 6-trinitrotoluene-contaminated soil in a reactor system. *Environ Sci Technol 32*:1676–1679.

Busscher HJ, Weerkamp AH, van der Mei H, van Pelt AWJ, de Jong HP, and Arends J (1984) Measurement of the surface free energy of bacterial cell surfaces and its relevance for adhesion. *Appl Environ Microbiol 48*:980–983.

Caldwell EL and Parr LW (1937) Ground water pollution and the bored-hole latrine. *J Infect Dis 61*:148–183.

Camper AK, Hayes JT, Sturman PJ, Jones WL, and Cunningham AB (1993) Effects of motility and adsorption rate coefficient on transport of bacteria through saturated porous media. *Appl Environ Microbiol 59*:3455–3462.

Champ DR and Schroeter J (1988) Bacterial transport in fractured rock: A field-scale tracer test at the Chalk River Nuclear Laboratories. *Water Sci Technol 20*(11/12):81–87.

Cuperus PL, van der Mei HC, Reid G, Bruce AW, Khoury AE, Rouxhet PG, and Busscher HJ (1992) The effects of serial passaging of lactobacilli in liquid medium on their physico-chemical and structural surface characteristics. *Cells and Materials 2*:271–280.

Dagostino L, Goodman AE, and Marshall KC (1991) Physiological responses induced in bacteria adhering to surfaces. *Biofouling 4*:113–119.

Dalton HM, Poulsen LK, Halasz P, Angles ML, Goodman AE, and Marshall KC (1994) Substratum-induced morphological changes in a marine bacterium and their relevance to biofilm structure. *J Bacteriol 176*:6900–6906.

Danielsson A, Norkrans B, and Björnsson A (1977) On bacterial adhesion—the effect of certain enzymes on adhered cells of a marine *Pseudomonas* sp. *Botan Marine 20*:13–17.

Davies DG and Geesey GG (1995) Regulation of the alginate biosynthesis gene *alg*C in *Pseudomonas aeruginosa* during biofilm development in continuous culture. *Appl Environ Microbiol 61*:860–867.

Davies DG, Parsek MR, Pearson JP, Iglewski BH, Costerton JW, and Greenberg EP (1998) The involvement of cell-to-cell signals in the development of a bacterial biofilm. *Science 280*:295–298.

Davis SN, Campbell DJ, Bentley HW, and Flynn TJ (1985) *Ground Water Tracers*. National Water Well Association, Robert S. Kerr Environmental Research Laboratory, Ada, Oklahoma.

de Bruin Wo, Kotterman MJJ, Posthumus MA, Schraa G, and Zehnder AJB (1992) Complete biological reductive transformation of tetrachloroethene to ethane. *Appl Environ Microbiol 58*:1996–2000.

DeFlaun M, Tanzer AS, McAteer AL, Marshall B, and Levy S (1990) Development of an adhesion assay and characterization of an adhesion-deficient mutant of *Pseudomonas fluorescens*. *Appl Environ Microbiol 56*:112–119.

DeFlaun MF, Marshall BM, Kulle E-P, and Levy SB (1994) Tn5 insertion mutants of *Pseudomonas fluorescens* defective in adhesion to soil and seeds. *Appl Environ Microbiol 60*:2637–2642.

DeFlaun MF, Oppenheimer SR, Streger S, Conee CW, and Fletcher M (1999) Alterations in adhesion, transport, and membrane characteristics in an adhesion-deficient pseudomonad. *Appl Environ Microbiol 65*:759–765.

de Marsily G (1986) *Quantitative Hydrogeology*. Academic Press, San Diego, California.

Derjaguin BV and Landau L (1941) Theory of stability of strongly charged lyophobic sols and of the adhesion of strongly charged particles in solutions of electrolytes. *Acta Physicochim URSS 14*:633–662.

DiStefano TD, Gossett JM, and Zinder SH (1991) Reductive dechlorination of high concentrations of tetrachloroethene to ethene by an anaerobic enrichment culture in the absence of methanogenesis. *Appl Environ Microbiol 57*:2287–2292.

El Fantroussi S, Naveau H, and Agathos SN (1998) Anaerobic dechlorinating bacteria. *Biotechnol Prog 14*:167–188.

Ellwood DC, Hunter JR, and Longyear VMC (1974) Growth of *Streptococcus mutans* in a chemostat. *Arch Oral Biol 19*:659–664.

Essaid HI, Bekins BA, Godsy EM, Warren E, Baedecker MJ, and Cozzarelli IM (1995) Simulation of aerobic and anaerobic biodegradation processes at a crude oil spill site. *Water Resources Res 31*(12):3309–3327.

Fletcher M (1976) The effects of proteins on bacterial attachment to polystyrene. *J Gen Microbiol 94*:400–404.

Fletcher M (1980) The attachment of bacteria to surfaces in aquatic environment. In Ellwood DC, Melling J, and Rutter PR (eds): *Adhesion of Microorganisms to Surfaces*. Academic Press, London, pp 87–108.

Fletcher M (1983) The effects of methanol, ethanol, propanol and butanol on bacterial attachment to surfaces. *J Gen Microbiol 129*:633-641.

Fletcher M (1996) Bacterial attachment in aquatic environments: A diversity of surfaces and adhesion strategies. In Fletcher M (ed): *Bacterial Adhesion: Molecular and Ecological Diversity*. New York: John Wiley & Sons, pp 1–24.

Fletcher M and Marshall KC (1982) Bubble contact angle method for evaluating substratum interfacial characteristics and its relevance to bacterial attachment. *Appl Environ Microbiol 44*:184–192.

Fontes DE, Mills AL, Hornberger GM, and Sherman JS (1991) Physical and chemical factors influencing transport of microorganisms in porous media. *Appl Environ Microbiol 57*:2473–2481.

Freeze RA and Cherry JA (1979) *Groundwater*. Upper Saddle River, NJ: Prentice-Hall.

Gannon JT, Manilal VB, and Alexander M (1991a) Relationship between cell surface properties and transport of bacteria through soil. *Appl Environ Microbiol 57*:190–193.

Gannon JT, Tan Y, Baveye P, and Alexander M (1991b) Effect of sodium chloride on transport of bacteria in a saturated aquifer material. *Appl Environ Microbiol 57*:2497–2501.

Gerba CP (1984) Microorganisms as groundwater tracers. In Bitton C and Gerba CP (eds): *Groundwater Pollution Microbiology*. New York: John Wiley & Sons, pp 225–233.

Glynn JR Jr, Belongia BM, Arnold RG, Ogden KL, and Baygents JC (1998) Capillary electrophoresis measurement of electrophoretic mobility for colloidal particles of biological interest. *Appl Environ Microbiol 64*:2572–2577.

Greenberg PE (1997) Quorum sensing in Gram negative bacteria. *ASM News 63*:371–377.

Harvey RW (1997) Microorganisms as tracers in groundwater injection and recovery experiments: A review. *FEMS Microbiol Rev 20*:461–472.

Harvey RW and Garabedian SP (1991) Use of colloid filtration theory in modeling movement of bacteria through a contaminated sandy aquifer. *Environ Sci Technol 25*:178–185.

Harvey RW, George LH, Smith RL, and LeBlanc DR (1989) Transport of microspheres and indigenous bacteria through a sandy aquifer: Results of natural- and forced-gradient tracer experiments. *Environ Sci Technol 23*:51–56.

Harvey RW, Kinner NE, MacDonald D, Metge DW, and Bunn A (1993) Role of physical heterogeneity in the interpretation of small-scale laboratory and field observations of bacteria, microbial-sized microsphere, and bromide transport through aquifer sediments. *Water Resources Res 29*:2713–2721.

Hermansson M, Kjelleberg S, Korhonen TK, and Stenström TA (1982) Hydrophobic and electrostatic characterization of surface structures of bacteria and its relationship to adhesion to an air-water interface. *Arch Microbiol 131*:308–312.

Huysman F and Verstraete W (1993) Water-facilitated transport of bacteria in unsaturated soil columns: Influence of cell surface hydrophobicity and soil properties. *Soil Biol Biochem* *25*(1):83–90.

Istok JD and Humphrey MD (1995) Laboratory investigation of buoyancy-induced flow (plume sinking) during two-well tracer tests. *Ground Water 33*(4):597–604.

James AM (1991) Charge properties of microbial cell surfaces. In Mozes N, Handley PS, Busscher HJ, and Rouxhet PG (eds): *Microbial Cell Surface Analysis.* New York: VCH, pp 221–262.

Jansen B and Kohnen W (1995) Prevention of biofilm formation by polymer modification. *J Industrial Microbiol 4*:191–196.

Jenkins MB and Lion LW (1993) Mobile bacteria and transport of polynuclear aromatic hydrocarbons in porous media. *Appl Environ Microbiol 59*:3306–3313.

Jenneman GE, McInerney MJ, Crocker MF, and Knapp RM (1986) Effect of sterilization by dry heat or autoclaving on bacterial penetration through Berea sandstone. *Appl Environ Microbiol 51*:39–43.

Jenneman GE, McInerney MJ, and Knapp RM (1985) Microbial penetration through nutrient-saturated berea sandstone. *Appl Environ Microbiol 50*:383–391.

Keswick BH, Wang D-S, and Gerba CP (1992) The use of microorganisms as ground-water tracers: A review. *Ground Water 120*:142–149.

Knox KW, Hardy LN, Markevics LJ, Evans JD, and Wicken AJ (1985) Comparative studies on the effect of growth conditions on adhesion, hydrophobicity, and extracellular protein profile of *Streptococcus sanguis* G9B. *Infect Immun 50*:545–554.

Korber DR, Lawrence JR, Sutton B, and Caldwell DE (1989) Effect of laminar flow velocity on the kinetics of surface recolonization by mot$^+$ and mot$^-$ *Pseudomonas fluorescens. Microbial Ecol 18*:1–19.

Lawrence JR, Korber DR, and Caldwell DE (1992) Behavioral analysis of *Vibrio parahaemolyticus* variants in high- and low-viscosity microenvironments by use of digital image processing. *J Bacteriol 174*:5732–5739.

Lyngkilde J and Christensen TH (1992) Redox zones of a landfill leachate pollution plume (Vejen, Denmark). *J Contam Hydrol 10*:273–289.

Makin SA and Beveridge TJ (1996) The influence of A-band and B-band lipopolysaccharide on the surface characteristics and adhesion of *Pseudomonas aeruginosa* to surfaces. *Microbiology 142*:299–307.

Marshall KC, Stout R, and Mitchell R (1971) Mechanism of the initial events in the sorption of marine bacteria to surfaces. *J Gen Microbiol 68*:337–348.

Marshall PA, Loeb GI, Cowan MM, and Fletcher M (1989) Response of microbial adhesives and biofilm matrix polymers to chemical treatments as determined by interference reflection microscopy and light section microscopy. *Appl Environ Microbiol 55*:2827–2831.

McDowell-Boyer LM, Hunt JR, and Sitar N (1986) Particle transport through porous media. *Water Resources Res 22*:1901–1921.

McEldowney S and Fletcher M (1986) Effect of growth conditions and surface characteristics of aquatic bacteria on their attachment to solid surfaces. *J Gen Microbiol 132*:513–523.

McKay LD, Cherry JA, Bales RC, Yahya MT, and Gerba CP (1993) A field example of bacteriophage as tracers of fracture flow. *Environ Sci Technol 27*(6):1075–1079.

Meier-Schneiders M, Busch C, and Diekert G (1993) The attachment of bacterial cells to surfaces under anaerobic conditions. *Appl Environ Microbiol 38*:667–673.

Mills AL, Herman JS, Hornberger GM, and DeJesús TH (1994) Effect of solution ionic strength and iron coatings on mineral grains on the sorption of bacterial cells to quartz sand. *Appl Environ Microbiol 60*:3300–3306.

Molin G, Nilsson I, and Stenson-Holst L (1982) Biofilm build-up of *Pseudomonas putida* in a chemostat at different dilution rates. *Eur J Appl Microbiol Biotechnol 15*:218–222.

Murphy EM, Ginn TR, Chilakapati A, Resch CT, Phillips JL, Wietsma TW, and Spadoni CM (1997) The influence of physical heterogeneity on microbial degradation and distribution in porous media. *Water Resources Res 33*(5):1087–1103.

Murphy EM, Schramke JA, Fredrickson JK, Bledsoe HW, Francis AJ, Sklarew DS, and Linehan JC (1992) The influence of microbial activity and sedimentary organic carbon on the isotope geochemistry of the Middendorf aquifer. *Water Resources Res 28*(3):723–740.

Natsch A, Keel C, Troxler J, Zala M, Albertini NV, and Defago G (1996) Importance of preferential flow and soil management in vertical transport of a biocontrol strain of *Pseudomonas fluorescens* in structured field soil. *Appl Environ Microbiol 62*(1):33–40.

Neumann AW, Good RJ, Hope CJ, and Sejpal M (1974) An equation of state approach to determine surface tensions of low energy solids from contact angles. *J Colloid Interface Sci 49*:291–304.

Norkrans B (1980) Surface microlayers in aquatic environments. *Adv Microbial Ecol 4*:51–81.

Oetzel S, Kass W, Hahn T, Reichert B, and Botzenhart K (1991) Field experiments with microbiological tracers in a pore aquifer. *Water Sci Tech 24*:305–308.

Paul JH and Jeffrey WH (1985) Evidence for separate adhesion mechanisms for hydrophilic and hydrophobic surfaces in *Vibrio proteolytica*. *Appl Environ Microbiol 50*:431–437.

Phelps TJ, Murphy EM, Pfiffner SM, and White DC (1994) Comparison between geochemical and biological estimates of subsurface microbial activities. *Microbial Ecol 28*:335–349.

Pieper AP, Ryan JN, Harvey RW, Amy GL, Illangasekare TH, and Metge DW (1997) Transport and recovery of bacteriophage PRD1 in a sand and gravel aquifer: Effect of sewage-derived organic matter. *Environ Sci Technol 31*:1163–1170.

Pitt WG, McBride MO, Barton AJ, and Sagers RD (1993) Air-water interface displaces adsorbed bacteria. *Biomaterials 14*:605–608.

Powell MS and Slater NKH (1982) Removal rates of bacterial cells from glass surfaces by fluid shear. *Biotech Bioeng 24*:2527–2537.

Powelson DK, Simpson JR, and Gerba CP (1990) Virus transport and survival in saturated and unsaturated flow through soil columns. *J Environ Qual 19*:396–401.

Price M (1996) *Introducing Groundwater*. New York: Chapman and Hall.

Prigogine I and Stengers I (1984) *Order Out of Chaos*. New York: Bantam.

Pringle JH and Fletcher M (1986) Influence of substratum hydration and adsorbed macromolecules on bacterial attachment to surfaces. *J Appl Environ Microbiol 6*:1321–1325.

Ragout A, Sineriz F, Kaul R, Guoqiang D, and Mattiasson B (1996) Selection of an adhesive phenotype of *Streptococcus salivarius* subsp. *thermophilus* for use in fixed-bed reactors. *Appl Environ Microbiol 46*:126–131.

Read RR and Costerton JW (1987) Purification and characterization of adhesive exopolysaccharides from *Pseudomonas putida* and *Pseudomonas fluorescens*. *Can J Microbiol 33*:1080–1090.

Reynolds PJ, Sharma P, Jenneman GE, and McInerney MJ (1989) Mechanisms of microbial movement in subsurface materials. *Appl Environ Microbiol 55*:2280–2286.

Rijnaarts HHM, Norde W, Bouwer EJ, Lyklema J, and Zehnder AJB (1993) Bacterial adhesion under static and dynamic conditions. *Appl Environ Microbiol 59*:3255–3265.

Roberson EB and Firestone MK (1992) Relationship between desiccation and exopolysaccharide production in a soil *Pseudomonas* sp. *Appl Environ Microbiol 52*:1284–1291.

Rosenberg M and Rosenberg E (1985) Bacterial adherence at the hydrocarbon-water interface. *Oil Petrochem Poll 2*:155–162.

Rosenberg M and Kjelleberg S (1986) Hydrophobic interactions: Role in bacterial adhesion. *Adv Microbial Ecol 9*:353–393.

Schafer A, Harms H, and Zehnder AJB (1998) Bacterial accumulation at the air-water interface. *Environ Sci Technol 32*(23):3704–3712.

Shamir UY and Harleman DRF (1967) Dispersion in layered porous media. *J Hydraul Div ASCE 93*(HY5):238–260.

Sharma PK, McInerney MJ, and Knapp RM (1993) *In situ* growth and activity and modes of penetration of *Escherichia coli* in unconsolidated porous materials. *Appl Environ Microbiol 59*:3686–3694.

Shea C, Nunley JW, Williamson JC, and Smith-Somerville HE (1991) Comparison of the adhesion properties of *Deleya marina* and the exopolysaccharide-defective mutant strain DMR. *Appl Environ Microbiol 57*:3107–3113.

Shonnard DR, Taylor RT, Hanna ML, Boro CO, and Duba AG (1994) Injection-attachment of *Methylosinus trichosporium* OB3b in a two-dimensional miniature sand-filled aquifer simulator. *Water Resources Res 30*:25–36.

Sieburth JM (1976) Dissolved organic matter and heterotrophic microneuston in the surface microlayers of the North Atlantic. *Science 194*:1415–1418.

Silliman SE, Konikow LF, and Voss CI (1987) Laboratory investigation of longitudinal disperison in anisotropic porous media. *Water Resources Res 23*(11):2145–2151.

Silliman SE and Simpson ES (1987) Laboratory evidence of the scale effect in dispersion of solutes in porous media. *Water Resources Res 23*:1667–1673.

Smith MS, Thomas GW, White RE, and Ritonga D (1985) Transport of *Escherichia coli* through intact and disturbed soil columns. *J Environ Qual 14*:87–91.

Starr RC, Gillham RW, and Sudicky EA (1985) Experimental investigation of solute transport in stratified porous media, 2. The reactive case. *Water Resources Res 21*:1043–1050.

Stenström TA (1989) Bacterial hydrophobicity, an overall parameter for the measurement of adhesion potential to soil particles. *Appl Environ Microbiol 55*:142–147.

Sudicky EA, Gillham RW, and Frind EO (1985) Experimental investigation of solute transport in stratified porous media, 1. The nonreactive case. *Water Resources Res 21*:1035–1041.

Sutherland IW (1983) Microbial exopolysaccharides—their role in microbial adhesion in aqueous systems. *CRC Crit Rev Microbiol 10*:173–201.

Szecsody JE, Brockman FJ, Wood BD, Streile GP, and Truex MJ (1994) Transport and biodegradation of quinoline in horizontally stratified porous media. *J Contam Hydrol 15*:277–304.

Taylor GT, Zheng D, Lee M, Troy PJ, Gyananath G, and Sharma SK (1997) Influence of surface properties on accumulation of conditioning films and marine bacteria on substrata exposed to oligotrophic waters. *Biofouling 11*:31–57.

van der Mei HC, Rosenberg M, and Busscher JH (1991) Assessment of microbial cell surface hydrophobicity. In Mozes N, Handley PS, Busscher JH, and Rouxhet PG (eds): *Microbial Cell Surface Analysis*. New York: VCH, pp 263–288.

van Loosdrecht MCM, Lyklema J, Norde W, Schraa G, and Zehnder AJB (1987) Electrophoretic mobility and hydrophobicity as a measure to predict the initial steps on bacterial adhesion. *Appl Environ Microbiol 53*:1898–1901.

van Loosdrecht MCM, Lyklema J, Norde W, and Zehnder AJB (1989) Bacterial adhesion: A physicochemical approach. *Microbial Ecol 17*:11–16.

van Pelt AWJ, Weerkamp AH, Uyen MHWJ, Busscher HJ, de-Jong HP, and Arends J (1985) Adhesion of *Streptococcus sanguis* CH3 to polymers with different surface free energies. *Appl Environ Microbiol 49*:1270–1275.

van Schie PM and Fletcher M (1999) Adhesion of biodegradative anaerobic bacteria to solid surfaces. *Appl Environ Microbiol 65*:5082–5088.

Verwey EJW and Overbeek JTG (1948) *Theory of the Solubility of Lyophobic Colloids.* Amsterdam: Elsevier.

Vesper SJ (1987) Production of pili (fimbriae) by *Pseudomonas fluorescens* and correlation with attachment to corn roots. *Appl Environ Microbiol 53*:1397–1405.

Vroblesky DA and Chapelle FH (1994) Temporal and spatial changes of terminal electron-accepting processes in a petroleum hydrocarbon-contaminated aquifer and the significance for contaminant biodegradation. *Water Resources Res 30*(5):1561–1570.

Wan J and Wilson JL (1994a) Colloid transport in unsaturated porous media. *Water Resources Res 30*(4):857–864.

Wan J and Wilson JL (1994b) Visualization of the role of the gas-water interface on the fate and transport of colloids in porous media. *Water Resources Res 30*(1):11–23.

Wan J, Wilson JL, and Kieft TL (1994) Influence of the gas-water interface on transport of microorganisms through unsaturated porous media. *Appl Environ Microbiol 60*(2): 508–516.

Weerkamp AH, Handley PS, Baars A, and Slot JW (1986) Negative staining and immunoelectron microscopy of adhesion-deficient mutants of *Streptococcus salivarius* reveal that the adhesive protein antigens are separate classes of cell surface fibril. *J Bacteriol 165*:746–755.

Weerkamp AH, Uyen HM, and Busscher HJ (1988) Effect of zeta potential and surface energy on bacterial adhesion to uncoated and saliva-coated human enamel and dentin. *J Dental Res 67*:1483–1487.

Weiss TH, Mills AL, Hornberger GM, and Herman J (1995) Effect of bacterial cell shape on transport of bacteria in porous media. *Environ Sci Technol 29*:1737–1740.

Wiencek KM and Fletcher M (1995) Bacterial adhesion to hydroxyl- and methyl-terminated alkanethiol self-assembled monolayers. *J Bacteriol 177*:1959–1966.

Wiencek KM and Fletcher M (1997) Effects of substratum wettability and molecular topography on the initial adhesion of bacteria to chemically defined substrata. *Biofouling 11*:293–311.

Williams V and Fletcher M (1996) *Pseudomonas fluorescens* adhesion and transport through porous media are affected by lipopolysaccharide composition. *Appl Environ Microbiol 62*:100–104.

Wood BD, Dawson CN, Szecsody JE, and Streile GP (1994) Modeling contaminant transport and biodegradation in a layered porous media system. *Water Resources Res 30*(6):1833–1845.

Wood WW and Ehrlich GG (1978) Use of baker's yeast to trace microbial movement in ground water. *Ground Water 16*:398–403.

Wrangstadh M, Szewzyk U, Östling J, and Kjelleberg S (1990) Starvation-specific formation of a peripheral exopolyssacharide by a marine *Pseudomonas* sp., strain S9. *Appl Environ Microbiol 56*:2065–2072.

ENVIRONMENTS AND MICROORGANISMS

3

CONSTRAINTS ON THE DISTRIBUTION OF MICROORGANISMS IN SUBSURFACE ENVIRONMENTS

FREDERICK S. COLWELL

Biotechnology Department, Idaho National Engineering and Environmental Laboratory, Idaho Falls, Idaho

1 INTRODUCTION

The earliest evidence for the presence of microbial communities in the terrestrial subsurface was obtained in the early part of this century and consisted of microbiological data acquired by culturing water from deep wells (e.g., Bastin et al., 1926). Among the studies that indicate a deep earth presence of microorganisms, some still depend on the collection of subterranean fluids in order to evaluate the presence of life at depth (Stetter et al., 1993). Because these samples are collected through an established well, the communities may not be completely representative of those indigenous to the habitat. However, a growing number of studies cite evidence for the presence of microbial communities in extreme geological subsurface environments through characterization of cores (Colwell et al., 1997; Fredrickson et al., 1991; Onstott et al., 1998; Parkes et al., 1994; Pedersen, 1993) or materials derived from mines or tunnels (Amy et al., 1992; Moser et al., 1999). Cores from

Subsurface Microgeobiology and Biogeochemistry, Edited by James K. Fredrickson and Madilyn Fletcher. ISBN 0-471-31577-X. Copyright 2001 by Wiley-Liss, Inc.

these newly explored geological strata are collected so that they are not contaminated and so they present an opportunity to assess the *in situ* conditions under which indigenous subsurface microbial communities exist. Also, unlike water samples, these cores allow scientists to consider the factors inherent to the sediments or rocks that limit the existence of microorganisms. Recently, these data have allowed researchers to place boundaries on the limits of the biosphere, based primarily on the perceived limits of life at high temperatures (Onstott et al., 1999; Whitman et al., 1998). Such studies are important for expanding our understanding of the significance of subterranean life with respect to the total living terrestrial biomass and for providing insight into the prospects for life on other planets (Gaidos et al., 1999).

One of the earliest careful microbiological investigations of the subsurface found virtually no limits to the presence of microbial life within samples that were acquired at depths to several hundred meters (Balkwill, 1989; Balkwill et al., 1989; Fredrickson et al., 1991). At the time, this seemed to be a confirmation in a newly discovered microbial habitat of Beijerinck's principle: "Everything is everywhere; the environment selects" (Atlas and Bartha, 1993). With ten additional years of experience, it would seem that this early deep sampling location may not have been conducted in a sufficiently extreme environment and that the most extreme habitats would, in fact, yield no microorganisms. A corollary statement for those zones in the earth where microorganisms cannot penetrate might be "don't go there." For example, although there is some debate about the depth of the biosphere prompted by claims of microorganisms capable of survival at 150–300°C (Baross and Deming, 1983; Gold, 1992), it is generally considered unlikely that life persists at these extremes. However, the notions held by microbiologists of the thermal extremes that prohibit life have been successfully challenged over the last two decades. It would be premature to consider the current consensus that 120°C is the thermal limit to life as the final word.

The environmental attributes of a given subsurface medium that constrain life can be divided into categories according to the type of constraint that they exert on microorganisms. These may be physical, chemical, or temporal attributes of the environment of interest. Temperature, pressure, water availability, radiation, and available space are the primary physical conditions that determine distribution of microorganisms in subsurface environments. The chemical constraints of an environment are summarized in the context of the flux and concentrations of required electron donors, terminal electron acceptors, and micronutrients, as well as the presence of toxic substances. The effects of these physical and chemical conditions are determined by the time intervals over which these factors are exerted on a cell. Thus, in reference to microbial survival in the subsurface, temporal attributes reflect the degree of stability of a habitat and thus the length of exposure to the particular stressors of that habitat. Of course, the physiology and the adaptive mechanisms of individual cells that find their way into the subsurface play important roles in determining their distribution and survival.

An understanding of how these phenomena exert an influence on microbial survival in the subsurface is of practical importance. The success of *in situ* bioremediation processes generally hinges on the effectiveness of microorganisms whose physiological activity is required to accomplish the desired cleanup task. Currently, many of these engineered solutions to underground contamination are conducted at shallow depths or in locations where the geological factors are not expected to pose significant constraints on the viable microbial community. However, there are exceptions. The minimal pore space requirements of microorganisms living in subsurface media may exclude them from locations where waste has been sequestered. Under these conditions, if direct contact between the active cells and the waste is required for degradation, then the rate of cleanup may be seriously impaired. Long-term storage of nuclear waste also requires some understanding of the limits of microbial survival. Several investigations have considered microbial resistance to the extreme environmental conditions that are imposed by such deep waste repositories (Stroes-Gascoyne and West, 1997; White and Ringelberg, 1996). By understanding the constraints on microbial survival in the subsurface, important insight can also be gained into the distribution, quantity, and regeneration rates of microbially derived products such as the methane that exists in many of the natural gas hydrates in subsurface marine sediments and represents a potential fuel resource (Ginsburg and Soloviev, 1998; Kvenvolden, 1999). Finally, as scientists consider the possibility of life on other planets, the limits of life on our own planet should be studied more thoroughly. Life on many other celestial bodies may only be possible in the subsurface, and therefore the phenomena associated with microbial survival in the earth's subsurface constitute important simulations of the conditions on other planets (Stevens, 1997).

This chapter will explore the different environmental factors that are inherent to the deep earth and how these factors are likely to limit microbial life. Although there have been some studies aimed at defining the dimensions of the biosphere based on collecting samples from the deep subsurface where extreme conditions are known to exist, this area of geomicrobiology research is in its infancy. Estimates of subsurface biomass and the ability of microorganisms to survive the brutal conditions of deep strata are qualified by the limited number of studies that have accessed the depths of our planet and therefore the limited types of environments that have been sampled. A further qualification of these studies are the detection limits of our methods for detecting microbial presence and activity and the accompanying scarcity of biomass in deep strata. Anywhere at the extreme physical and chemical boundaries of life, viable cells become rare and a conclusion that cells are absent is always dependent on the methods used to interrogate the sample. Fundamentally, low densities of biomass represent a significant challenge to detection. Currently, both classical methods of microorganism cultivation and molecular methods are incomplete in their ability to survey naturally occurring microbial communities (Amann et al., 1995; Siering, 1998). So, this challenging effort remains a process of discovery guided by our preexisting understanding of microbial subsistence under controlled experiments and our ability to confirm the presence of life in samples that contain living cells at the threshold of detection.

2 TEMPERATURE

Considerable research has been dedicated to understanding the maximum temperatures that allow microbial survival (Lowe et al., 1993). In this arena, many early experiments focused on the microbiology of foods or methods of sterilization (see, e.g. Frazier and Westhoff, 1978). Recently, high-temperature research efforts have been largely driven by interest in new industrial or environmental applications of microorganisms or their enzymes (Blochl et al., 1995; Herbert, 1992). Studies of deep marine hydrothermal vents and hot springs have generated ecological data and the ensuing studies of isolates have improved our comprehension of the physiological and molecular characteristics unique to these microorganisms.

An interesting point of reference for biosynthesis and life at high temperatures comes from thermodynamic calculations of the assembly of amino acids. Such considerations indicate that autotrophic synthesis of amino acids is more favorable under hydrothermal vent conditions of $100°C$ than under cooler seawater conditions (Amend and Shock, 1998). Although this may be the case under these hyperthermophilic conditions, fundamental attributes of microbial survival appear to breakdown at higher temperatures. The generally accepted maximum temperature for growth of microorganisms is somewhat above $110°C$ (Blochl et al., 1995), with microorganisms reported to grow at temperatures up to $113°C$ (Stetter, 1998, 1999). Because the covalent bonds in proteins and nucleic acids are disrupted at $250°C$ and the tertiary structure of many of these biological molecules is altered at even lower temperatures, there is a necessary limit to the temperatures that microorganisms can survive (White, 1984). The endospores of thermophilic bacteria that are especially tolerant to wet heat are only able to survive for minutes at $121°C$ (Brown, 1994). For now, this temperature serves as a rough guide to the upper thermal limit of life, even though microorganisms have now been reported to survive autoclaving (Stetter, 1999). This knowledge suggests that beyond a certain depth in the earth, high temperatures will ultimately prevent life.

Given that high temperature poses a severe constraint on the distribution of microorganisms in the subsurface, the rate of temperature increase in the subsurface (geothermal gradient) as a rough determinant of the maximum depth where microbial life can exist should be considered. Worldwide, the geothermal gradient varies according to the heat flux to the surface and the ability of geological material to conduct heat. In some of the subsurface environments that have been studied, the geothermal gradient varies markedly, providing evidence that the depths to which life can exist, as controlled by temperature, will vary also (Fig. 3.1). One set of end-member habitats having extremely high temperatures with respect to survival of microbial life are geothermal hot spots where life-limiting temperatures rise to the earth's surface. Yellowstone National Park, Iceland, and other thermal regions around the world are examples. At the opposite end of this upper thermal limit boundary are the sediments in the marine trenches that are overlain by $>10,000\,m$ of cold ocean water. In these locations, whatever the geothermal gradient, extreme depths from the earth's surface (as measured from sea level) must be reached in the sediments before life is prevented by high temperature.

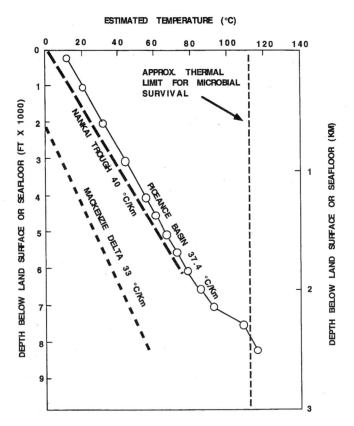

FIGURE 3.1 Depiction of geothermal gradients from the Piceance Basin western Colorado (Spencer, 1990), the Mackenzie River Delta, Northwest Territories (Dallimore and Collett, 1999), and the Nankai Trough off the coast of Japan (Tsuji et al., 1998), various deep subsurface environments that have been sampled for microorganisms. The vertical line indicates the approximate upper temperature limit for known microorganisms. Data for the Mackenzie Delta are shown from where the temperature exceeds 0°C below 650 m, the depth of the ice-bonded permafrost. The estimated geothermal gradient (°C/km) for each of the subsurface environments is shown on the plots.

   In polar regions where surface temperatures are low enough to allow permafrost to form, microorganisms encounter thermal conditions that range upward from several degrees below zero. Permafrost is any ground that remains below 0°C for more than 2 years, although many permafrost environments also contain taliks where water is not frozen (Harris, 1986). Permafrost may extend as deep as several hundred meters below land surface. Microorganisms have been found to persist in areas where the earth is permanently frozen, with estimates of microbial survival in the range of millions of years (Vorobyova et al., 1997). In terms of supporting microbial communities, the consistent low temperatures are believed to make permafrost the most stable deep subsurface environment for sustaining life (Vorobyova et al., 1997).

Recent microbial studies conducted as a part of the coring of a gas hydrates research well in the Canadian Arctic suggest that microbial life exists well below zones of ice-bonded permafrost (Colwell et al., 1999). These habitats represent areas where the normal geothermal gradient is suppressed by the cold atmospheric conditions, and so the maximum depth to which life may exist as delimited by temperature can be extended by as much as a kilometer into the earth relative to locations with more moderate surface temperatures.

3 PRESSURE

Recently, interest in the pressure limits of microorganisms has risen along with the study of thermal limits of microorganisms because of potential industrial applications (Kato and Bartlett, 1997). Investigations of microorganisms that live at high pressures have been conducted primarily on samples collected from deep marine waters (Marquis, 1982; Yayanos, 1995) and more recently from deep marine sediments (Horikoshi, 1998; Takami et al., 1997). A general summary of microbial responses to hydrostatic pressure indicates a linear decrease in activity with increasing pressure or depth (Jannasch and Taylor, 1984). However, variations to this trend exist.

Three general classifications of microorganisms with respect to pressure tolerance have been developed. Varying by degree and range of tolerance, barotolerant microbes exhibit a maximal rate of growth at atmospheric pressure but some lower rate of growth at higher hydrostatic pressures. In general, it is believed that most microorganisms are insensitive to hydrostatic pressure in the range of 10.1–60.8 MPa and therefore qualify as barotolerant (Jannasch and Taylor, 1984). On the other hand, barophiles have a maximal rate of growth at some elevated hydrostatic pressure but can grow at atmospheric pressure. Obligate barophiles require elevated hydrostatic pressure to grow and these bacteria may die if they are decompressed (Chastain and Yayanos, 1991). Although most of the microorganisms characterized with respect to pressure are marine microorganisms, one investigation determined that a deep terrestrial subsurface isolate is barophilic, growing faster at 20 MPa than at atmospheric pressure (Delwiche et al., 1996). It is reasonable to assume that all classes of pressure tolerant microorganisms may be found in the deep terrestrial subsurface and this portion of the earth's habitable realm is notably unstudied in this regard (Yayanos, 1986, 1995).

In many studies of pressure-tolerant marine microorganisms, attention has been directed toward the preservation of the *in situ* pressures to minimize the impact of decompression on the microbial communities in the samples (Chastain and Yayanos, 1991; Jannasch et al., 1976; Yayanos, 1995). Maintenance of the *in situ* pressure of a deeply acquired seawater sample during transit to the surface and subsequent maintenance of this pressure regime as the sample is evaluated in the laboratory is a complex, but relatively straightforward, task of fluid exchange under pressure. In contrast, far more complicated techniques and hardware must be used in order to maintain a rock core under continuous pressure. Recovery of obligate barophiles that are exquisitely sensitive to decompression from deep terrestrial hard rock cores therefore will rarely be accomplished and such studies have yet to be performed. The superficial layers of soft marine sediments located beneath considerable oceanic depths (to 11,000 m) and consequently subject to high hydrostatic pressures have been sampled with brief decompression and investigated for microorganisms (Kato et al., 1997, 1998; Takami et al., 1997). These studies indicate the presence of obligate barophiles in the sediment community, along with barotolerant microorganisms. Recently, samples from marine sediments where the water depth is 6,292 m (65 MPa) were obtained at pressure and split, with half of the samples maintained and cultured under *in situ* pressures (not decompressed) and the other half cultured at atmospheric pressure (Yanagibayashi et al., 1999). Barophiles became dominant in the cultures that were not decompressed, whereas barotolerant microorganisms became dominant in the decompressed cultures. This research demonstrates that some of the microbial components of a community from a high-pressure environment likely escape detection if *in situ* pressure is not maintained through sample handling and incubations.

In terrestrial subsurface environments, hydrostatic pressure that is based on the depth of the groundwater column above the location of microbial communities should be considered an important constraint on microbial activity and survival in extremely deep strata. The subsurface depth at which the hydrostatic pressure begins to exert its influence depends on the depth to groundwater in a given system. As noted earlier, microorganisms are generally tolerant to hydrostatic pressures in the range of 10.1–60.8 MPa (Jannasch and Taylor, 1984). Based on a simple consideration of hydrostatic head as the key determinant of pressure, a range of 10–60 MPa corresponds to depths of approximately 1000 to 6000 m if the water table is near the surface. Where the depth to groundwater is deeper, as for example, several hundred meters below ground surface at Rainier Mesa in Nevada (Amy et al., 1992), then greater depths will be achieved before the average microorganism is inhibited by hydrostatic pressure.

To our knowledge, cores from as deep as 6000 m have rarely if ever been collected for microbiological analysis. Based on a mean continental geothermal gradient of 19°C/km (Sclater et al., 1980), and a mean surface temperature of 15°C, 6000 m corresponds to a temperature of 129°C. Although the geothermal gradient varies dramatically depending on local conditions, these general trends suggest that in many locations a thermal boundary is a more significant limit to subsurface microbial life than pressure (Onstott et al., 1999). If microbial life anywhere on the planet is limited by pressure, then it would have to be in the sediments of the deep ocean trenches.

In subsurface systems that are hydraulically open (connected to the surface), the bacteria within these deep materials do not necessarily "feel" the pressure exerted by the mass above them. This is because they exist within the framework of the geological strata in the interstices or attached to the rock itself, and the cells themselves do not bear the lithostatic pressure of the formation. However, in hydrologically closed systems, such as where an aquifer is confined by impermeable zones or within small fluid inclusions, deep earth microorganisms experience a lithostatic pressure gradient. The lithostatic pressure gradient depends on the density of the overburden that generally ranges between 1.7 and 2.7 g/cc (Keys, 1997). Studies that have considered this aspect of the pressure burden for cells caught within the superstructure of the overlying strata have not been conducted.

Because of the difficulty of working with high-pressure gases, few studies have been conducted on microbial tolerance to these conditions. Still, deep subsurface microorganisms must contend with or may require various biogenic and thermogenic gases. If a free gas is present, then within the liquid film where microbial cells exist, the concentration of these gases (e.g., hydrogen, methane, and carbon dioxide) is determined by the concentrations of the gases in the free gas phase (Nelson et al., 1992a, 1992b). Gases that are hydrophobic will partition into hydrophobic portions of cells (e.g., the cell membrane) and have the potential to alter the structure and function of these cells (Thom and Marquis, 1984). Hyperbaric carbon dioxide may become dissolved in pore water and can significantly decrease the pH of a cell's immediate environment. High concentrations of free methane in strata underlying gas hydrates in marine sediments likely translate to high concentrations of dissolved methane in adjacent saturated zones. Thus, hyperbaric methane has the potential to inhibit methanogenic activity in these locations due to thermodynamic constraints (Paull and Ussler, 1999). Furthermore, as will be discussed under transport-related limitations, subsurface strata that are charged with free gas at pressure may represent a boundary to microbial movement in these geological strata.

4   WATER ACTIVITY

Similar to the early studies of temperature tolerance of microorganisms, research in food preservation provided the first insight into the survivability of microorganisms under exceedingly dry conditions (Frazier and Westhoff, 1978). Although microorganisms have an absolute demand for water to sustain their growth, nonvegetative cells are resistant to desiccation and may even survive in such states for millions of years (Cano and Borucki, 1995). Thus, while microbial activity or cell growth may be prevented in dry subsurface locations, viable microorganisms in the form of spores and other resting stages may still occur. Where the levels of moisture in a vadose system become low, microorganisms are stressed in several ways. Such stresses include those that are inside the cell and related to the accumulation of solutes and the low internal water potential as well as those that are outside the cell and occur because of diminished diffusion of essential nutrients and substrates as water becomes more scarce (Stark and Firestone, 1995).

For microorganisms, total water in the environment is less important than the liquid water that is available for biochemical processes (Atlas and Bartha, 1993). For microbial studies, such available water is expressed either as water activity or water potential. Water activity a_w depends on the amount of water (in moles) relative to the amount of solute and the activity coefficients for water and the solute (Frazier and Westhoff, 1978). The a_w for pure water is 1.0, and the osmotic and matric forces of a subsurface vadose environment decrease the a_w to some lower fraction. Whereas most microorganisms are metabolically active only at a_w values greater than 0.96, some fungi can be active at values as low as 0.60 (Atlas and Bartha, 1993).

Water potential is a thermodynamic term related to a_w that establishes the difference in free energy between a soil sample and pure water at the same temperature (Ehrlich, 1990). Water potential is expressed in negative MPa and is the suction that must be applied to a sample to withdraw water from it. Water potential is the measurement most frequently used to describe the water status of soil environments. Microorganisms have been cultured from samples obtained in arid vadose zones where the water potential is low (Kieft et al., 1993). In this study, culturable cells were detected in Miocene-aged vitric tuff of Rainier Mesa in Nevada where the water potentials ranged from -46 to -62 MPa. These water potential values, which are not uncommon in arid climate subsurface environments, are considered low relative to those used in typical soil studies (Stark and Firestone, 1995). It is interesting to note that for some microorganisms, a slow water potential decrease ensures better survival of the population than when the water potential is decreased rapidly in a stepwise fashion (Poirier et al., 1997). Unsaturated subsurface strata that undergo gradual changes in water potential may contain cells that are hardened to survive under low water potentials in contrast to subsurface environments where frequent wetting events cause extreme water potential fluctuations. A thorough treatment of microorganisms in the vadose zone considering the research already conducted in this area is provided elsewhere in this volume (see Chapter 5).

Subsurface microbial communities are surely constrained by the unavailability of water in arid land vadose zones and where solute concentrations become high in porewaters. Among the published subsurface investigations where microorganisms were either never or rarely recovered or detected, arid land vadose zones stand out as being relatively austere habitats (Kieft et al., 1993; Palumbo et al., 1994). Estimates of microbial activity using geochemical modeling indicate that deep vadose zones rank as low as deep consolidated high-temperature rocks, producing as little as 10^{-13} moles of carbon dioxide/L/year (Kieft and Phelps, 1997). These values are as much as 10 orders of magnitude lower than similar determinations for microbial communities in subsurface aquifers and perhaps 14 orders of magnitude lower than communities in surface soils. Through similar modeling approaches, deep arid vadose zone microorganisms are thought to have doubling times that are as long as 3×10^5 years (Onstott et al., 1999).

5   GEOLOGIC CONSTRAINTS

In shallow subsurface environments, the notion that microorganisms can be transported through porous media is not disputed and has evolved into a key area of research. The earliest investigations of groundwater microbiology established that disease-causing bacteria were capable of survival and transport in aquifers, such that these organisms were able to infect new hosts due to mixing of sewage contaminated waters with drinking water (Atlas and Bartha, 1993). Recent studies have characterized the transport of microorganisms in the subsurface with a focus on bioremediation processes (Harvey et al., 1989), microbial enhanced oil recovery via porous media plugging schemes (Cusack et al., 1990; Lappin-Scott et al., 1988), or the use of microbes as tracers (Harvey, 1997). Relative to the deep subsurface, shallow systems are typically readily inoculated with microorganisms from the surface either by natural or artificial means and are more likely to be influenced by human manipulations. Much research is still needed in order to place transport-related constraints on life at greater depths or in more extreme subsurface habitats. Additional experiments that take into account cell size, shape, hydrophobicity, and electrostatic charge, as well as the characteristics of the subsurface strata, solutes, and hydrogeology, need to be conducted to catalog fully the properties that influence microbial movement in the subsurface (Lawrence and Hendry, 1996). Although there is a lack of field-scale investigations of microbial transport in the deep earth where transport limitations may be significant, some speculation is possible based on what is known of these deep strata.

Constraints on microbial colonization of the subsurface due to transport limitations are probably most significant in deep subsurface systems. In general, the transition from shallow to deep systems in the context of available space for microbial transport can be considered a transition from an interconnected porous medium near the surface to one where interconnected porosity, pore throat size, and the number and size of open fractures diminish with depth. Of course, exceptions can occur, most notably where there may be a surface outcrop of low-porosity, low-permeability material exposed due to erosion or uplift. Still, even in these cases, the loss of overburden mass may help to open fractures or microfractures in the geologic strata that allow microbial colonization. In contrast, deep burial of strata represents a long-term process of compaction and the consequent loss of available pore and fracture space where microorganisms can move. Infrequent seismic events may open space at depth, but such openings should not last long under the persistent lithostatic pressure. In addition, the process of compaction acts in concert with processes of mineralization that also narrow the pathways in which microorganisms can be transported (Tissot and Welte, 1983). Ultimately, the space for microbial survival is limited to austere microniches, perhaps fluid inclusions, that resist further compaction yet are isolated from an essential supply of nutrients or sink for waste products.

Approaches that combine microbial analysis of cores or water samples from specific depths with noninvasive geophysical tools to characterize the bulk environment from which the organisms are derived have real potential to advance our understanding of underground life (Ghiorse, 1997). An example of a geophysical tool that can be used to collect data pertaining to porosity is the acoustic-velocity log. Acoustic logs use a transducer to transmit acoustic energy through the fluid in a borehole into the surrounding geologic materials and then back to an acoustic receiver. This tool records the travel time of pulsed acoustic waves through the geologic medium, the velocity of the signal being related to the lithology and porosity of the rocks (Keys, 1997). The compressional-wave velocity varies for different geologic media (e.g., sandstone: 4600–5200 m/sec; granite: 5800–6100 m/sec; freshwater: 1500 m/sec) and is related to porosity and density as well as the size, distribution, and types of grains and pore spaces. An acoustic log obtained from INEL-1, a deep borehole at the Idaho National Engineering and Environmental Laboratory, indicates that with increasing depth in the borehole the velocity of the acoustic signal increases (Fig. 3.2). Overall, the increase in compressional velocity with increasing depth suggests that the signal is traveling through material that is less porous and therefore less likely to provide space for microbial colonization. Interestingly, the high geothermal gradient (ca. 50°C/km) at this location also probably limits microbial survival to relatively shallow zones of the subsurface. The direct inspection of cores and cuttings allows interpretation of the acoustic velocity data through an understanding of the lithology of the particular subsurface environment. Although these acoustic data provide accurate information regarding the trend in porosity reduction, it should be noted that even such a trend has exceptions. With an acoustic televiewer tool (Keys, 1997) in the INEL-1

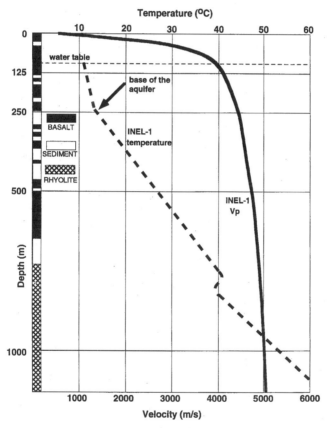

FIGURE 3.2 Temperature and acoustic velocity (Vp) profiles plotted versus depth and stratigraphy in the INEL-1 deep borehole. At the total depth of the borehole (3000 m), the temperature is approximately 140°C. The acoustic velocity, a function of the geologic material through which the signal passes, also generally increases with depth as the porosity of the geologic strata decreases. Figure modified from Smith et al. (1994).

borehole, a 2.3-m-wide fracture was detected at 2272 m (R.P. Smith, personal communication). Intriguingly, the presence of such a fracture where water temperatures exceed 100°C may represent a conduit for transporting hyperthermophiles in the subsurface.

Collection of cuttings from drilling fluids and "mud logs," which may include measurements of gases that exsolve from the circulating drilling fluids, are used by drillers and geologists to understand the subsurface and can also be helpful in interpreting geological strata with reference to the likelihood of microbial transport and colonization. Gas concentrations in the drilling fluids directly reflect the concentrations in the formation at depth. Mud logs provide a depth-oriented documentation of borehole and formation characteristics such as rate of drill bit penetration, interpretive lithology (inferred from the cuttings and the rate of penetration), evidence of oil or coal, and concentration of gases including carbon dioxide, methane, ethane, and propane. Specific events associated with the drilling or coring process are also noted on the mud log at the corresponding time and depth at which the event occurred.

A model of how such logs can aid in understanding constraints on microbial transport in subsurface systems is provided by research conducted in the Piceance Basin of western Colorado. The Piceance Basin is a hydrocarbon-bearing province with natural gas-producing reservoirs that vary in age from Late Jurassic to Eocene (Johnson and Rice, 1990; Lorenz, 1987). As part of a research project aimed at understanding the limits of microbial colonization in subsurface environments, in 1994 the Department of Energy Subsurface Science Program piggybacked on a drilling effort aimed at developing the natural gas resource at several thousand meter depths. Three cores were obtained for microbiological analysis, from depths of 856–862, 1996–1997, and 2091–2096 m. Formation gases in the drilling muds were determined on-site by using gas chromatography with a flame ionization detector (Fig. 3.3). These data suggest that although methane and other gases were present throughout the borehole, the real peak in methane concentrations did not occur until the borehole had exceeded 2000 m. Concentrations of methane below the 2000-m boundary averaged 100-fold higher than those above the boundary and these higher levels of gas indicated that the rock was overpressured. The transition to strata that contained high concentrations of natural gas corresponded to data that indicated the absence of microorganisms in the deepest core. Microorganisms were evident in the two uppermost cores, but the core from 2091–2096 m was deemed to be hydrologically isolated (Colwell et al., 1997). Thus, the apparent limit to microbial colonization occurred over a 100-m section where hyperbaric methane or the rock conditions that trapped it appeared to block microbial transport.

Although it is difficult to sort through all the factors that may influence the ability of microorganisms to colonize a specific deep environment, at least in the case of the Piceance Basin it appears that the stratigraphic transition to exceedingly high concentrations of gas serves as a possible barrier to microbial transport. It seems likely that if a stratum has porosity and permeability characteristics that allow gas entrapment and potential overpressurization, then microbial transport should be severely impeded. Such a barrier to advective microbial migration may occur in

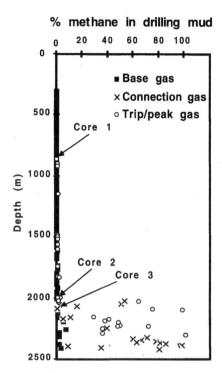

FIGURE 3.3 Methane concentration measured in drilling fluids circulating in the deep borehole 1-M-18 in the Piceance Basin as the borehole increased in depth. Core 1, Core 2, and Core 3 designate depths where cores were obtained for microbiological analysis. "Base gas" represents measurements made continuously on the circulating fluids. "Connection gas" and "trip" or "peak gas" measurements were made when the drilling fluids were temporarily not circulating, therefore allowing a potential bleeding of gas from the surrounding geological medium before fluid circulation was started again. Methane was determined by gas chromatography with a flame ionization detector by Rocky Mountain Geo-Engineering Corp. (Grand Junction, CO).

other geologic systems such as in marine sediments where a bottom-simulating reflector (BSR) has been detected. BSRs are believed to indicate the presence of free gas, often beneath gas hydrate-bearing sediments. If methane is trapped as free gas at these locations, then it seems likely that such a zone would be impermeable to microorganisms as well. This presents the possibility that the only way microorganisms can be incorporated in deeper levels below the free gas zone is by the incremental burial of new sediments and gradual deposition of sediments upon which the microorganisms are attached. Laboratory experiments have demonstrated that when dissolved gases in porous media come out of solution, they may form a relative permeability seal (Benzing and Shook, 1996; Benzing et al., 1996). Such a seal, which is like a "vapor lock" seal, can induce an increase in pressure and prevent the measurable flow of water past the free gas. The exsolution of gas and the formation of the pressure seal do not necessarily conform to bedding planes, but rather are dependent on the isotherm and the tendency of the gas to come out of solution according to Henry's law.

6 CHEMICAL AND ENERGY RESOURCES

The geological formation in which microorganisms reside has the potential to determine their survival because of its origins as well as its recent geological history. The fundamental characteristics of a given formation will determine the types, quantities, and distribution of the sources of energy and nutrients for cells. Different mineralogical and geochemical environments determine the availability of dissolved chemical species, which include electron donors and terminal electron acceptors (Krumholz et al., 1997). The physical attributes of a formation that determine whether it is porous or massive can likewise determine whether microorganisms are able to access the dissolved chemical species that they require for survival. Microorganisms in the subsurface are most likely to survive where there is an ample supply of energy-yielding substrates as well as appropriate electron acceptors. Such conditions occur where organic-rich reduced sediments provide a steady supply of low-molecular-weight organic compounds or reduced inorganic chemicals that provide energy to sustain trapped microorganisms (Fredrickson et al., 1997; Krumholz et al., 1997). In contrast, organic poor, oxidized minerals (e.g., igneous rocks) offer little for microorganisms to thrive or even survive (Palumbo et al., 1994). In either case, in order for microorganisms to survive in an environment for any length of time, they need to receive relatively high-energy electrons and shuttle these electrons to lower-energy states. The time scale over which these processes must occur in order to sustain some form of life is disputed. Indeed, even the question of what actually constitutes microbial viability is debated (Phelps, 1998).

Some of the most intense speculation in this regard is currently occurring in the arena of astrobiology. The subsurface of planets or moons within our solar system is believed to be the location where near-earth extraterrestrial life is most likely to occur and these environments may bear some similarity to the earth's subsurface. Besides requiring water and certain elements, microbial life requires energy for metabolism, growth, maintenance, and reproduction. Our current knowledge about life and its requirements, based on the study of earth-bound ecology, suggests that the fundamental factor determining the presence of life on nonterrestrial planets or moons would be the ability to take advantage of the energy supplied by the available oxidation–reduction reaction couples (Gaidos et al., 1999). The overall tendency toward chemical equilibrium is the threat to survival, and without the external input of energy via the chemical products of sunlight, radioactive decay, or geothermal energy, life will not last long. Although high levels of appropriate reductants and oxidants are available on the earth's surface, the deep biosphere can become limited by the supply of these resources. The origin and history of the geologic formation can determine the flux rates of appropriate redox couples to specific locations and thereby determine whether life can be sustained.

In a different sense, high concentrations of toxic chemical species have the potential to limit the viability of microorganisms. These toxins may result from a situation where limited flux through the system causes a buildup of microbial metabolic byproducts and subsequently inhibits further metabolic activities. Alternatively, high concentrations of external toxins can accumulate in subsurface microbial niches, for example, porphyry intrusive ore deposits, in which past hydrothermal alteration of minerals has been followed by an increased concentration of heavy metals in newly mineralized fractures and veins compared to the surrounding rock matrix (Wilkinson et al., 1994). Although many microorganisms may exhibit significant heavy metal resistance (Kuenen and Tuovinen, 1981), at some point these toxic elements would be lethal to the cells. The fractures through which hydrothermal waters move and where the metals may be deposited are also the most likely location for the transport and deposition of cells.

Perhaps the most extreme conditions exist where heavy metals and radioactivity coexist. These conditions are met in the natural nuclear fission reactor at Oklo, Gabon, Africa where sedimentary strata contain high concentrations of uranium ore (Gauthier-Lafaye et al., 1997). Similarly, the gold in the Carbon Leader of the Witwatersrand Super Group in the gold mines in South Africa coincides with high concentrations of uranium (Moser et al., 1999). Lower microbial diversity as measured by phylogenetic methods exists in the groundwater of the Oklo site than is present in the granitic environment of the Äspö, Sweden site where radioactivity is insignificant (Crozier et al., 1999).

7 FLUX OF CHEMICAL AND ENERGY RESOURCES

Consideration of how environments change geologically and the time scale for such change can determine the survival of the microorganisms trapped therein because change directs the flux of the resources that these communities require. The redistribution of nutrients, microorganisms, or favorable growth conditions is possible when geological events such as hydrothermal activity, fluid and gas cycling, and tectonism are imposed on the microbial communities within subsurface strata. At mid-oceanic ridges and spreading centers, which are perceived to be porous due to the basalts that dominate in the system, high flux rates of water and nutrients move through subsurface strata that are made even more porous by proximal tectonic events (Delaney et al., 1998; Jannasch and Taylor, 1984). These conditions are believed to be responsible for the sudden increases in bacterial numbers that have been observed in marine waters proximal to a hydrothermal eruption. Hydrothermal events are frequently coupled to a high efflux of reduced microbial energy sources (Brock, 1978) and the decrepitation and redistribution of minerals within consolidated rock masses. The rapid alteration and weathering of rocks in the subsurface have the potential to cause changes in both the overall porosity of the system and localized porosities.

Another high-energy subsurface system that has the potential to conduct large quantities of microorganisms as well as the electron donors and acceptors that are required for their activity is the marine sediments that consist of gas hydrates. These are becoming the focus of increased attention because of their potential as a large source of methane for both fuel and contributions to climate change (Kvenvolden, 1999). Thus far, studies of these mineral deposits that contain huge quantities of biogenic and thermogenic methane and whose stability is dependent on high pressure and low temperature are still in the earliest phases of investigation. Methane seeps that are inferred to be associated with gas hydrates, such as the Haakon Mosby mud volcano off the coast of Norway (Vogt et al., 1997) and seeps that may originate from deep sources of natural gas such as on the continental shelf in the Gulf of Mexico (Sasson et al., 1998), represent important newly discovered features that link subsurface environments of high chemical and physical energy with the surface or at least the seafloor. Like the hydrothermal systems at the mid-oceanic ridges, there are often dense communities of macrofauna that exist on the seafloor where the methane is released into the water column (Kuramoto et al., 1998; Naganuma et al., 1999; Sasson et al., 1998; Vogt et al., 1997). Although not yet thoroughly studied, these features may be analogous to the high-temperature vents in that microbial biomass that is supported by the flow of a reductant (in this case, methane) from the subsurface is the food supply for invertebrate species.

Typically, the presence of a marine gas hydrate zone is inferred by the existence of a suitable temperature/pressure regime as well as a BSR in the sediments. BSRs are measured using a seismic signal generated in the overlying water column and indicate the location of free gas bubbles beneath the hydrate stability zone (Dickens et al., 1997; Holbrook et al., 1996). This seismic reflectance line is usually laterally contiguous in the sediments; however, in some notable exceptions there are breaks in the BSR (Wood and Gettrust, 1999). Although considerably more research is needed to determine the nature of these features, these breaks in the BSR may represent areas where methane has escaped from the sediments and the free gas that is characteristic of the BSR no longer exists. Often where the BSR is discontinuous the topography of the seafloor suggests the past existence of a mud volcano-type feature (Wood and Gettrust, 1999). If these seafloor features are linked to breaks in the BSR, then these sites are the location of considerable methane release and probably an associated microbial plume. Large numbers of defunct mud volcanoes and the associated evidence of dead macrofauna are relics of these events (Kuramoto et al., 1998).

In a similar manner, some evidence exists for the large-scale movement of fluids in porous media associated with terrestrial Gulf Coast sediments (Hanor and Sassen, 1990). It is believed that the dissolution of salt diapirs at depth causes density inversions in the pore fluids, which are significant enough to force lateral and vertical movement of these fluids in subsurface sediments. Some of this fluid movement may occur through fractures or faults as opposed to the matrix porosity. Recent evidence suggests that these strata also contain active anaerobic microbial communities (Krumholz et al., 1999).

At the opposite extreme of the high-flux environments that are typified by hydrothermal vents and methane seeps, deep cratonic environments in the center of continental land masses would seem to be a subsurface habitat that is slow to exhibit change. The only significant flow of water in these massifs occurs in fractures and this is where microbial communities of any significance are found. The best reference sites in this regard are the two well-studied underground laboratories in Sweden: Stripa and Äspö (Pedersen, 1993, 1997). Access to these locations via tunnels has allowed information to be gained on the presence and distribution of microorganisms in granite and iron ore. Although large fractions of these subsurface media are likely to be inaccessible to microorganisms due to prohibitively small pore throat sizes, considerable microbial biomass exists within the fractures and faults that conduct water through these geologic features (Pedersen, 1997). Given adequate flux of groundwater through these fractures, it seems likely that the microorganisms within the fractures play key roles in the biogeochemistry of the aquifer. Important microbial processes include the mobilization of iron through the metabolism of iron reducing bacteria, the reduction of sulfate to sulfide by sulfate reducers, and the reduction of carbon dioxide to organic compounds by hydrogen utilizers (Pedersen, 1997).

8 COMBINED FACTORS

Superimposed on all considerations of the abiotic factors that control the survival and access of microorganisms in deep subsurface strata are biological factors inherent to the cells themselves. The types of microorganisms that can exist within and transit through the deep earth may be considerably different than those that successfully colonize the more hospitable strata of the upper earth. Besides requiring the obvious attributes of tolerance to high temperature, high pressure, and low hydraulic flux, these microorganisms may benefit by being metabolically versatile. A survey of the characteristics of thermophiles and hyperthermophiles suggests that they are broadly adapted to extract energy from the simple reduced compounds that are typical of these environments. Many of these microorganisms are anaerobes that can use sulfides, manganese, hydrogen, methane, carbon monoxide, and other inorganic compounds (Lowe et al., 1993). Reports describing microbial enrichments or isolates from extremely deep and hot subsurface locations often suggest that certain physiological traits (e.g., ability to use multiple, simple electron donors and acceptors) may aid microbial persistence in those environments (Colwell et al., 1997; Liu et al., 1997; Stetter et al., 1993). In a specific case, *Bacillus infernus*, a thermophilic, strictly anaerobic metal-reducing bacterium was isolated from the Taylorsville Triassic Basin at a depth of 2700 m below land surface where the *in situ* temperature is 76°C. This microorganism grows on lactate or formate with Fe(III), Mn(IV), trimethylamine oxide, or nitrate as electron acceptors and it can also ferment glucose (Boone et al., 1995).

In addition to having multiple metabolic capabilities, many extremophiles are multiply tolerant of the physical challenges that they likely encounter in the deep subsurface. For example, in thermophiles high hydrostatic pressures can increase the growth rates of bacteria (Nelson et al., 1992b), increase the upper temperature limits of growth (Erauso et al., 1993; Holden and Baross, 1995), and stabilize proteins (Hei and Clark, 1994). Also, there is evidence that hyperthermophiles are capable of repairing damage to their chromosomes induced by ionizing radiation (DiRuggiero et al., 1997). Microorganisms at depth that are exposed to both high temperatures and high levels of radiation may be uniquely adapted to withstand both stressors. In addition to these physiological responses to combined factors of life at depth, some research even points to the possibility that microbial behavior allows cells to avoid locations that are too hot. The extremophile *Sulfolobus acidocaldarius* responds to excessively high temperatures by swimming away from such habitats (Lewus and Ford, 1999). So, in subsurface strata where there is sufficient energy for cells to be motile and fissures or fractures in which to move, cells may optimize their location with respect to temperature.

9 SUMMARY

Based on the known constraints to microbial life in all environments and the understanding of the types of factors that exist in the subsurface, estimates can be made for the key environmental parameters that are most likely to limit microbial life in the subsurface. Certainly, at some depth within the earth, temperature can become the single factor preventing microbial survival. Therefore, the geothermal gradient establishes the innermost possible boundary to life. However, other factors can pull the boundary closer to the surface. High pressure may extend the thermal limits for microorganisms, but in the cases studied this only results in extension of the maximum temperature by a few degrees. Research on the other factors that control where life exists in the deep earth has been minimal to date. The flux of electron acceptors, electron donors, other nutrients, waste products, and toxic compounds is generally tied to the flux of water. Microorganisms may be severely limited in strata where the permeability is sufficiently low to prevent movement of fluids or in vadose zones where the water potential is too low to sustain life.

Depending on the geothermal gradient and the climate conditions near the surface at any given location, it should be possible to determine a topographic isotherm within the earth where life is restricted. Such a profile will appear at the surface in locations where hot springs occur and will dip deep into sediments in the ocean's deep trenches where the overlying water column maintains low temperatures well into the earth's surface. This depiction of the biosphere must include hypothetical or empirical limits that are placed on microbial communities because of the absence of water or the presence of transport-limiting phenomena such as mineralization, compaction, and pockets of high-pressure gas. It might also include areas where thermodynamic constraints prevent niches from being occupied. In addition to areas where life may be at the verge of extinction, these plots should indicate locations where life in the subsurface is believed to be comparatively healthy. These are characterized by fluid circulation resulting from hydrothermal influence, the presence of density gradients in porewaters, or large-scale release of gases. Each of these processes is likely sufficient to drive the flux of resources that microorganisms require in order to remain active, if only on an intermittent basis. Such a conceptual model should aid in the preparation of a more accurate inventory of the total biomass that exists deep in the earth and a more accurate assessment of the importance of this biomass to geochemical cycling (Whitman et al., 1998). Furthermore, the development of such a model will help to identify gaps in our understanding of what determines the boundaries of life on our planet.

Research on the existence of microbial communities in specific subsurface environments will see advances in the coming years, largely due to application of new interdisciplinary approaches for characterizing these environments. Focus on understanding the distribution of microorganisms on a micron scale will provide direct information on where these organisms prefer to live with respect to the minerals in their habitat (Holman et al., 1998; Siering, 1998; Tobin et al., 1999). Both molecular biology and advanced microscopy techniques will be applied to this question. At the other end of the scale range, the use of geophysical tools for describing the environment in the context of microbial requirements will provide insights on how the bulk properties of geological media control microbial presence in the subsurface (Ghiorse, 1997). By applying these new approaches to the fundamental question of geological constraints on microbial survival in the subsurface, we will obtain important data that will improve our ability to design or understand *in situ* bioremediation, select sites for waste repositories, evaluate economically significant biogeochemical processes, and determine the potential for life on other planets. The ability to recognize all aspects of the environment within which cells exist will be a significant step toward considering microbial distribution in these subsurface environments and realizing where life can become established and where it cannot.

Acknowledgments   The author's research at the Idaho National Engineering and Environmental Laboratory is funded through Contract DE-AC07-76IDO1570. The support of F. J. Wobber (U.S. Department of Energy, Office of Science), H. Guthrie (U.S. Department of Energy, Fossil Energy Office), and T. Uchida (Japanese Petroleum Exploration Company) is gratefully acknowledged. Y. Fujita and T. C. Onstott provided helpful reviews of the manuscript.

REFERENCES

Amann RI, Ludwig W, and Schliefer K-H (1995) Phylogenetic identification and *in situ* detection of individual microbial cells without cultivation. *Microbiol Rev 59*:143–169.

Amend JP and Shock EL (1998) Energetics of amino acid synthesis in hydrothermal ecosystems. *Science 281*:1659–1662.

Amy P, Haldeman D, Ringelberg D, Hall D, and Russell C (1992) Comparison of identification systems for classification of bacteria isolated from water and endolithic habitats within the deep subsurface. *Appl Environ Microbiol 58*:3367–3373.

Atlas RM and Bartha R (1993) *Microbial Ecology: Fundamentals and Applications*, 3rd ed. Redwood City, CA: Benjamin Cummings.

Balkwill D (1989) Numbers, diversity, and morphological characteristics of aerobic, chemoheterotrophic bacteria in deep subsurface sediments from a site in South Carolina. *Geomicrobiol J 7*:33–52.

Balkwill DL, Fredrickson JK, and Thomas JM (1989) Vertical and horizontal variations in the physiological diversity of the aerobic chemoheterotrophic bacterial microflora in deep southeast coastal plain subsurface sediments. *Appl Environ Microbiol 55*:1058–1065.

Baross JA and Deming JW (1983) Growth of "black smoker" bacteria at temperatures of at least 250°C. *Nature 303*:423–426.

Bastin ES, Greer FE, Merritt CA, and Moulton G (1926) Bacteria in oil field waters. *Science 63*:21–24.

Benzing WM and Shook GM (1996) Study advances view of geopressure seals. *Oil and Gas J* Volume 94, 62–66.

Benzing WM, Shook GM, and Leroy SD (1996) The formation and behavior of "vapor lock" pressure seals and associated hydrocarbon accumulations in geologically young basins. *Trans Gulf Coast Assoc Geolog Soc XLVI*:25–39.

Blochl E, Burggraf S, Fiala G, Lauerer G, Huber G, Huber R, Rachel R, Segerer A, Stetter KO, and Volkl P (1995) Isolation, taxonomy and phylogeny of hyperthermophilic microorganisms. *World J Microbiol Biotechnol 11*:9–16.

Boone DR, Liu Y, Zhao Z, Balkwill DL, Drake GR, Stevens TO, and Aldrich HC (1995) *Bacillus infernus* sp. nov., an Fe(III)- and Mn(IV)-reducing anaerobe from the deep terrestrial subsurface. *Int J Syst Bacteriol 45*:441–448.

Brock T (1978) In Starr MP (ed): *Thermophilic Microorganisms and Life at High Temperatures*. New York: Springer-Verlag.

Brown KL (1994) Spore resistance and ultraheat treatment and processes. *J Appl Bacteriol Symp Suppl 76*:67S-80S.

Cano RJ and Borucki MK (1995) Revival and identification of bacterial spores in 25- to 40-million-year-old Dominican amber. *Science 268*:1060–1064.

Chastain R and Yayanos A (1991) Ultrastructural changes in an obligately barophilic marine bacterium after decompression. *Appl Environ Microbiol 57*:1489–1497.

Colwell F, Delwiche ME, Blackwelder D, Wilson M, Lehman RM, and Uchida T (1999) Microbial communities from core intervals, JAPEX/JNOC/GSC Mallik 2L-38 gas hydrate research well. *Geolog Surv Can Bull 544*:189–195.

Colwell FS, Onstott TC, Delwiche ME, Chandler D, Fredrickson JK, Yao Q-J, McKinley JP, Boone DR, Griffiths R, Phelps TJ, Ringelberg D, White DC, LaFreniere L, Balkwill D, Lehman RM, Konisky J, and Long PE (1997) Microorganisms from deep, high temperature sandstones: Constraints on microbial colonization. *FEMS Microbiol Rev 20*:425–435.

Crozier RH, Agapow PM, and Pedersen K (1999) Towards complete biodiversity assessment: An evaluation of the subterranean bacterial communities in the Oklo region of the sole surviving natural nuclear reactor. *FEMS Microbiol Ecol 28*:325–334.

Cusack F, Lappin-Scott H, Singh S, Rocco M, and Costerton W (1990) Advances in microbiology to enhance oil recovery. *Appl Biochem Biotech 24/25*:885–898.

Dallimore SR and Collett TS (1999) Regional gas hydrate occurrences, permafrost conditions, and Cenozoic geology, Mackenzie Delta area. *Geolog Surv Canad Bull 544*:31–43.

Delaney JR, Kelley DS, Lilley MD, Butterfield DA, Baross JA, Wilcock WSD, Embley RW, and Summit M (1998) The quantum event of oceanic crustal accretion: impacts of diking at mid-ocean ridges. *Science 281*:222–230.

Delwiche ME, Colwell FS, Tseng H-Y, Yao Q-J, and Onstott TC (1996) Pressure and temperature adaptation of a bacterium from 2 kilometers beneath the surface of the earth. 96th General Meeting of the American Society for Microbiology, New Orleans, LA.

Dickens GR, Paull CK, Wallace P, and the ODP Leg 164 Scientific Party (1997) Direct measurements of *in situ* methane quantities in a large gas-hydrate reservoir. *Nature* *385*:426–428.

DiRuggiero J, Santangelo N, Nackerdien Z, Ravel J, and Robb FT (1997) Repair of extensive ionizing-radiation DNA damage at 95 degrees C in the hyperthermophilic archaeon *Pyrococcus furiosus*. *J Bacteriol 179*:4643–4645.

Ehrlich HL (1990) *Geomicrobiology*, 2nd ed. New York: Marcel Dekker, Inc.

Erauso G, Reysenbach A-L, Godfroy A, Meunier J, Crump B, Partensky F, Baross JA, Marteinsson V, Barbier G, Pace NR, and Prieur D (1993) *Pyrococcus abyssi* sp. nov., a new hyperthermophilic archaeon isolated from a deep-sea hydrothermal vent. *Arch Microbiol 160*:338–349.

Frazier WC and Westhoff DC (1978) *Food Microbiology*, 3rd ed. New Delhi: Tata McGraw-Hill Publ. Co.

Fredrickson J, Balkwill D, Zachara J, Li S, Brockman F, and Simmons M (1991) Physiological diversity and distributions of heterotrophic bacteria in deep Cretaceous sediments of the Atlantic coastal plain. *Appl Environ Microbiol 57*:402–411.

Fredrickson JK, McKinley JP, Bjornstad BN, Long PE, Ringelberg DB, White DC, Suflita JM, Krumholz L, Colwell FS, Lehman RM, and Phelps TJ (1997) Pore-size constraints on the activity and survival of subsurface bacteria in a Late Cretaceous shale-sandstone sequence, northwestern, New Mexico. *Geomicrobiol J 14*:183–202.

Gaidos EJ, Nealson KH, and Kirschvink JL (1999) Life in ice-covered oceans. *Science 284*:1631–1632.

Gauthier-Lafaye F, Blanc PL, Bruno J, Griffault L, Ledoux E, Louvat D, Michaud V, Montoto M, Oversby V, Delvillar LP, and Smellie J (1997) The last natural nuclear fission reactor. *Nature 387*:337.

Ghiorse WC (1997) Subterranean life. *Science 275*:789.

Ginsburg GD and Soloviev VA (1998) *Submarine Gas Hydrates*. St. Petersburg: VNIIOkean-geologia.

Gold T (1992) The deep, hot biosphere. *Proc Nat Acad Sci USA 89*:6045–6049.

Hanor J and Sassen R (1990) Evidence for large-scale vertical and lateral migration of formation waters, dissolved salt, and crude oil in the Louisiana Gulf Coast. In *Gulf Coast oils and gases: their characteristics, origin, distribution, and exploration and production significance*. In Schumacher D and Perkins B (eds): Proceedings of the Annual Research Conference, Gulf Coast Section, Society of Economic Paleontologists and Mineralogists Foundation, published by Society of Economic Paleontologists and Mineralogists, Austin, TX, pp 283–296.

Harris SA (1986) *The Permafrost Environment*. Totowa, NJ: Barnes and Noble Books.

Harvey RW (1997) Microorganisms as tracers in groundwater injection and recovery experiments: A review. *FEMS Microbiol Rev 20*:461–472.

Harvey RW, George LH, Smith RL, and LeBlanc DR (1989) Transport of microspheres and indigenous bacteria through a sandy aquifer: Results of natural- and forced-gradient tracer experiments. *Environ Sci Technol 23*:51–56.

Hei DJ and Clark DS (1994) Pressure stabilization of proteins from extreme thermophiles. *Appl Environ Microbiol 60*:932–939.

Herbert R (1992) A perspective on the biotechnological potential of extremophiles. *Trends Biotechnol 10*:395–402.

Holbrook SW, Hoskins H, Wood WT, Stephen RA, Lizarralde D, and Leg 164 Science Party (1996) Methane hydrate and free gas on the Blake Ridge from vertical seismic profiling. *Science 273*:1840–1843.

Holden JF and Baross JA (1995) Enhanced thermotolerance by hydrostatic pressure in the deep-sea hyperthermophile *Pyrococcus* strain ES4. *FEMS Microbiol Ecol 18*:27–34.

Holman H-YN, Perry DL, and Hunter-Cevera JC (1998) Surface-enhanced infrared absorption-reflectance (SEIRA) microspectroscopy for bacteria localization on geologic material surfaces. *J Microbiol Meth 34*:59–71.

Horikoshi K (1998) Barophiles: Deep-sea microorganisms adapted to an extreme environment. *Curr Opin Microbiol 1*(3):291–295.

Jannasch HJ, Wirsen CO, and Taylor CD (1976) Undecompressed microbial populations from the deep sea. *Appl Environ Microbiol 32*:360–367.

Jannasch HW and Taylor CD (1984) Deep-sea microbiology. *Ann Rev Microbiol 38*:487–514.

Johnson RC and Rice DD (1990) Occurrence and geochemistry of natural gases, Piceance Basin, Northwest Colorado. *Amer Assoc Petrol Geol Bull 74*:805–829.

Kato C and Bartlett DH (1997) The molecular biology of barophilic bacteria. *Extremophiles 1*:111–116.

Kato C, Li L, Nogi Y, Nakamura Y, Tamaoka J, and Horikoshi K (1998) Extremely barophilic bacteria isolated from the Mariana Trench, Challenger Deep, at a depth of 11,000 meters. *Appl Environ Microbiol 64*(4):1510–1513.

Kato C, Li L, Tamaoka J, and Horikoshi K (1997) Molecular analyses of the sediment of the 11,000-m deep Mariana Trench. *Extremophiles 1*(3):117–123.

Keys WS (1997) *A Practical Guide to Borehole Geophysics in Environmental Investigations.* New York: CRC Lewis Publishers.

Kieft T and Brockman F (2000) Vadose zone microbiology. In Fredrickson JK and Fletcher M (eds): *Subsurface Microbiology and Biogeochemistry.* New York: John Wiley & Sons.

Kieft TL, Amy PS, Brockman FJ, Fredrickson JK, Bjornstad BN, and Rosacker LL (1993) Microbial abundance and activities in relation to water potential in the vadose zones of arid and semiarid sites. *Microbial Ecol 26*:59–78.

Kieft TL and Phelps TJ (1997) Life in the slow lane: Activities of microorganisms in the subsurface. In Amy PS and Haldeman DL (eds): *The Microbiology of the Terrestrial Deep Subsurface.* Boca Raton, FL: CRC Press.

Krumholz LR, McKinley JP, Ulrich FA, and Suflita JM (1997) Confined subsurface microbial communities in Cretaceous rock. *Nature 386*:64–66.

Krumholz LR, Patel S, Tanner RS, Elias DA, and Carson DB (1999) Anaerobic microbial activities in deep Gulf coastal sediments. 4th International Symposium on Subsurface Microbiology, Vail, CO.

Kuenen JG and Tuovinen OH (1981) In Starr MP, Stok H, Truper HG, Balows A, and Schlegel HG (eds): *The Genera Thiobacillus and Thiomicrospira (The Prokaryotes).* New York: Springer-Verlag, pp 1023–1036.

Kuramoto S, Hiramura J, Joshima M, and Okuda Y (1998) Nature of gas hydrates in the Nankai accretionary prism. The International Symposium on Methane Hydrates, Resources in the Near Future, Chiba City, Japan.

Kvenvolden KA (1999) Potential effects of gas hydrate on human welfare. *Proc Natl Acad Sci USA 96*:3420–3426.

Lappin-Scott HM, Cusack F, and Costerton JW (1988) Nutrient resuscitation and growth of starved cells in sandstone cores: A novel approach to enhanced oil recovery. *Appl Environ Microbiol 54*:1373–1382.

Lawrence JR and Hendry MJ (1996) Transport of bacteria through geologic media. *Can J Microbiol 42*:410–422.

Lewus P and Ford RM (1999) Temperature-sensitive motility of *Sulfolobus acidocaldarius* influences population distribution in extreme environments. *J Bacteriol 181*:4020–4025.

Liu SV, Zhou J, Zhang C, Cole DR, Gajdarziska-Josifovska M, and Phelps TJ (1997) Thermophilic Fe(III)-reducing bacteria from the deep subsurface: The evolutionary implications. *Science 277*:1106–1108.

Lorenz JC (1987) Reservoir sedimentology of rocks of the Mesaverde Group, Multiwell Experiment Site and East-Central Piceance Basin, Northwest Colorado. In Johnson RC (ed): *Geologic History and Hydrocarbon Potential of Late-Cretaceous-Age, Low Permeability Reservoirs, Piceance Basin, Colorado.* Morgantown, WV: Department of Energy.

Lowe SE, Jain MK, and Zeikus JG (1993) Biology, ecology, and biotechnological applications of anaerobic bacteria adapted to environmental stresses in temperature, pH, salinity, or substrates. *Microbiol Rev 57*:451–509.

Marquis RE (1982) Microbial barobiology. *BioScience 32*:267–272.

Moser D, Onstott TC, Pfiffner S, White DC, Peacock A, Phelps T, Deflaun M, Hoek J, Ghiorse WC, Colwell F, Kieft T, Reysenbach A-L, Fredrickson JK, Southam G, Kotelnikova S, Slater G, Omar G, Pratt L, Boone D, Pedersen K, and Sher W (1999) The Witwatersrand deep microbiology project: A window into the extreme environment of deep subsurface microbial communities. 4th International Symposium on Subsurface Microbiology, Vail, CO.

Naganuma T, Meisel CJ, Wada H, Kato Y, Takeuchi A, Fujikura K, Naka J, and Fujioka K (1999) Sea-floor fissures, biological communities, and sediment fatty acids of the Northern Okushiri Ridge, Japan Sea: Implications for possible methane seepage. *The Island Arc 8*:232–244.

Nelson CM, Schuppenhauer MR, and Clark DS (1992a) Effects of hyperbaric pressure on a deep-sea archaebacterium in stainless steel and glass-lined vessels. *Appl Environ Microbiol 57*:3576–3580.

Nelson CM, Schuppenhauer MR, and Clark DS (1992b) High-pressure, high-temperature bioreactor for comparing effects of hyperbaric and hydrostatic pressure on bacterial growth. *Appl Environ Microbiol 58*:1789–1793.

Onstott TC, Phelps TJ, Colwell FS, Ringelberg D, White DC, Boone DR, McKinley JP, Stevens TO, Long PE, Balkwill DL, Griffin WT, and Kieft T (1998) Observations pertaining to the origin and ecology of microorganisms recovered from the deep subsurface of Taylorsville Basin, Virginia. *Geomicrobiology 15*:353–385.

Onstott TC, Phelps TJ, Kieft T, Colwell FS, Balkwill DL, Fredrickson JK, and Brockman F (1999) Microbial abundance and activity in the deep subsurface. In Seckbach J (ed): *Enigmatic Microorganisms and Life in Extreme Environments.* Norwell, MA: Kluwer.

Palumbo AV, McCarthy J, Parker A, Pfiffner S, Colwell FS, and Phelps TJ (1994) Potential for microbial growth in arid subsurface sediments. *Appl Biochem Biotechnol 45/46*:823–834.

Parkes RJ, Cragg BA, Bale SJ, Getliff JM, Goodman K, Rochelle PA, Fry JC, Weightman AJ, and Harvey SM (1994) Deep bacterial biosphere in Pacific Ocean sediments. *Nature 371*:410–413:.

Paull CK and Ussler W (1999) Blake Ridge microbial methane accumulations: *In situ* production or products of migration and concentration? *Amer Geophys Union, Fall Meeting*. San Francisco, CA.

Pedersen K (1993) The deep subterranean biosphere. *Earth-Sci Rev 34*:243–260.

Pedersen K (1997) Microbial life in deep granitic rock. *FEMS Microbiol Rev 20*:399–414.

Phelps TJ (1998) Metabolic strategies for survival in the deep underworld. 98th General Meeting, American Society for Microbiology, Atlanta, GA.

Poirier I, Marechal PA, and Gervais P (1997) Effects of the kinetics of water potential variation on bacteria viability. *J Appl Microbiol 82*:101–106.

Sasson R, MacDonald IR, Guinasso NL, Joye S, Requejo AG, Sweet ST, Alcala-Herrera J, DeFreitas DA, and Schink DR (1998) Bacterial methane oxidation in sea-floor gas hydrate: Significance to life in extreme environments. *Geology 26*:851–854.

Sclater JG, Jaupart C, and Galson D (1980) The heat flow through oceanic and continental crust and the heat loss of the earth. *Rev Geophys Space Phys 18*:269–311.

Siering PL (1998) The double helix meets the crystal lattice: The power and pitfalls of nucleic acid approaches for biomineralogical investigations. *Amer Mineralog 83*:1593–1607.

Smith RP, Josten NE, and Hackett WR (1994) Upper crustal seismic and geologic structure of the eastern Snake River Plain: Evidence from drill holes and regional geophysics at the Idaho National Engineering Laboratory. *EOS 75*(44):685.

Spencer CW (1990) Comparison of overpressuring at the Pinedale Anticline Area, Wyoming, and the Multiwell Experiment Site, Colorado. *US Geolog Surv Bull 1886*:C1–C16.

Stark JM and Firestone MK (1995) Mechanisms for soil-moisture effects on activity of nitrifying bacteria. *Appl Environ Microbiol 61*:218–221.

Stetter KO (1998) Hyperthermophiles: Isolation, classification, and properties. In Horikoshi K and Grant WD (eds): *Extremophiles*. New York: John Wiley & Sons.

Stetter KO (1999) Extremophiles and their adaptation to hot environments. *FEBS Lett 452*:22–25.

Stetter KO, Huber R, Blochl E, Kurr M, Eden RD, Fielder M, Cash H, and Vance I (1993) Hyperthermophilic archaea are thriving in deep North Sea and Alaskan oil reservoirs. *Nature 365*:743–745.

Stevens TO (1997) Subsurface microbiology and the evolution of the biosphere. In Amy PS and Haldeman DL (eds): *The Microbiology of the Terrestrial Deep Subsurface*. New York: CRC Press.

Stroes-Gascoyne S and West JM (1997) Microbial studies in the Canadian nuclear fuel waste management program. *FEMS Microbiol Rev 20*:573–590.

Takami H, Inoue A, Fuji F, and Horikoshi K (1997) Microbial flora in the deepest sea mud of the Mariana Trench. *FEMS Microbiol Lett 152*(2):279–285.

Thom S and Marquis R (1984) Microbial growth modification by compressed gases and hydrostatic pressure. *Appl Environ Microbiol 47*:780–787.

Tissot BP and Welte DH (1983) *Petroleum Formation and Occurrence*. New York: Springer-Verlag.

Tobin KJ, Onstott TC, Deflaun M, Colwell F, and Fredrickson J (1999) *In situ* imaging of microorganisms in geologic material. *J Microbiol Meth 37*:201–213.

Tsuji Y, Furutani A, Matsuura S, and Kanamori K (1998) Exploratory surveys for evaluation of methane hydrates in the Nankai Trough Area, offshore central Japan. The International Symposium on Methane Hydrates, Resources in the Near Future, Chiba City, Japan.

Vogt PR, Cherkashev G, Ginsburg G, Ivanov G, Milkov A, Crane K, Lein A, Sundvor E, Pimenov N, and Egorov A (1997) Haakon Mosby mud volcano provides unusual example of venting. *EOS 78*:549–557.

Vorobyova E, Soina V, Gorlenko M, Minkovskaya N, Zalinova N, Mamukelashvili A, Gilichinsky D, Rivkina E, and Vishnivetskaya T (1997) The deep cold biosphere: Facts and hypotheses. *FEMS Microbiol Rev 20*:277–290.

White D and Ringelberg D (1996) Monitoring deep subsurface microbiota for assessment of safe long-term nuclear waste disposal. *Can J Microbiol 42*:375–381.

White RH (1984) Hydrolytic stability of biomolecules at high temperatures and its implication for life at 250 °C. *Nature 310*:430–432.

Whitman WB, Coleman DC, and Wiebe WJ (1998) Prokaryotes: The unseen majority. *Proc Natl Acad Sci USA 95*:6578–6583.

Wilkinson WH, Vega LA, and Titley SR (1994) Geology and ore deposits at Mineral Park. In Titley SR (ed): *Advances in Geology of the Porphyry Copper Deposits, Southwestern North America*. Tucson, AZ: University of Arizona Press.

Wood W and Gettrust JF (1999) Laterally varying seismic reflection strengths in the Cascadia Margin. Boston, MA: American Geophysical Union.

Yanagibayashi M, Nogi Y, Li L, and Kato C (1999) Changes in the microbial community in Japan Trench sediment from a depth of 6292 m during cultivation without decompression. *FEMS Microbiol Lett 170*:271–279.

Yayanos A (1986) Evolutional and ecological implications of the properties of deep-sea barophilic bacteria. *Proc Natl Acad Sci USA 83*:9542–9546.

Yayanos AA (1995) Microbiology to 10,500 meters in the deep sea. *Ann Rev Microbiol 49*:777–805.

4

DIVERSITY AND ACTIVITY OF MICROORGANISMS IN DEEP IGNEOUS ROCK AQUIFERS OF THE FENNOSCANDIAN SHIELD

K. PEDERSEN

Göteborg University, Department of Cell and Molecular Biology, Microbiology, Göteborg, Sweden

1   INTRODUCTION

Igneous rocks are the predominant solid constituents of the Earth, formed through cooling of molten or partly molten material at or beneath the Earth's surface. Such rocks are penetrated by man for a variety of purposes, such as mining for metals, boreholes for the extraction of groundwater and rock heat, and tunnels and vaults for communication, transport, defense, storage, and deposition. The concept of underground storage or deposition of hazardous wastes is commonly applied to materials that cannot be destroyed. One group of such wastes is metals, especially mercury and radionuclides. Disposal concepts vary from country to country, but igneous rocks are commonly in target in countries with access to these types of geological structures. Deposition of long-lived hazardous wastes requires extremely good knowledge of the igneous rock to be used as a host. The radioactive waste disposal industry invests large sums in research on the safe underground disposal of all types of radioactive wastes. In the beginning of the Swedish long-term radioactive waste disposal

Subsurface Microgeobiology and Biogeochemistry, Edited by James K. Fredrickson and Madilyn Fletcher. ISBN 0-471-31577-X. Copyright 2001 by Wiley-Liss, Inc.

research program on subterranean microbiology, previously unknown microbial ecosystems were revealed at depths exceeding 1000 m (Pedersen and Ekendahl, 1990). This discovery triggered a thorough exploration of the subterranean biosphere in the igneous rock aquifers of the Fennoscandian Shield (Pedersen, 1997a, 2000).

The Deep Biosphere Laboratory at Göteborg University (DBL) has, since 1987, performed active research at 11 subterranean sites in Sweden and Finland (Fig. 4.1), exploring depths to 1700 m (Table 4.1). The DBL has also investigated and is still investigating a number of other subterranean sites. They are (1) the natural nuclear reactors of Oklo, in Gabon (Crozier et al., 1999; Pedersen et al., 1996b), (2) the natural hyperalkaline springs of Jordan (Pedersen et al., 1997a), (3) the underground research laboratory at Pinawa, in Canada (Stroes-Gascoyne et al., 1997), (4) the ultra-deep gold mines of Witwatersrand in South Africa (Barnicoat et al., 1997), and (5) subsea floor basement rock environments of the Pacific Ocean (Smith et al., 2000). This chapter will, however, deal mainly with findings obtained by the DBL during 12 years of investigation of microorganisms in Finnish and Swedish igneous rock aquifers. A total of 75 specific borehole depths in 53 different boreholes (Table 4.1) have been investigated for geology, chemistry, numbers of microorganisms, and microbial diversity and activity.

Early investigations by the DBL sought to determine how the potential risk of radionuclide migration could be affected by microorganisms that were thought able

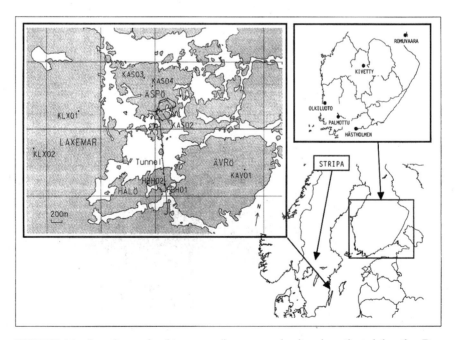

FIGURE 4.1 Locations of subterranean igneous rock sites investigated by the Deep Biosphere Laboratory at Göteborg University. Detailed information about the sites is presented in Table 4.1.

TABLE 4.1 Site Data Comprising All Boreholes in Igneous Rock Investigated for Microbiology by the Deep Biosphere Laboratory at Göteborg University during 1987–1999.

Site	Year	Borehole	Depths (m)	Site Characteristics	Original Scientific Publications
Hålö (S)	1992–1996	HBH01	45	Hålö is an island above a part of the Äspö HRL tunnel (Fig. 4.1). The rock is mainly a red to gray porphyritic monzo–granite belonging to the vast Transscandinavian granite-Porphyry belt (Gàal and Gorbatschev, 1987) with intrusion ages (U–Pb) between 1760 and 1840 Myr. Major fractures and fracture zones control recharge, discharge, and groundwater flow through the island (Banwart et al., 1996). The groundwater is dilute shallow groundwater, brackish Baltic seawater, and saline native groundwater (11–$4890 \, mg \, Cl \, L^{-1}$). (See the Äspö site below for more details).	Banwart et al. (1996) Pedersen et al. (1996a) Kotelnikova and Pedersen (1998)
	1992–1996	HBH02	10		
Hätholmen (F)	1997	HH-KR1	938–948	The rock consists of Rapakivi-type granite. The main fracture minerals include calcite, dolomite, iron hydroxides, and clay minerals. Iron sulfides occur only sporadically. Iron and iron hydroxides are very common, and thus iron is important in controlling E_h processes. The groundwater ranges from fresh to brackish to saline and is composed of several different end-members (Haveman et al., 1998).	Haveman et al. (1999)
	1997	HH-KR2	907–912		
	1998	HH-KR3	211–216		
	1998	HH-KR4	683–688		
	1999	HH-KR5	382–387		
	1998	HH-KR6	62–66		
Kivetty (F)	1997	KI-KR5	717–725	The main rock types are granodiorite and granite. Major fracture minerals include calcite, iron sulfides, iron oxyhydroxides, and clay minerals. Iron hydroxides (goethite and limonite, FeOOH) and quartz are also present in some places. The iron oxyhydroxides are mostly found at shallow depths, down to about 130–170 m. Calcites and iron sulfides are found below 50–100 m depth. All groundwater is freshwater, with a regional maximum of total dissolved solids of less than $200 \, mg \, L^{-1}$. Sulfur compounds are thought to control redox processes, buffering the redox to below $-300 \, mV$ at depth (Haveman et al., 1998).	Haveman et al. (1999)
	1998	KI-KR13	494–499		

(continued)

TABLE 4.1 (*continued*)

Site	Year	Borehole	Depths (m)	Site Characteristics	Original Scientific Publications
Laxemar (S)	1988	KLX01	272–277 456–461 680–702	Laxemar is a part of the Precambrian bedrock where Småland granites predominate the older, Sveocarelian complexes. The borehole KLX01 was situated in the center of a major block. The groundwater is brackish with an E_h at or below -220 mV. A second borehole, KLX02, was drilled to 1700-m depth. The groundwater ranges from fresh to brackish to very saline at depth (approximately 10% total dissolved solids) and is composed of several different end members (Gustafsson et al., 1988).	Pedersen and Ekendahl (1990) Pedersen and Ekendahl (1992b)
	1990	KLX01	830–841 910–921 999–1078		
	1999[a]	KLX02	100–1700		
Olkiluoto (F)	1997–1998	OL-KR3	243–253 438–443	The main rock types include gneisses, schists, granodiorites, and granites. The major fracture minerals include calcite, pyrite, chlorite, and other clay minerals. Sulfur species have an important role in controlling redox processes. The groundwater ranges from fresh to brackish to saline and is composed of several different end members (Haveman et al., 1998).	Haveman et al. (1999)
	1998	OL-KR4	861–866		
	1997	OL-KR8	302–310		
	1997–1999	OL-KR9	563–571 470–475 324–332		
Palmottu (F)	1997	OL-KR10	80–132	The Palmottu area occurs within a zone of metamorphosed supracrustal volcanic and sedimentary rocks that extends from southwest Finland into central Sweden (Blomqvist et al., 1995). The site has an U-Th mineralization with a total length of about 400 m that is related to the latest stages of orogenic events in Finland about 1800–1700 Myr ago. The surroundings of Palmottu are characterized by granites and highly metamorphosed migmatitic rocks. The main types of groundwater comprise fresh Ca-HCO₃ waters, dilute Ca-HCO₃-SO₄ waters, and slightly saline waters of the Na-Cl or Na-SO₄ type.	
	1998	R302	80–100		
	1998	R337	32–38		
	1998–1999	R387	119–127 304–309		

Site	Year	Borehole	Depth (m)	Description	Reference
Romuvaara (F)	1998	RO-KR10	561–566	The main rock types include different types of gneisses, intersected by granodiorite and metadiabase dykes. The major fracture minerals are calcite, iron sulfides, iron oxyhydroxides (goethite and limonite, FeOOH), and clay minerals. The iron oxyhydroxides are found at shallow depths down to 150 m. Iron sulfides, including pyrite, are found at depths of greater than 100 m. All groundwater is fresh, with maximum Cl^- concentrations of less than 300 mg L^{-1} at depth. No effects of glacial melt waters have been detected. Sulfur compounds are thought to control redox processes, buffering E_h below -200 mV (Haveman et al., 1998).	Haveman et al. (1999)
	1998	RO-KR11	540–545		
Stripa (S)	1987–1991	V1	865–866	The Stripa mine was an underground research facility between 1976 and 1991. The ore consisted of a quartz-banded haematite and occurred in a lepatite formation. Adjacent to the lepatite formation is a large body of medium-grained granite through which the boreholes run from drifts in the mine at 360 (V1) and 410 m (V2). Most fractures were partly or fully sealed with chlorite, epidote, or calcite. Because of silica weathering, the pH of the groundwater approached 10. All groundwater is fresh to slightly brackish, with maximum Cl^- concentrations of less than 700 mg L^{-1} at depth (Nordstrom et al., 1985).	Pedersen and Ekendahl (1992a)
		V2	799–807		Ekendahl et al. (1994)
			812–820		Ekendahl and Pedersen (1994)
			970–1240		
Ävrö (S)	1987	KAV01	420–425	See Äspö for geology, hydrogeology, and geochemistry.	Pedersen and Ekendahl (1990)
			522–531		
			558–563		
			635–924		

(continued)

TABLE 4.1 (*continued*)

Site	Year	Borehole	Depths (m)	Site Characteristics	Original Scientific Publications
Äspö (S)	1988–1989	KAS02	202–214	The geology is characterized by a red to gray porphyritic granite-granodiorite (known regionally as the "Småland type") dated to 1760–1850 Myr. The rock area belongs to the vast Transscandinavian granite-Porphyry belt (Gäal and Gorbatschev, 1987). The main fracture-filling minerals at Äspö are, in decreasing order, chlorite, calcite, epidote, fluorite, quartz, hematite, Fe-oxyhydroxides, pyrite, and clay minerals (Smellie et al., 1995). The hydrological situation at Äspö is characterized by a low hydraulic gradient. The recharge–discharge is mainly controlled by tectonic zones and major fractures. The Äspö groundwaters are shown to be mainly reducing (available E_h data record values of -250 to -350 mV), nearly all total iron is in the ferrous state, and sulfide is detectable in small quantities up to approximately 1 mg L^{-1}. The waters are moderately alkaline, generally between 7.3–8.3. Approximate depth trends show a change from an Na-Ca(Mg)HCO$_3$-Cl-type groundwater at near surface (0–150 m), through an Na-Ca(Mg)Cl-SO$_4$ type at depths of 300–800 m, and finally the deepest most saline waters are of the Ca-Na(Mg)Cl-SO$_4$ type occurring below approximately 800 m. Increasing with depth are Cl, Br, Na, Ca, SO$_4$, Sr, and Li, and decreasing are HCO$_3$, Mn, Mg, Fe$_{(tot)}$, Fe$_{(+II)}$, and TOC (Smellie et al., 1995).	Pedersen and Ekendahl (1990) Pedersen et al. (1997b) Pedersen et al. (1996a)
			314–319		
			463–468		
			860–924		
	1996	KAS02	207–208		
	1989	KAS03	129–134		
			860–1002		
	1992	KAS03	533–626		
	1987–1989	KAS04	195–205		
			290–300		
			380–415		

Äspö Hard Rock Laboratory (HRL) tunnel (S)

See Äspö above for geology, hydrogeology, and geochemistry. The Äspö HRL has been constructed as part of the development of the Swedish concept for deep geological disposal of spent nuclear fuel and the work has been divided into three phases; preinvestigation (1986–1990), construction (1990–1995), and operation (1995–). The Äspö HRL is situated on the island of Äspö, adjacent to the Baltic coast of Sweden approximately 400 km south of Stockholm. The access tunnel to the HRL descends with a declination of 14% from the Baltic shoreline for a distance of approximately 1700 m under the seafloor, where it spirals down and terminates 460 m below the island of Äspö (Fig. 4.4). An extensive geoscientific evaluation and detailed site characterization were executed during all three work phases described above and the results of the work thus far completed have been published in a series of reports, summarized by Stanfors et al. (1997) and Rhén et al. (1997).

References:
Pedersen et al. (1996a)
Pedersen et al. (1997c)
Kotelnikova et al. (1998)
Kotelnikova and Pedersen (1998)
Motamedi and Pedersen (1998)
Banwart et al. (1996)

Borehole	Years	Depth
KR0012	1992–1996	68
KR0013	1992–1996	68
KR0015	1992–1996	68
SA813B	1992–1996	112
SA923A	1992–1996	134
SA1062A	1992–1996	143
HA1327B	1992–1996	179
SA14320A	1992–1996	192
KA2511A	1994–1996	345
KA2512A	1994–1996	345
KA2858A	1995–1996	380
KA2862A	1994–1996	380
KA3005A	1994–1996	400
KA3010A	1994–1996	400
KA3067A	1994–1996	409
KA3105A	1996–1996	414
KA3110A	1994–1996	414
HD0025A	1994–1996	420
KA3385A	1994–1996	446
KA3539G	1999–[a]	446
KA3548A01	1999–[a]	446
KA3600F	1999–[a]	446
KJ0050F01	1999–[a]	448
KJ0052F02	1999–[a]	448
KJ0052F03	1999–[a]	448

See Fig. 4.1 for the respective location of each site. F = Finland, S = Sweden.
[a] Ongoing investigations that are planned to continue for up to 20 years.

to possibly survive in deep groundwater (Birch and Bachofen, 1990; Pedersen and Albinsson, 1991, 1992). It soon became apparent that microbial populations can be detected in any deep aquifer studied (Pedersen and Ekendahl, 1990), and more attention was paid to assaying the activity of these microorganisms with radiotracer methods (Ekendahl and Pedersen, 1994; Pedersen and Ekendahl, 1992a, 1992b). The results indicated a notable metabolic and species diversity and motivated the introduction of 16/18S rDNA sequencing for assessment of subterranean microbial diversity (Ekendahl et al., 1994; Pedersen et al., 1996a, 1997c). Nucleic acid probes for 16/18S rRNA are, at present, being applied for *in situ* identification of phylogenetic groups of microorganisms (Pedersen, 1997a). Work has also been performed to describe new species adapted to life in igneous rock aquifers (Kalyuzhnaya et al., 1999; Kotelnikova et al., 1998; Motamedi and Pedersen, 1998). The finding of many hydrogen-utilizing chemolithotrophs invoked the model of a subterranean hydrogen-driven biosphere in the igneous rock aquifers of the Fennoscandian Shield (Pedersen, 1993b, 1997a; Pedersen and Albinsson, 1992). All vital necessary components in the model have now been proven to exist at depth, but a mass balance analysis remains.

2   DRILLING AND EXCAVATION METHODS USED FOR THE EXPLORATION OF MICROORGANISMS IN DEEP IGNEOUS ROCK AQUIFERS

2.1   Drilling and Sampling of Aquifer Rock Surfaces

All sampling of igneous hard rock aquifer material and groundwater requires penetration of the rock to reach the target aquifers. There is only one principal method of achieving this goal, and that is via drilling (Fig. 4.2). The detailed procedure can be modified in many ways, but there are two main drilling approaches, namely from the surface of land and sea, or from an underground tunnel. Drilling in hard rock can be done with either percussion or core drilling. The percussion drill does not recover any rock and it may introduce air into intersected aquifers. The debris created during drilling, and groundwater once an open fracture has been intersected, are forced to the surface by the air. Drilling is done using compressed air with a pressure higher than that of the groundwater. This method is limited to about 100–150 m depth, after which the pressure needed will be too high for standard percussion-drilling machines. Deeper boreholes must instead be drilled by core drilling, producing a core that can be used to map the geological strata penetrated. The retrieval of rock aquifer material during drilling always requires core drilling. Triple-tube drilling (Fig. 4.3) is the best available method for obtaining cores with the smallest disturbance. This type of drilling has recently been used with good results by the DBL in the Äspö Hard Rock Laboratory (HRL) tunnel.

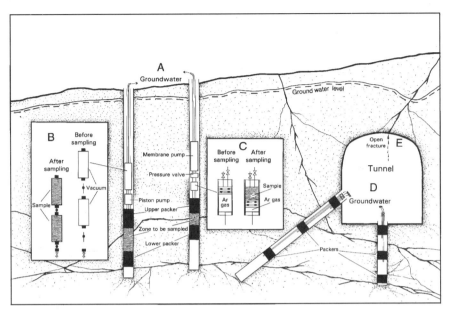

FIGURE 4.2 Access to aquifer material and groundwater occurs via drilling of boreholes from the ground surface or tunnels. After retrieval of drill core material, the boreholes are packed off in one or several sections, each of which isolates one or more specific aquifers. (*a*) Down hole pumps of various types force groundwater from the aquifer to the ground surface for subsampling. (*b*) Borehole BAT sampler that can be opened and closed from the surface and is designed for gas sampling (Torstensson, 1984). (*c*) The PAVE borehole sampler that can be opened and closed from the surface and is designed for gas and microbiological sampling (Haveman et al., 1999). One or more sample vessels can be used simultaneously. (*d*) Tunnel boreholes do not require pumps when the tunnel is below the groundwater table. Aquifers can be packed off and connected to sampling devices in the tunnel with pressure-resistant tubes. It is important to understand the potential danger and technical difficulties associated with the high hydrostatic pressure occurring at depth. For example, there is 42–45 atmospheres of pressure in the boreholes drilled at the bottom of the Äspö HRL. (*e*) Open fractures in tunnels can be sampled directly and represent groundwater with minimal disturbance, except for the pressure decrease due to water entering the tunnel.

2.2 Evaluation of the Contamination Risk during Drilling and Excavation

The risk for contamination of intersected aquifers during drilling is great. Drill water may penetrate aquifers. Therefore, drill water contamination is controlled by the addition of one or more tracers directly to the drill water. Fluorescent stains have been employed on a routine basis to drill the boreholes listed in Table 4.1. The stains used were fluorescein or uranine. The drill water contamination of aquifers penetrated by surface-drilled boreholes decreased from some 2.6–13% in KAV01 to 0.06–0.8% in KAS02-03-04 and KLX01 (Pedersen and Ekendahl, 1990). This decrease was due to a new drilling technique that was employed for KAS02 and later

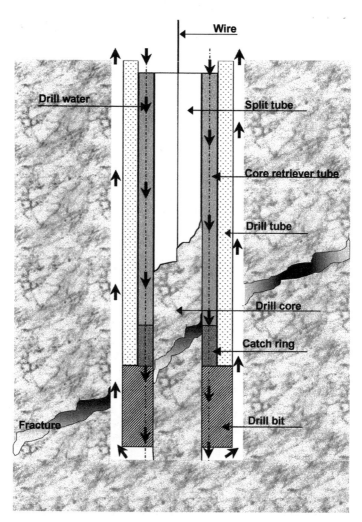

FIGURE 4.3 The triple-tube drilling principle. The use of a core retriever minimizes the exposure of the core to drill water and delivers the core intact to the surface. This is especially important when multifractured rock is penetrated. The drill tube protects the drill core retriever form contact with aquifer systems intersected during drilling. This is of great importance when fractures with contaminants, for example, hydrocarbons, are penetrated. The split tube keeps intersected fractures intact with small rock pieces and loose fracture fillings in their original places throughout the retrieval of the core from the rock to the subsampling occasion.

boreholes, but not for KAV01: the so-called telescope-type drilling (Almén and Zellman, 1991; Pedersen, 1993a). Drilling in tunnels reduces the risk of drill water contamination when the drilling is done at tunnel depths deeper than about 60–70 m. The risk decreases with depth because of an increase in aquifer pressure that will be higher than that used for the drill water, at least in water-conducting fractures.

Pumping or flushing of boreholes to measure the maximum hydraulic water capacity will concurrently clean aquifers and the borehole of drill water, mud, and cuttings, provided there is enough water inflow for intense pumping.

The potential risk for microbial contamination of hard rock aquifers was examined, using 16/18S rDNA sequencing and culturing methods, during drilling of the Äspö HRL SELECT boreholes (Pedersen et al., 1997c). It was found that the tubing used for the drill water supply constituted a source of bacterial contamination to the aquifers via the drilling equipment. The sequencing results showed that although large numbers of contaminating bacteria were introduced into the boreholes during drilling, they did not become established at detectable levels in the aquifers. The number of microorganisms varied from 10^5–10^6 cells mL^{-1} in the drill water introduced into the boreholes. The drill water contamination of the studied boreholes was less than one part per 1000; in other words, fewer than 10^3 cells mL^{-1} in the aquifers of the drilled boreholes could be expected to originate from contamination during the drilling operation. This number was more than one order of magnitude smaller than the number found directly after drilling and could therefore not explain the origin of the observed total populations in the new boreholes. Microorganisms must therefore have been present in the aquifers before the drilling operation began.

2.3 Sampling of Groundwater from Hard Rock Aquifers

Surface boreholes generally require pumping to retrieve groundwater while tunneling under the groundwater level accesses artesian boreholes that can be sampled without pumps; this reduces the risk of contamination. With tunneling, it is also easy to collect samples at *in situ* pressure. Most tunnels used for igneous rock microbiology research are found in mines at various stages of mining. Some, such as the Stripa mine in Sweden (Table 4.1), and the Kamaishi mine (Ishii et al., 2000) and Tono uranium mine (Murakami et al., 1999) in Japan, have been closed for mining and are only used for research. Others, such as the Witwatersrand gold mines in South Africa (Barnicoat et al., 1997), are used for research during active mining. Mines generally penetrate large bodies of rock, depending on the age of the mine and the rock formation of interest. This fact, together with the extensive pumping needed to keep the mines dry, will in time alter the surrounding groundwater system in a way that makes it difficult to assure that the observations made reflect pristine groundwater. Such disturbances can be decreased significantly when underground facilities are built solely for research. Tunnels such as the Swedish Äspö HRL (Fig. 4.4), Grimsel in the Swiss Alps (Frieg et al., 1998) and the Underground Research Laboratory in Pinawa, Canada (Jain et al., 1997) are examples. Data on changes of the hydrodynamic situation and the biogeochemical record obtained during construction can be used as a baseline when interpreting information.

Pumping from the surface and withdrawal of groundwater via tubes may produce unwanted effects. These effects include degassing due to a pressure decrease and the possibility of microbial biofilm formation on tube walls during prolonged pumping through more than 1000-m-long tubing in very deep boreholes. Such effects can be avoided by the application of down-hole samplers. The DBL has tested the

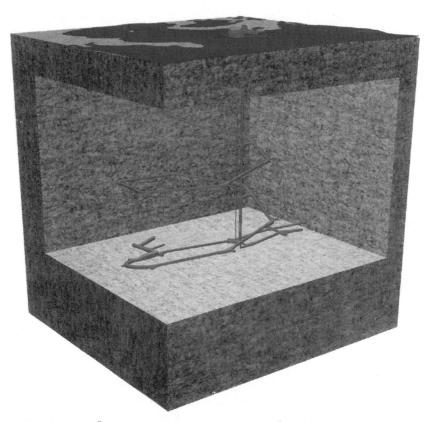

FIGURE 4.4 The Äspö HRL is situated on the island of Äspö, adjacent to the Baltic coast of Sweden approximately 400 km south of Stockholm. The access tunnel to the HRL descends with a declination of 14% from the Baltic shoreline for a distance of approximately 1700 m under the seafloor, where it spirals down and terminates 460 m below the island of Äspö (Fig. 4.1). The total length of the tunnel is 3600 m. Äspö itself, characteristic for the surrounding islands, comprises a slightly undulating topography (10 m above sea level) of well-exposed rock.

suitability of two types of down-hole samplers, in addition to sampling via pumping. The samplers were the BAT (Torstensson, 1984) and the PAVE (Haveman et al., 1999) samplers (see Figs. 4.2b and 4.2c). The BAT sampler was constructed with gas sampling as a major aim and consisted of two sterile cylindrical tubes (or one larger tube) made of stainless steel. The tubes were supplied with nitrile rubber stoppers and evacuated. They were opened and closed at the sampling depths by penetration of the stoppers with hypodermic needles using a mechanical device controlled from the ground level. For several reasons, however, this sampler was not a reliable tool for microbiology sampling. The piston pump, which was very difficult to clean, was placed before the sample containers and the sudden decrease in pressure when the sample containers opened seemed to shear biofilms from the pump. Additionally, the pressure difference between the evacuated sample cylinders and the groundwater at depth (up to 100 atmospheres) may have caused a "French press" effect on cells in the groundwater. This may have disrupted some cells by a sudden drop in pressure when they passed out of the narrow hypodermic needle orifice penetrating the sampling containers. The BAT sampler (Fig. 4.2b) never came into routine use for microbiology sampling. The boreholes at ground surface at Hålö, Laxemar, Palmottu, Ävrö and Äspö (Table 4.1) were sampled with the borehole pump technique described under "A" in Fig. 4.2. A mobile chemistry laboratory was used on the ground surface for sample retrieval and preparations (Grenthe et al., 1992).

The PAVE system was constructed with both gas and microbiological sampling as major objectives. The system consists of a rubber membrane pump placed above a sample container with two sterile, evacuated, and closed pressure vessels filled with argon gas so that the movable piston can move to the top of the pressure vessel (Fig. 4.2c). The argon pressure is set just below the hydrostatic pressure at the sample depth, which makes the drop in pressure during sampling negligible. Prior to sampling, the complete PAVE system is disinfected by rinsing for 30 min with a 10 mg L^{-1} chlorine dioxide water solution (Freebact, Wecantech AB, Märsta, Sweden), then flushed with sterile water for 10 min. To ensure the efficiency of the sterilization, control samples were analyzed and growth was not detected (Haveman et al., 1999). The section of the borehole to be sampled with PAVE is packed off with inflatable rubber packers as was done with the BAT system (Fig. 4.2). Groundwater is pumped from the packed-off zone past the closed pressure vessels and out of the borehole. Groundwater parameters (i.e., pH, E_h, conductivity, and temperature) are monitored in nitrogen gas shielded flow through cells in the field lab on the surface until they stabilize. At this point, samples for field and laboratory analysis for hydrogeochemical characterization are collected. After this phase, the pressure valve of the PAVE is opened. Groundwater pressure pushes the piston in the sampler down to fill the sampler with groundwater. The valve is left open for several hours to allow water to pump through the sampler, and then the PAVE sampler is closed and raised out of the borehole. This system was employed for sampling of the Hästholmen, Kivetty, Olkiluoto, and Romuvaara sites listed in Table 4.1.

Sampling from tunnels under the groundwater table (Fig. 4.2d) reduces sampling difficulties significantly. Boreholes become artesian, and packers and pressure-resistant tubes with valves are all that is needed for successful sampling. The high pressure encountered with increasing tunnel depths requires very robust anchoring and packer systems but the installation of such equipment is routine. This sampling method was used for all the Äspö HRL tunnel boreholes described in Table 4.1. A final possibility is to sample open fractures that enter tunnels underground (Fig. 4.2e). These are free of drilling effects, but may have encountered some disturbance from the blasting operations used for tunnel construction. Parts of the Äspö tunnel were drilled with a tunnel-boring machine (diameter of the drill = 5.5 m). The disturbance on the surrounding rock mass is minimal in those (lower) parts of the tunnel.

3   THE DEEP IGNEOUS ROCK AQUIFER ENVIRONMENT

3.1   The Fennoscandian Shield

The Fennoscandian Shield (also called the Baltic Shield) is the slightly vaulted Precambrian rock area comprising parts of Norway, most of Sweden, all of Finland, and most of Karelen and the Kola Peninsula. The Fennoscandian Shield mostly consists of granite and gneiss rocks. Apart from these, it is covered with platforms of sedimentary rock in the north, east, and south and it borders on the younger Scandinavian mountains to the west (Norway). The age of the Fennoscandian Shield rock ranges from 3100 million years in the north to 1700 million years in the south and west (Gàal and Gorbatschev, 1987). The rock composition varies significantly between the sites discussed in this chapter, but most of the rock types found (Table 4.1) are characterized as igneous. When formed, igneous rocks are too hot to host life. Therefore, observed life in these rocks must have entered after cooling and fracturing of the rock mass. The rock considered has generally been of granitic composition with quartz, feldspars, and mica as the bulk rock minerals. In addition to these, there are accessory minerals that influence the hydrochemical conditions (Table 4.1). Calcite may influence pH and HCO_3^-, pyrite dissolution and precipitation affect redox, apatite may act as a source of HPO_4^{2-}, and fluorite is involved in F^- exchange between dissolved and solid phases. Clay minerals may act as ion exchangers. Clays are common fracture-filling minerals and some of them have been formed in fractures because of weathering reactions elsewhere in the aquifer system. Minor amounts of iron oxyhydroxide minerals, such as goethite and limonite, are found in the fractures, especially in the shallow (< 100 m) part of the rock. Old fractures and vaults in igneous rock commonly contain ferric iron because of the historical oxidizing action of deep, very hot water. Disintegration of this water to oxygen and hydrogen occurs when the water comes under the very high pressure and temperature, which prevail in contact with magmatic layers under the crust (Apps and van de Kamp, 1993). During its flow, the oxygen in this hydrothermally altered water oxidizes ferrous iron in the rock closest to the transport paths to ferric iron, giving the rock bordering to the fractures a red color up to a distance of several tens of centimeters from the aquifer.

3.2   Groundwater Flow in Igneous Rock Aquifers

The distribution of flow in hard rock aquifers has an influence on groundwater composition. The hydraulic conductivity varies considerably between different locations. Structures such as fracture zones may act as conductors and have a dominating influence. Vertical conductive zones are important for groundwater recharge at depth. Horizontal zones may act as hydraulic shields and separate groundwater with different composition. Deep groundwater with a relatively high salinity will have a higher density, which helps to stabilize the layering (Smellie and Wikberg, 1991). The openings in rock fractures are potential channels for groundwater. Model studies have been done of flow and transport in fractures with variable apertures (Moreno et al., 1985). The results suggest that considerable channeling is to be expected in such fractures, and that there is a tendency for some pathways to carry much more water than others. In a limited mass of rock, one or a few channels will dominate the flow and transport of nutrients and microorganisms. Hydraulic conductivities have been measured in boreholes at different depths. This information together with the groundwater surface topography (which in Sweden approximately corresponds to the ground surface topography) is used to calculate the groundwater flowfield. Groundwater flow at about 500-m depth is calculated to be in the range of $0.01-1 \, \mathrm{L \, m^{-2} \, year^{-1}}$ (Pedersen and Karlsson, 1995). Near the surface, there is an increase in hydraulic conductivity and flow. At or below sea level, the hydraulic gradient evens out and, because of this, the flow rate is very small here. The hydraulic gradients increase considerably near a tunnel because the flow pattern is different from what it was before tunneling. Note that a significant part of the inflow comes from aquifers situated deeper than the tunnel position concerned. This situation is considered when data from the Äspö HRL are interpreted.

3.3 Geochemistry of Igneous Rock Groundwater

In general, groundwater under land in Sweden is of meteoric origin. The infiltrated water is almost "pure water" derived from rain or melting snow, with dissolved air as a significant component. The processes in the biologically active soil zone are very important in determining the composition of the recharge water. Oxygen will be consumed and carbon dioxide produced. The carbonic acid will react with minerals such as calcite and feldspars to form carbonate ions. Calcium and alkali ions will be released to the water. Ion exchange with clay minerals may affect the proportions of cations. Organic materials, such as humic and fulvic acids and other substances, will be added to the water from the soil. Biological processes will also have an influence on the groundwater if seawater infiltrates through organic rich seabottom sediments.

At great depths or under the sea floor, saline water is found where chloride is the dominating anion (Table 4.2). The most common cation in saline groundwater is either sodium or calcium. The saline water may have a marine origin, but other end-members are also possible, depending on location and other factors. Very deep, at depths of 1000–1500 m or more, the salinity can be very high, 8% or more (e.g., KLX02, Table 4.2), exceeding that of seawater. It is also common that in coastal regions saline groundwater is found at shallower depths than further inland. This may, of course, be relic seawater that infiltrated several thousands of years ago, when land near the Swedish east coast was covered by the sea due to the glacial depression (land displaced downward by the weight of the ice cover). The infiltration of seawater continued until land was reclaimed by the land uplift, which is continuing in Sweden. However, an alternative explanation could be the lack of a driving hydraulic force under the "flat" surface of the sea. With little or no hydraulic gradient in the groundwater beneath the bottom of the sea, fossil saline conditions can be preserved for very long periods of time. They need not always be the result of a relatively recent infiltration of seawater. In other words, deep saline water may have originated long before the last glaciation some 10,000 years ago.

TABLE 4.2 Selected Chemical Parameters of the Stripa, Laxemar, and Äspö HRL Sites (Pedersen, 1997b) and Four of the Finland Sites (Haveman et al., 1999).

Borehole	Depth (m)	pH	Temp. (°C)	E_h (mV)	HCO_3^- (µM)	DOC (µM)	Fe_{tot} (µM)	Fe^{2+} (µM)	S^{2-} (µM)	SO_4^{2-} (mM)	Na^+ (mM)	Ca^{2+} (mM)	Cl^- (mM)	TDS (g/l)
Hästholmen														
HH-KR1	937–947	7.0	13.8	−46	380	310	28.6	25	<0.3	1.44	213	92.3	433	24.7
HH-KR2	907–912	7.3	10.0	−214	450	300	23.3	17	<0.3	0.35	231	75.9	420	23.9
HH-KR23	211–216	7.8	19.6	—[a]	2220	340	21.5	12	1.9	6.70	109	16.2	149	9.54
Kivetty														
KI-KR5	717–725	8.1	5.6	—	1440	180	1.8	1.6	0.3	0.02	0.43	0.40	0.1	0.144
Laxemar														
KLX01	830–841	8.2	19.5	−270	104	—	3.9	3.8	2.3	7.10	120	77	259	15.7
KLX01	910–921	8.4	21.2	—	98	—	0.94	0.92	11	8.10	135	97	315	19.0
KLX01	999–1078	8.5	22.9	−220	190	—	6.5	6.4	5.6	7.20	146	115	355	22.1
KLX02	1420–1705	7.9	35.0	−334	230	75	6.1	6.1	<0.3	11.30	370	495	1330	76
Olkiluoto														
OL-KR3	243–253	8.3	—	−180	380	140	1.8	1.1	11	0.01	63	7.0	77.8	4.57
OL-KR8	302–310	7.8	—	—	730	130	0.3	0.5	0.6	4.90	88	25.5	135	8.51
OL-KR10	324–332	7.9	10.4	—	360	120	6.6	6.1	0.5	0.09	84	30.9	152	8.73
OL-KR9	563–571	8.2	—	—	220	130	0.2	02	<03	0.01	183	81.1	324	19.2

	561–566	8.4	14.4	−418	1800	1050	0.9	0.2	<0.3	0.03	1.09	0.26	0.1	0.172
Romuvaara														
RO-KR10	561–566	8.4	14.4	−418	1800	1050	0.9	0.2	<0.3	0.03	1.09	0.26	0.1	0.172
Stripa														
V2	799–807	9.5	18	205	158	—	0.3	0.27	<0.3	0.052	3.78	0.80	5.07	0.30
V2	812–821	9.4	18	199	50	—	0.14	0.09	4.4	0.86	9.10	5.60	19.7	1.23
V2	970–1240	10.2	26	−3	57	—	0.07	0.07	100	0.38	9.10	2.82	14.4	0.88
Äspö HRL														
KR0012	68	7.7	9.3	—	4980	920	3.5	3.5	—	1.18	17.7	3.2	23.7	1.75
KR0013	68	7.5	8.9	—	4900	920	6.4	6.0	—	1.20	33.4	8.9	50.5	3.38
KR0015	68	7.5	8.9	—	4960	1500	6.6	6.3	—	1.15	23.0	4.8	22.3	2.07
KA3005/2[b]	400	7.6	14.3	—	1300	170	11	10.5	—	3.16	75	33.6	152	8.96
KA3010/2	400	7.6	14.3	—	910	210	13.9	12.7	—	3.49	82	46.8	186	10.9
KA3110/1	414	7.6	13.4	—	2700	340	19.8	18.9	—	2.84	69	15.0	108	6.63

[a]No data.
[b]Number after slash denotes sampled borehole section.

Typical groundwater compositions at different depths and locations encountered in the course of the research and exploration of the sites described in this chapter are given in Tables 4.2 and 4.3. It is obvious that concentrations of major constituents, such as the cations sodium and calcium and the anions bicarbonate and chloride, may vary considerably in concentration depending on where and at which depth the samples have been taken. Chloride behaves conservatively, but many other ions obviously interact with the minerals. This is particularly evident in groundwater of marine origin. An example is the ion exchange of calcium for sodium and vice versa. A further observation is that ions such as potassium and magnesium, which are common in seawater, are evidently suppressed in groundwater, presumably by reactions with the minerals (not shown). In addition, sulfate is partly consumed, probably by sulfate-reducing bacteria (SRB). Carbonate is less common at depth, possibly because slow reactions with the rock minerals cause precipitation of carbonate as calcite and that autotrophic microbial activity produces organic carbon and methane.

The pH of granitic groundwater in Sweden is buffered by the carbonate system and is slightly alkaline. Calcite is an abundant mineral in the rocks and can, together with available feldspars, react with acids. Therefore, "acid rain" or any similar disturbance of pH does not propagate very far down underground. Measurements of redox potential with E_h electrodes commonly give values of between -100 and $-400\,mV$. E_h is dependent in part on pH and the ferrous iron concentration, but the low concentrations of redox active species in groundwater make the measurement of E_h a delicate operation. *In situ* measurement has been found to offer the most accurate results (Grenthe et al., 1992).

3.4   Gases Dissolved in Igneous Rock Groundwater

The studied groundwater contains dissolved gases such as nitrogen, hydrogen, carbon dioxide, methane, some ethane, propane, and the noble gases helium, neon, argon, krypton, and radon (Table 4.3). Oxygen is only found, if at all, at very shallow depths. The amount of gas dissolved varied from $18-340\,mL$ gas L^{-1} groundwater. The Finnish groundwater generally contained more gas than the Swedish groundwater. Local and depth variations are common within and between the studied sites for many of the observed gases. Nitrogen is the dominating gas in most samples examined. Some of the nitrogen may have been dissolved from air in rain and surface waters that infiltrate as groundwater over time. However, the solubility of nitrogen at $10°C$ and atmospheric pressure is $870\,\mu M$. Most of the nitrogen values in Table 4.3 exceed this solubility limit so other sources of dissolved nitrogen to groundwater must exist. Nitrogen gas can be used by nitrogen-fixing microorganisms as a source of nitrogen. Many microorganisms produce nitrogen gas from nitrate during the anaerobic respiration process termed denitrification. Microbial processes could then contribute to the pool of dissolved nitrogen gas in groundwater through denitrification of nitrogen dissolved from deep mantle rocks (Apps and van de Kamp, 1993), but it is not known whether this process occurs at the studied sites.

TABLE 4.3 The Content of Nitrogen, Hydrogen, Helium, Argon, and Carbon-containing Gases and the Total Volumes of Gas Extracted from Groundwater Samples of the Stripa, Laxemar, and Äspö HRL Sites and Four of the Finland Sites.

Borehole	Depth (m)	N_2	H_2	He	Ar	CO	CO_2	CH_4	C_2H_2	C_2H_4	C_2H_6	C_3H_6	C_3H_8	Volume gas (mL)
							μM gas							
Hästholmen														
HH-KR1	937–947	4800	78	600	400	<0.3	30	4	0.002	0.009	0.14	0.010	0.01	158
HH-KR2	907–912	6500	13	900	900	<0.5	10	9	0.004	0.009	0.12	<0.009	0.02	228
HH-KR3	211–216	2800	0.6	9	800	<0.2	20	2	<0.002	0.003	0.02	<0.003	<0.003	77
Kivetty														
KI-KR5	717–725	4900	0.06	0.7	400	<0.2	50	6	<0.02	0.02	0.2	<0.05	0.02	147
Laxemar														
KLX01	830–841	2074	—ᵃ	205	—	0.02	21	1.2	<0.004	<0.004	<0.004	—	—	52
KLX01	910–921	1650	—	156	—	0.004	22	1.2	<0.004	<0.004	<0.004	—	—	41
KLX01	999–1078	803	—	109	—	0.03	71	1.4	<0.004	<0.004	<0.004	—	—	22
Olkiluoto														
OL-KR3	243–253	1870	2.9	49	140	<0.2	2	1170	<0.02	0.002	4.1	<0.04	<0.04	82
OL-KR8	302–310	1660	<0.08	40	30	<0.08	6	2	0.002	0.004	0.008	<0.02	<0.02	47
OL-KR10	324–332	2240	0.2	180	50	<0.2	6	3330	<0.02	<0.02	16	<0.05	0.3	145
OL-KR9	563–571	2290	15	480	40	<0.6	6	10,500	<0.08	0.007	69.4	0.01	0.5	340

(*continued*)

TABLE 4.3 *(continued)*

Borehole	Depth (m)	N_2	H_2	He	Ar	CO	CO_2	CH_4	C_2H_2	C_2H_4	C_2H_6	C_3H_6	C_3H_8	Volume gas (mL)
							μM gas							
Romuvaara														
RO-KR10	561–566	1100	0.08	0.2	140	<0.06	4	3	<0.006	0.001	0.07	<0.01	0.01	33
Stripa														
V2	799–807	1115	—	<10	—	<0.04	1.4	11	<0.004	<0.004	0.013	—	—	25
V2	812–821	1383	—	<10	—	<0.04	0.49	7.6	<0.004	<0.004	0.027	—	—	31
V2	970–1240	1093	—	<10	—	<0.04	0.5	13	<0.004	<0.004	0.129	—	—	25
Äspö HRL														
KR0012	68	981	—	1.8	—	0.004	270	46	<0.004	<0.004	<0.004	—	—	29
KR0013	68	1115		4.9	—	0.008	430	88	<0.004	<0.004	<0.004	—	—	37
KR0015	68	981	—	2.9	—	0.004	670	182	<0.004	<0.004	<0.004	—	—	41
KA3005/2[b]	400	1157	0.075	78	—	<0.04	48	77	<0.004	<0.004	<0.004	—	—	32
KA3005/4	400	1190	0.005	170	—	<0.04	94	82	<0.004	<0.004	<0.004	—	—	34
KA3010/2	400	1812	1.4	354	—	0.06	6.3	2.5	<0.004	<0.004	<0.004	—	—	49
KA3110/1	414	663	0.65	20	—	<0.04	82	41	<0.004	<0.004	<0.004	—	—	18

[a]Not examined.
[b]Number after slash denotes sampled borehole section.

Carbon dioxide concentration generally decreases with depth. Active organisms expel carbon dioxide from degradation of organic material, and many autotrophic microorganisms transform carbon dioxide to organic carbon. The concentration of this gas may, therefore, be influenced by microorganisms in addition to the effects it may have on the carbonate system, pH, and mineral precipitation and dissolution. One obvious example of this process was recorded when the intrusion of shallow groundwater to the Äspö HRL tunnel was followed. Microorganisms degraded organic carbon to carbon dioxide, which resulted in a significant increase in alkalinity of the groundwater and presumably also influenced mineral precipitation on the aquifer surfaces (Banwart et al., 1996).

The content of hydrogen and methane varies considerably among the studied sites. Hästholmen had the highest hydrogen values, and Olkiluoto showed some very high methane values. Data on hydrogen in hard rock have been published previously. Between 2 and 1600 µM hydrogen were found in groundwater from Canadian Shield and Fennoscandian Shield rocks (Sherwood Lollar et al., 1993a, 1993b). Methane occurs frequently in subterranean environments globally, not just in hard rock environments. Evidence of an ongoing methane-generating process in deep Swedish granite has been published (Flodén and Söderberg, 1994; Söderberg and Flodén, 1991, 1992). Pockmarks in Baltic sea sediments were found, indicating gas eruption from fracture systems in the underlying granite, mainly methane. Between 1 and 18,600 µM of methane in groundwater from Canadian Shield and Fennoscandian Shield rocks have been registered (Sherwood Lollar et al., 1993a, 1993b). Data indicate that levels of up to 720 µM of methane exist at depths to 440 m at Äspö HRL (Kotelnikova and Pedersen, 1997). The origin of methane at the sites listed in Table 4.3 has not been well researched. The significant content of $C_{2-3}H_{2-8}$ suggests that most of the methane found at the Finnish sites is of inorganic origin. The lack of these compounds at Äspö and Laxemar would suggest a biological origin (Des Marais, 1996). Some results on the $^{13}C/^{12}C$ signatures indicate a biogenic origin for the Äspö methane (Banwart et al., 1996).

4 FOSSILS OF MICROORGANISMS IN A FRACTURE CALCITE MINERAL

Many old fractures in hard rock are no longer open, because they have been filled with precipitating minerals such as calcite, dolomite, pyrite, epidote, or chlorite (Table 4.1). This fact poses interesting possibilities in the search for microbial fossils that have been captured in fluid inclusions in hard rock aquifers. It has been reported for halophilic bacteria (Vreeland et al., 1999). Two boreholes in the Äspö HRL tunnel carried groundwater that was oversaturated with carbonate and calcium (SA813B and SA923A, Table 4.1). These boreholes were explored for attaching and biofilm-forming microorganisms (Pedersen et al., 1996a). Glass surfaces exposed to flowing groundwater at flowrates of 8–14 mm sec^{-1} for 67 days were subsequently observed through light and scanning electron microscopes. Some 1.2×10^6 cells/cm^{-2} were found more or less buried in dense precipitates of

FIGURE 4.5 (A) A bacterial biofilm mixed with calcite precipitates on a surface exposed to slowly flowing groundwater from the borehole SA813B in the Äspö HRL tunnel. 16S rDNA analysis and culturing results demonstrated this biofilm to be predominated by an acetogenic bacterium and a SRB; see Chapter 4 (Pedersen) (scale bar = 1 μm). (B) Thin-section transmission electron microscopy of a calcite precipitate that coated a fracture in crystalline rock from 207 m below sea level in southeast Sweden. Fossil microorganisms (M) are arranged in a typical biofilm formation.

calcite (Fig. 4.5a). Although partly artificial, this experiment supports the idea that microbes attached to fracture minerals in deep hard rock may become buried in precipitates, or trapped in fluid inclusions, during growth of fracture minerals.

An unusually long fracture core sample from a 207-m borehole length in KAS02 (Table 4.1) was extensively investigated with respect to stable isotopes (Tullborg et al., 1999). Obvious signs of biological activity in the form of light $\delta^{13}C$ signatures in the calcite carbon were found. It was therefore assumed that fossil microorganisms may be detectable in the fracture minerals. Investigating the fracture with electron microscopy did indeed reveal bacterium-like structures in parts of it (Fig. 4.5b). Energy-dispersive X-ray analysis found these structures to be enriched with carbon (Pedersen et al., 1997b). The presence of fossil microorganisms in the deep igneous rock fracture mineral was an indication that microbial life was present deep under the island of Äspö long before drilling and excavation of the Äspö HRL. This observation, and the biofilm example discussed above, seem to reflect a past and a present situation of attached subterranean microorganisms that are neatly linked. The isotope and electron microscopy results strongly suggest that microbial activity was ongoing earlier in the deep granitic aquifers of Äspö. The presence of modern autotrophic and heterotrophic microbial life in aquifers in this rock volume has likewise been repeatedly demonstrated (Kotelnikova and Pedersen, 1998). Modeling historical and present geochemical processes in deep granitic aquifers should require biologically catalyzed reactions to be correct. However, it remains to be determined at which rates subterranean microorganisms shuttle carbon between various dissolved and solid phases in hard rock aquifers.

5 NUMBERS OF ATTACHED AND UNATTACHED MICROORGANISMS IN DEEP GROUNDWATER AND THEIR RELATIONSHIPS WITH THE GEOCHEMICAL RECORD

5.1 Total Number of Microorganisms

Total numbers of subsurface microorganisms vary notably depending on the site studied. Values in the range of 10^3–10^8/mL of groundwater or gram of sediment have been reported for deep environments (Fredrickson and Onstott, 1996; Ghiorse and Balkwill, 1983; Ghiorse and Wilson, 1988; Pedersen, 1993a). The total numbers of microorganisms in igneous rock groundwater samples have been evaluated by DBL ever since the first boreholes were examined in 1987 (Table 4.1). Unattached microorganisms have been counted with epifluorescent microscopy after filtration using 0.2-μm filters and staining with acridine orange (AO) and/or 4',6-diamidino-2-phenylindole (DAPI) (Fry, 1990; Herbert, 1990). The average total number of cells commonly registered in the Fennoscandian igneous rock aquifers is generally within the range of 10^5–10^6 cells mL^{-1}, although the range of single observations is from 1×10^3 to 5×10^6 cells mL^{-1} (Fig. 4.6). A large set of boreholes examined at a site results in a larger range of total numbers than a small set of boreholes. Such correlation can be expected if there are large variations between the aquifers examined at one site. This seems to be the case for Laxemar, Stripa, and Äspö HRL, showing ranges of total numbers of almost three orders of degree. The variability of the total number in specific boreholes has been studied extensively at Stripa and Äspö. It was found to be remarkably small. The Stripa borehole V2 delivered reproducible and nonvariable numbers for the whole period studied (4 years) (Ekendahl and Pedersen, 1994; Pedersen and Ekendahl, 1992a). Four newly drilled boreholes in the Äspö HRL tunnel were sampled three times during 1 year. Each borehole showed similar total numbers over this time period (Pedersen et al., 1997c), as did seven other boreholes at the Äspö HRL sampled three times during a period of 6 months (Pedersen et al., 1996a). The variability in total numbers between boreholes and the nonvariability in total numbers within boreholes are indicative of stable environments with little or no changes in the conditions for microbial life. Such conditions may, however, vary considerably between sites and boreholes intersecting the Fennoscandian Shield igneous rock aquifers. This observation compares well with data on the groundwater chemistry of boreholes. They may vary significantly between boreholes, depths, and sites (Table 4.2), but they are nonvariable within specific boreholes over time (Nilsson, 1995).

Correlation analyses of the total microbial numbers with several other groundwater parameters were performed early on during the DBL program (Pedersen and Ekendahl, 1990) and showed that the only measured parameter which correlated with the total number was the amount of total organic carbon (TOC). The same relation was later found for the shallow (0–105 m) groundwater of Bangombé, in

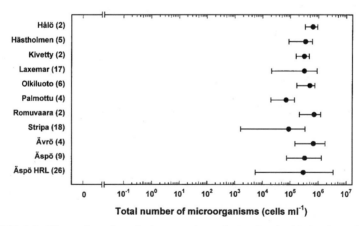

FIGURE 4.6 The total number of microorganisms observed at the 11 sites investigated. The data are extracted from publications according to Table 4.1, except for some recent, unpublished data obtained during spring 1999. The figure shows the average total number of microorganisms per site (•) and the bar gives the range of the data used to calculate the average number. Numbers in parentheses after the site names give the sum of observations for that site and they consist of a set of mean values based on two to six repetitions according to respective publications.

Oklo, Gabon (Pedersen et al., 1996b). It appears that the total number of microorganisms in groundwater does not correlate with any other measured parameter. Repeated attempts (not published) to find significant correlation between individual groundwater parameters, other than TOC and total numbers have failed.

5.2 Viable Counts of Microorganisms in Igneous Rock Groundwater

The plate count technique has been employed to determine the number of colony-forming units (CFU) in deep igneous rock groundwater. The percentages of total numbers that could be cultured from Äspö borehole groundwater with this method ranged from less than 0.1–10%, with an average of 1.7% (Pedersen and Ekendahl, 1990). The media used were general-purpose types for heterotrophic bacteria. Further characterization was required to obtain information regarding the kinds of microorganisms obtained. Typically, the genera *Pseudomonas* and *Shewanella* were found. More recent use of this method resulted in very low CFU percentages of the total numbers, less than 0.1% (Pedersen et al., 1997c). These low viable count numbers and the inability of the plate count method to reveal information about the metabolic diversity of the investigated microorganisms motivated adaptation of more selective media for different predominant physiological groups. A range of anaerobic selective culture media for various physiological groups were, therefore, developed and applied on Äspö HRL tunnel groundwater and at our five groundwater sites in Finland (Table 4.1). Anaerobic Hungate tubes and serum bottles with aluminum crimp sealed butyl rubber stoppers were used for most probable number (MPN) determinations with a set of different media (Haveman et al., 1999; Kotelnikova and Pedersen, 1998). Figure 4.7 summarizes the results of MPN enumeration of iron- and sulfate-reducing bacteria, heterotrophic and autotrophic acetogens, and heterotrophic and autotrophic methanogens.

Iron-reducing bacteria (IRB) and SRB were detected at all sites in Finland at most depths investigated, in the range of 10^0–10^4 cells mL^{-1} (Figs. 4.7a and 4.7b). An inverse relationship between IRB and SRB, which correlated with the predominating types of fracture minerals at the sites, became evident when the Hästholmen and Olkiluoto sites were compared (Haveman et al., 1999). Olkiluoto had a relatively high number of SRB and a low number of IRB, whereas the opposite was true for Hästholmen. Hästholmen groundwater is rich in iron, containing up to two orders of magnitude more total iron than in Olkiluoto (Table 4.1). Fracture minerals at Hästholmen include iron hydroxides, but pyrites are only sporadically present, while in the Olkiluoto fracture minerals, pyrite is one of the major components. This indicates that the presence or absence of pyrite as a fracture mineral correlates well with the presence or absence of SRB at the compared sites. Pyrite formation on fractures in these cold aquifers may reflect long-term SRB activity, which is not apparent from groundwater chemistry data. Examination of the concentrations of dissolved sulfate and sulfide was much less effective in predicting the presence or absence of SRB than was work with the MPN of SRB. No correlation could be found between sulfate or sulfide concentrations and SRB. The largest numbers of SRB in the samples from Finland were found in boreholes with either very low or

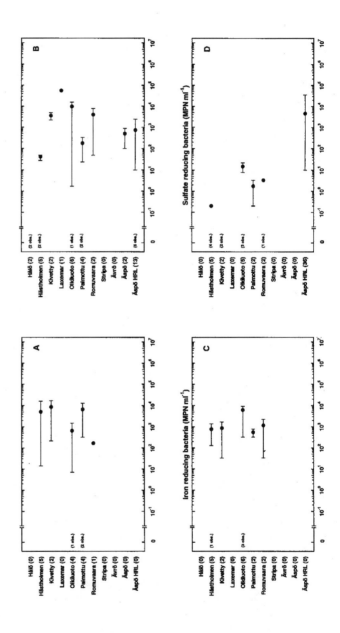

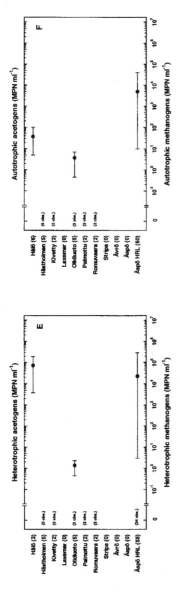

FIGURE 4.7 (*a–f*) Most probable numbers of physiological groups of microorganisms observed at the eleven sites investigated. The data are extracted from publications according to Table 4.1, except for recent, unpublished data obtained during spring 1999. The figures show the average number of microorganisms per site (●) and the bar gives the range of the data used to calculate the average number. Numbers in parentheses after the site names give the sum of observations for that site and each observation consists of a MPN determination with three or five parallel tubes in the dilution series. Numbers in parentheses above 0 on the *x*-axis give the number of negative observations.

very high sulfate and generally low sulfide concentrations. This example demonstrates that it is important to identify conservative indicators of microbial activity that are insensitive to transport processes, because the timeframes of various processes in hard rock may span millions of years at steady-state conditions. Even at the slowest transport rate, sulfate may replenish to SRB during such long periods, resulting in a steady-state concentration of sulfate and a buildup of pyrite precipitates. Long-term and very slow processes should be the focus when searching for evidence of subterranean microbial activity. Signatures in fracture minerals therefore seem to be reliable indicators of past and present microbial activity, especially if stable isotope ratios are added to the analysis protocol (Des Marais, 1996; Pedersen et al., 1997b).

In most hydrothermally altered fractures (see above), IRB have access to an almost unlimited source of ferric iron, provided they can access it. Humic and fulvic acids are common in most deep groundwaters and these complex compounds have been demonstrated to act as electron shuttles between ferric iron sources and IRB (Coates et al., 1998). The molecule size of these compounds is small enough to allow penetration of the rock matrix, which then enables iron reduction of parts of the rock that are not directly accessible to the IRB. Attempts to correlate numbers of IRB with amounts of ferric and ferrous iron have not been successful, because the iron redox couple is sensitive to inorganic processes much more than are sulfur redox couples, at least where reduction is concerned. Therefore, it is not possible to discriminate between biological and chemical iron redox reactions. The adaptation of mixing models has been more fruitful (Banwart et al., 1996). The effect of IRB on carbon dioxide and ferrous iron production has been demonstrated in a shallow groundwater intrusion system at the Äspö HRL tunnel. Organic carbon in the groundwater that reached the studied fracture zone was oxidized with ferric iron as the electron acceptor. This process rapidly reached a steady state that has been sustainable since the start of measurements in March 1991. A similar approach was taken at the Äspö HRL for determination of sulfate reduction along the tunnel. It was found that the MPN of SRB correlated well with geological, hydrological, and groundwater isotope data, indicative of ongoing sulfate reduction (Laaksoharju et al., 1995; Pedersen, 1997a).

Repeatedly obtained pure cultures and 16S rDNA sequences of acetogenic bacteria from Äspö HRL groundwater indicated that this physiological group of bacteria was important to the subterranean environment (Pedersen et al., 1996a). Later applications of enrichment media for heterotrophic and autotrophic acetogens support this hypothesis. Autotrophic acetogens form acetate from hydrogen and carbon dioxide, and the carbon may then be further transformed to methane by the acetoclastic methanogens. Heterotrophic acetogens were found at all the Finnish sites (Fig. 4.7c). Autotrophic acetogens frequently occurred in the Äspö HRL tunnel and at several of the Finland sites (Fig. 4.7d). The numbers of acetate-producing bacteria in the Äspö HRL environment correlated well with the numbers of heterotrophic methanogens (Fig. 4.7e), including acetoclastic ones (Kotelnikova and Pedersen, 1998).

TABLE 4.4 *In vitro* radiotracer and radiographic estimations of carbon transformations by unattached microorganisms in deep groundwater. The radiotracer data have been normalized to mole per liter of groundwater per hour to enable comparisons, although the time for incubation varies from 6–9.5 hours for Laxemar, Stripa, and Äspö up to 4–10 days for Äspö HRL. Acetate and leucine were labeled with ^{3}H and all other compounds were labeled with ^{14}C. See the following publications for details: Ekendahl and Pedersen (1994), Kotelnikova and Pedersen (1998), and Pedersen and Ekendahl (1990, 1992a, 1992b).

Borehole	Depth (m)	Transformation of Carbon Compounds (nMh^{-1})/(% active cells)								
		CO_2 to Cells	CO_2 to CH_4	CO_2 to Acetate	Formate to Cells	Acetate to Cells	Acetate to CH_4	Lactate to Cells	Glucose to Cells	Leucine to Cells
Laxemar										
KLX01	831–841	-[a]/-	[b]		-/-	0.001/29		0.230/21	0.072/2	0.005/56
KLX01	910–921	0.028/-			-/-	0.001/27		0.700/16	0.016/-	0.023/87
KLX01	999–1078	0.160/3			0.085/51	0.001/21		3.160/83	0.048/-	0.048/98
Stripa										
V2	799–807	0.520/5			0.018/4	·/·		0.043/16	0.031/5	0.002/55
V2	812–820	0.110/5			0.005/6	·/·		0.016/34	0.031/8	0.008/23
V2	970–1240	0.150/-			-/-	·/·		0.110/6	0.110/-	0.008/9

(*continued*)

TABLE 4.4 *(continued)*

Borehole	Depth (m)	Transformation of Carbon Compounds (nMh^{-1})/(% active cells)								
		CO$_2$ to Cells	CO$_2$ to CH$_4$	CO$_2$ to Acetate	Formate to Cells	Acetate to Cells	Acetate to CH$_4$	Lactate to Cells	Glucose to Cells	Leucine to Cells
Äspö HRL										
KR0012B	68	.	1170/·	9630/·	.	.	12,460/·	.	.	.
KR0013B	68	.	0/·	4620/·	.	.	3100/·	.	.	.
KR0015B	68	.	1470/·	15,300/·	.	.	12,370/·	.	.	.
SA813B	112	.	0/·	33,230/·	.	.	6300/·	.	.	.
SA1420A	192	.	0/·	3/·	.	.	120/·	.	.	.
KA2511A	345	.	530/·	2220/·	.	.	300/·	.	.	.
KA2512A	345	.	0/·	0/·	.	.	1290/·	.	.	.
KA2862A	380	.	7/·	0.0/·	.	.	0/	.	.	.
KA3005A	400	.	150/·	30/·	.	.	30/·	.	.	.
KA3010A	400	.	59/·	2/·	.	.	0/·	.	.	.
KA3067A	409	.	980/·	2/·	.	.	110/·	.	.	.
KA3105A	414	.	70/·	7/·	.	.	130/·	.	.	.
KA3110A	414	.	510/·	65/·	.	.	0/·	.	.	.
HB0025A	420	.	25/·	90/·	.	.	130/·	.	.	.
KA3385A	446	.	20/·	0/·	.	.	0/·	.	.	.

[a]Not detected.
[b]Not examined.

The presence of hydrogen and carbon dioxide in most deep groundwater examined (Table 4.3) indicates that autotrophic methanogenesis is possible. The MPN analyses, indeed, report significant numbers of this group at Hålö, the Äspö HRL, and in Olkiluoto (Fig. 4.7f). We have no obvious explanation for the lack of evidence of methanogens in most Finnish samples. At Äspö, the media used were developed and adjusted during repeated sampling in the tunnel. Alternatively, during the investigations in Finland, we only had one opportunity of sampling and adjustment of the media to the prevailing conditions at each borehole level. All samples from Finland were collected from the ground surface in boreholes with the PAVE method (Fig. 4.2), which offers one sample of 300-mL groundwater per level. Generally, it is not financially possible to repeat such sampling campaigns. However, it cannot be concluded, based on the lack of positive MPN enrichments, that there were in fact no, or very few, methanogens at the studied Finland sites.

6 ESTIMATIONS OF CARBON TRANSFORMATION ACTIVITIES IN IGNEOUS ROCK AQUIFERS

6.1 Methodology

Radiolabeled compounds for the estimation of microbial activity have been in use in microbial ecology for decades (Grigorova and Norris, 1990). With this technique, samples are incubated with the radiotracer of interest and examined. Cells, or products of cells, can be separated and examined for radioactivity using standard liquid scintillation counting techniques. This method will yield average activity results for the whole sample. The activity of individual cells can be examined using a microautoradiography (MARG) technique (Tabor and Neihof, 1982, 1984). Both the liquid scintillation and MARG techniques have been applied to microorganisms from the Laxemar, Stripa, and the Äspö HRL sites, with varying radiotracers and incubation times (Table 4.4). The advantages of the MARG technique are that individual cells can be examined and the method successfully combined with nucleic acid probing, offering specific information on selected metabolic activities (Amann et al., 1995). However, at very low metabolic rates, the MARG method is less applicable than liquid scintillation. In an aquatic sample supplemented with a radiolabeled substrate for a short period, an individual cell must have a minimum uptake rate to become sufficiently labeled to produce a positive microradiogram. The lowest radioactivity that resulted in cells radioactive enough to expose the film faster than the background radiation was found to be 10^{-3} disintegrations/min (Pedersen and Ekendahl, 1992a). This level corresponds to $0.1–1 \times 10^{-16}$ mole ^{14}C/bacterium and 2.1×10^{-19} mole ^{3}H/bacterium for the method used to generate the MARG data in Table 4.4. A prolonged incubation time, more than a couple of hours, will lower the detection limits, but will concurrently increase the background and also allow for growth on the added substrates. The data in Tables 4.4 and 4.5 were generated with short incubation times for Laxemar and Stripa, and extended incubation times for the Äspö HRL samples.

TABLE 4.5 The total number of bacteria in groundwater and on surfaces exposed to flowing groundwater, and the amounts of carbon dioxide and lactate transformed by these microorganisms. Data are from Pedersen and Ekendahl (1992a, 1992b).

		Unattached bacteria			Attached bacteria			Hypothetical Volume-to-Surface Ratios in a 0.1-mm-wide Fracture					
Borehole	Depth (m)	Cells $m^{-3} \times 10^{10}$	$\mu m\ CO_2$ $m^{-3}\ day^{-1}$	μm lactate $m^{-3}\ day^{-1}$	Cells $m^{-2} \times 10^{10}$	$\mu m\ CO_2$ $m^{-2}\ day^{-1}$	μm lactate $m^{-2}\ day^{-1}$	Cells $m^{-2} \times 10^{10}$	$\mu m\ CO_2$ $m^{-2}\ day^{-1}$	μm lactate $m^{-2}\ day^{-1}$	Cells $m^{-3} \times 10^{10}$	$\mu m\ CO_2$ $m^{-3}\ day^{-1}$	μm lactate $m^{-3}\ day^{-1}$
Laxemar													
KLX01	831–841	1.5	—[a]	5.6	0.09[b]	0.9	2.6				1200	—	9200
KLX01	910–921	2.1	0.68	17	0.12[b]	1.1	6.0				1100	32,000	7000
KLX01	999–1078	6.8	4	76	0.10[b]	1.5	0.14				300	7500	36
Stripa													
V2	799–807	0.54	12.5	1	1.2[c]	—	2.0				44,000	—	40,000
V2	812–820	0.18	2.6	0.4	7.1[c]	0.48	5.5				790,000	3600	280,000
V2	970–1240	12	3.5	2.7	5.9[c]	1.8	37				9800	10,000	270,000

[a] Not detected.
[b] 20 days of exposure to flowing groundwater.
[c] 120 days of exposure to flowing groundwater.

6.2 *In vitro* Activity of Unattached Cells

Both the Laxemar and Stripa populations assimilated all added compounds at varying rates. Carbon dioxide was assimilated at relatively low rates as were formate, acetate, and glucose. The fastest uptake was found to occur with lactate. Generally, ten times more lactate than other ^{14}C-labeled compounds was assimilated, and up to 83% of the deepest population at Laxemar actively assimilated lactate. Lactate seemed to be the preferred carbon source and it can be utilized by IRB, SRB, and heterotrophic acetogens at the anaerobic conditions prevailing in deep groundwater. A look at Figs. 4.7*b* and 4.7*d* will confirm that these three physiological groups of bacteria are indeed common. The incorporation of acetate was not high, but as a ^{3}H-labeled acetate was used, the sensitivity to the MARG technique was higher than for a corresponding ^{14}C-compound. MARG revealed that a large proportion of the cells assimilated this compound [see Pedersen and Ekendahl (1992a) for a detailed discussion of method sensitivity]. Acetate is used by many IRB, SRB, and heterotrophic methanogens, both of which are frequently represented in the MPN data (Figs. 4.7*b* and 4.7*e*). Likewise, acetoclastic methanogens have also been found to be active in many of the studied Äspö HRL boreholes (Table 4.4). Uptake of carbon dioxide by the Stripa and Laxemar populations points to the presence of autotrophic organisms. Later studies confirmed these assumptions, and autotrophic acetogens and methanogens have been enriched, enumerated (Figs. 4.7*d* and 4.7*f*), and isolated at many of the studied sites. Autotrophic methane production was common at the Äspö HRL, as shown by the radiotracer scintillation results (Table 4.4). Consequently, a relatively large data set describing carbon transformations has been obtained by DBL over a 12-year investigation period. When these data are compared with MPN data, it appears that most of the observed radio-labeled carbon transformations can be correlated with the culturing of physiological groups of microorganisms that can transform the corresponding carbon compounds.

Significant formation rates of methane and acetate were obtained *in vitro* from Äspö groundwater at a temperature (17°C) close to the *in situ* temperature (10–17°C) (Table 4.4). The general trends for heterotrophic methane formation and for acetate formation followed those observed with MPN and enrichments (Kotelnikova and Pedersen, 1998). The highest activity was found in the shallow boreholes (45–68 m), which also had the highest numbers of heterotrophic methanogens and homoacetogens. Autotrophic methane formation did not, however, follow the culturability trend. This may have resulted from the increasing difficulty with increasing depth to mimic *in situ* conditions *in vitro* for parameters such as pressure and dissolved gases. With this exception, three independent methods, MPN enumerations (Fig. 4.7), enrichments (Kotelnikova and Pedersen, 1998) and radio-tracer assays (Table 4.4), have all established the presence of active heterotrophic methanogens and homoacetogens in the examined groundwater.

6.3 *In vitro* Activity of Attached Cells

An aquifer in hard rock is made of two surfaces that are wavy and rough. They are in contact with each other at some points and are distanced from each other at others.

Aquifers generally expose a large surface to the contained groundwater and the surface-to-volume ratio increases as the aquifer width decreases. Microorganisms commonly have a strong tendency to attach to surfaces in aquatic environments (Marshall, 1984) and to form biofilms (Characklis and Marshall, 1990). The ability of deep aquifer microorganisms to form biofilms on artificial glass and rock surfaces introduced in aquifers has been studied with slowly flowing groundwater at three depths in Stripa (Ekendahl and Pedersen, 1994; Pedersen and Ekendahl, 1992a), four depths at Laxemar (Pedersen and Ekendahl, 1990, 1992b), and four depths in the Äspö HRL tunnel (Pedersen et al., 1996a). The time of exposure ranged from 20–161 days at flowrates from 1–31 mm sec^{-1}. All surfaces exhibited attached cells and in cases of prolonged exposure (> 25 days), growing colonies could also be found. This showed that the microorganisms in the aquifers not only attached passively, they were able to both attach and grow on the surfaces.

A hypothetical comparison of cell numbers and activities of attached and unattached bacteria in a 0.1-mm-wide fracture is shown in Table 4.5. It demonstrates the potential importance of attached microorganisms versus unattached microorganisms in subsurface environments. The attached bacteria have generally exhibited a higher activity per cell than have the unattached bacteria (not shown) and together they would be up to five orders of magnitude more active than all the unattached microorganisms suspended in the assumed 0.1-mm fracture. The natural flowrate of groundwater in deep aquifers depends on the existing hydraulic gradient. It is generally very slow, much slower than the flowrate used in the experiments described above. It is still an open question whether attached microorganisms are common and active on aquifer rock surfaces under pristine conditions. It would be necessary to drill and directly examine undisturbed aquifers to answer this question. The very high drilling cost and the relatively low probability of intersecting an open fracture during a drilling operation tie up this question with very high experimental costs before a statistically significant observation series would be obtained.

6.4 *In vitro* Viability of Unattached and Attached Cells

Leucine assimilation is virtually specific to bacteria if low (nM) concentrations are applied. This amino acid is used for protein synthesis by many bacteria during growth (Kirchman et al., 1985). It can also be used as a carbon and energy source and fermented by proteolytic clostridia via the Strickland reaction. High percentages, up to 98%, of most populations described in Tables 4.4 and 4.5, assimilated leucine. This showed that major portions of the studied communities were viable.

7 SPECIES DIVERSITY AND PHYLOGENY OF MICROORGANISMS IN IGNEOUS ROCK AQUIFERS

7.1 Molecular Investigations

The MPN assays and activity measurements described above have supplied extensive information about present and active physiological groups in the deep aquifers examined, but these methods do not reveal species diversity and phylogeny. Classical microbiology involves characterization and species affiliation based on large sets of phenotypic and genotypic data. It is very time-consuming and not a suitable technique for the screening of environmental samples for species diversity. The concept of microbial diversity has been changed by the growth of available sequence data from ribosomal 16/18S rDNA. The cloning and sequencing of RNA from microbes living in their natural environments have revealed a genetic diversity beyond the dreams of researchers whose tools were limited not so long ago to culturing and microscopy (Pace, 1997). An early strategy of DBL was to add 16/18S rDNA sequencing of environmental DNA to the investigations. The first site thus examined was the Stripa borehole V2 (Ekendahl et al., 1994). Attached microorganisms were studied. All sequences found among the 72 clones investigated belonged to the eubacterial branch Proteobacteria. Two of the major groups fell into the beta group and the third into the gamma group. The next site to be investigated was the Äspö HRL tunnel. In the first campaign, 155 clones of unattached and attached bacteria from nine boreholes were sequenced (Pedersen et al., 1996a). A comparison of the predominating 16S rRNA gene sequences with international databases revealed three clone groups that could be identified as bacteria on the genus level: the *Bacillus*, *Desulfovibrio*, and *Acinetobacter* genera. One of the clone groups could be identified as a fungus. A second campaign was executed during a contamination control investigation while drilling boreholes in the Äspö HRL tunnel (Pedersen et al., 1997c); 158 clones were sequenced. Several clones showed a high similarity with 16S rRNA genes from known and sequenced bacteria, such as *Bacillus*, *Desulfovibrio*, *Desulfomicrobium*, *Methylophilus*, *Acinetobacter*, *Shewanella*, and a yeast.

The diversity of the bacterial community detected in the Äspö HRL groundwater is not large compared with the numbers expected in surface soils. For example, 4000 species were found in 1 g of soil by Torsvik et al. (1990). Of the 385 sequenced clones from Stripa and the Äspö HRL, 122 constituted unique sequences. Each represented a possible species not entered in the database at the time of the comparison. On average, approximately one-third of the sequenced clones represented unique species. In other DBL investigations, identical to the Äspö HRL study, a similar molecular biodiversity per total number of sequenced clones was observed, namely, 44 specific clone groups out of 130 sequenced clones from five boreholes at the natural nuclear reactor in Bangombé, Gabon, Africa (Pedersen et al., 1996b). Twenty specific clone groups out of 67 sequenced clones from radioactive waste buffer material (Stroes-Gascoyne et al., 1997) and 23 specific clone groups out of 87 sequenced clones from alkaline spring water (Pedersen et al., 1997a). All these investigations clearly have not exhausted the sequences to be found, because new sequences were found in nearly every additional sample collected. The 16/18S rDNA sequence data collection clearly has to be scaled up significantly for the study of most groundwater sites, requiring an application of automated procedures. *Bacteria* and *Archaea* are likely to be the first organisms for which automated biodiversity assessment is practical (although other soil microbiota may also be surveyed in the same way). As major constituents of the global community, *Bacteria* and *Archaea* deserve assessment in their own right in addition to their value as indicators (Crozier et al., 1999).

7.2 Characterization and Description of New Species

The molecular work described above has provided extensive insights into the phylogenetic diversity of igneous rock aquifer microorganisms. However, it does not reveal species-specific information unless 100% identity of the 16S rDNA gene with a known and described microorganism is obtained. The huge diversity of the microbial world makes the probability of such a hit very small and none of the 122 specific sequences mentioned above had 100% 16S rDNA identities with described species. Still, if 100% identity is obtained, there may yet be strain-specific differences in some characteristics unrevealed by the 16S rDNA information (Fuhrman and Campbell, 1998). DNA reassociation studies are required before a final judgment about species identity can be made (Stackebrandt and Goebel, 1994). If species information is required, time-consuming methods in systematic micro-biology must be applied to pure cultures. Known genera or species can be identified through these methods. Several isolates from the Laxemar, Äspö, and Äspö HRL sites have been identified as *Shewanella putrefaciens*, *Pseudomonas vesicularis*, and *Pseudomonas fluorescens* (Pedersen and Ekendahl, 1990; Pedersen et al., 1996a). An isolate that does not match a known genus or species obviously provides the opportunity to describe a new species.

Three new subterranean species from deep igneous rock aquifers have been described and reported by DBL. Sulfate-reducing bacteria are common in the aquifers studied (Fig. 4.7b) and three SRB species, based on their different 16S rDNA sequences, were repeatedly isolated from different boreholes in the Äspö HRL tunnel [KAS03 at 533–626 m, and KR0013, SA1062, and HA1327 (Pedersen et al., 1996a)]. One of them, Aspo-1, had a 16S rDNA identity greater than 99% with *Desulfomicrobium baculatum*; it was not studied further. This genus seems to be very common at Äspö since its 16S rDNA sequence was repeatedly retrieved from other tunnel boreholes at Äspö (KA2858 and KA3105) (Pedersen et al., 1997c). The isolate Aspo-2 was characterized in detail and described as a new species, *Desulfovibrio aespoeensis* (Motamedi and Pedersen, 1998). It is a meso-philic species with growth characteristics that appear well adapted for life in the aquifers from which it was isolated. Three autotrophic methane-producing strains of *Archaea* were isolated from the Äspö HRL tunnel boreholes at depths of 68, 409, and 420 m, respectively. These organisms were nonmotile, small, thin rods, 0.1–0.15 μm in diameter, and able to utilize $H_2 + CO_2$ or formate as substrates for growth and methanogenesis. One of the isolates, denoted A8p, was studied in detail. Phylogenetic characterization based on 16S rRNA gene sequence comparisons placed this isolate in the genus *Methanobacterium* (Kotelnikova et al., 1998). Phenotypic and phylogenetic characteristics indicate that the alkaliphilic, halo-tolerant strain A8p represents a new species and we proposed the name *Methano-bacterium subterraneum*. It grew with a doubling time of 2.5 hours under optimal conditions (20–40°C, pH 7.8–8.8, and 0.2–1.2 *M* NaCl). *M. subterraneum* is eurythermic since it can grow at a wide range of temperatures from 3.6–45°C.

Methane is common in most groundwater studied (Table 4.3) and there has been a growing interest in methanotrophs at DBL. Their consumption of oxygen with methane as electron donor is beneficial to repositories for high-level radioactive waste. Their activities have, therefore, been studied in detail (Kotelnikova and Pedersen, 1999). During the DBL investigations of microbial methane oxidation in the Äspö HRL tunnel, several oxygen-dependent methanotrophic isolates were obtained. One of them has been successfully described in close collaboration with Russian experts on methylotrophic bacteria (Kalyuzhnaya et al., 1999). Methane-utilizing bacteria were first enriched from deep granitic rock environments and affiliated by amplification of functional and phylogenetic genes. Type I methano-trophs belonging to the genera *Methylomonas* and *Methylobacter* dominated in the enrichment cultures from depths below 400 m. A pure culture of an obligate methanotroph, strain SR5, was then isolated and characterized. Based on phenotypic and genotypic characteristics, we proposed that the strain SR5 is a new species, *Methylomonas scandinavica*. Pigmented motile rods of this new species contained intracytoplasmic membranes as stacks of vesicles, assimilated methane via the ribulose monophosphate pathway, and had an incomplete tricarboxylic acid cycle. *M. scandinavica* grows at temperatures of 5–30°C with an optimum of 15°C, close to the *in situ* temperature. Whole cell protein, enzymatic, and physiological analyses of *M. scandinavica* revealed significant differences between this and the other representatives of Type I methanotrophs. The prospect of anaerobic methane oxidation is an intriguing possibility that has been approached in different environments (Hindrichs et al., 1999). However, absolute evidence of a laboratory culture of an anaerobic methane-consuming species or consortium is still lacking.

8   IS THERE A DEEP HYDROGEN-DRIVEN BIOSPHERE IN DEEP IGNEOUS ROCK AQUIFERS?

The repeated observations of autotrophic, hydrogen-dependent microorganisms in the deep aquifers studied by DBL (Figs. 4.7*d* and 4.7*f,* Table 4.4) imply that hydrogen is an important electron and energy source, and that carbon dioxide is an important carbon source for the subsurface biosphere. Hydrogen and carbon dioxide have been found in μM concentrations at most of the sites investigated (Table 4.3) together with methane. Methane is a major product of autotrophic methanogens, shown to be very active *in vitro* at Äspö HRL (Table 4.4). Therefore, a model of a hydrogen-driven biosphere in deep Fennoscandian Shield igneous rock aquifers has been suggested by DBL during the course of its subsurface microbiology program (Kotelnikova and Pedersen, 1998; Pedersen, 1993b, 1997a, 1999; Pedersen and Albinsson, 1992; Pedersen, 2000). The organism base for this biosphere is suggested to be composed of autotrophic acetogens that have the capability of oxidizing hydrogen and fixing carbon dioxide to produce acetate, and methanogens that yield methane from hydrogen and carbon dioxide (autotrophic methanogens) or from acetate produced by autotrophic acetogens (acetoclastic methanogens) (Fig. 4.8). All components needed for the life cycle in Fig. 4.8 have been shown to occur in deep igneous rock aquifers and the microbial activities expected have been demonstrated at significant rates *in vitro*. The model has, consequently, convincing support from the qualitative data obtained.

The theory of a deep biosphere driven by hydrogen generated in deep geological strata requires more research on possible sources of hydrogen. There are at least six possible processes by which crustal hydrogen is generated (Apps and van de Kamp, 1993): (1) reaction between dissolved gases in the system of carbon-hydrogen-oxygen-sulfur in magmas, especially in those with basaltic affinities; (2) decomposition of methane to carbon (graphite) and hydrogen at temperatures above 600°C; (3) reaction between carbon dioxide, water, and methane at elevated temperatures in vapors; (4) radiolysis of water by radioactive isotopes of uranium, thorium and their daughters, and potassium; (5) cataclasis of silicates under stress in the presence of water; and (6) hydrolysis by ferrous minerals in mafic and ultramafic rocks. It is important to explore the scale of these processes and the rates at which the produced hydrogen is becoming available for deep microbial ecosystems.

It also remains to examine *in situ* metabolic rates for the model in Fig. 4.8, which will require meticulous experimental conditions, because of very slow metabolic rates under nondisturbed conditions. The central question to address during such an

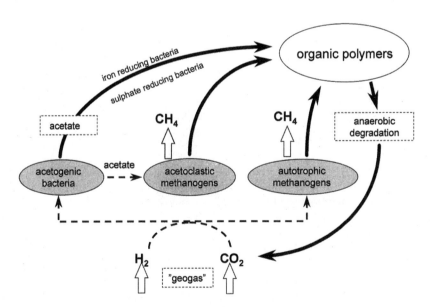

FIGURE 4.8 The deep hydrogen-driven biosphere hypothesis, illustrated by its carbon cycle. At relevant temperature and water availability conditions, intraterrestrial microorganisms are capable of performing a life cycle that is independent of sun-driven ecosystems. Hydrogen and carbon dioxide from the deep crust of earth are used as energy and carbon sources.

endeavour is whether hydrogen-driven microbial chemolithotrophic *in situ* activities at depth are in balance with possible renewal rates of hydrogen. An indisputable yes on this question is crucial for the unequivocal confirmation of a deep hydrogen-driven biosphere in the deep igneous rock aquifers of the Fennoscandian Shield.

Acknowledgments This work was supported by the Swedish Nuclear Fuel and Waste Management Company and the Swedish Natural Science Research Council. The author is grateful to numerous colleagues at the Deep Biosphere laboratory at Göteborg University and Äspö hard rock laboratory, and Lotta Hallbeck and Shelley Haveman for critical comments on the manuscript.

REFERENCES

Almén K-E and Zellman O (1991) *Field Investigation Methodology and Instruments Used in the Preinvestigation Phase, 1986–1990*. Stockholm: Swedish Nuclear Fuel and Waste Management Co., SKB TR 92-21.

Amann RI, Ludwig W, and Schleifer K-H (1995) Phylogenetic identification and *in situ* detection of individual microbial cells without cultivation. *Microbiol Rev 59*:143–169.

Apps JA and van de Kamp PC (1993) Energy gases of abiogenic origin in the Earth's crust. The future of energy gases. U.S. Geologic Survey Professional Paper 1570 81–132.

Banwart S, Tullborg E-L, Pedersen K, Gustafsson E, Laaksoharju M, Nilsson A-C, Wallin B, and Wikberg P (1996) Organic carbon oxidation induced by largescale shallow water intrusion into a vertical fracture zone at the Äspö Hard Rock Laboratory (Sweden). *J Contam Hydrol 21*:115–125.

Barnicoat AC, Henderson IHC, Knipe RJ, Yardley BWD, Napier RW, Fox NPC, Keney AK, Muntingh DJ, Strydom D, Winkler KS, Lawrence SR, and Cornford C (1997) Hydrothermal gold mineralization in the Witwatersrand basin. *Nature 386*:820–824.

Birch L and Bachofen R (1990) Complexing agents from microorganisms. *Experientia 46*:827–834.

Blomqvist R, Suksi J, Ruskeeniemi T, Ahonen L, Niini H, Vuorinen U, and Jakobsson K (1995) *The Palmottu Natural Analogue Project. Summary Report 1992–1994*. Espoo: Geological Survey of Finland, Nuclear Waste Disposal research, YST-88.

Characklis WG and Marshall KC (1990) *Biofilms*. New York: John Wiley & Sons.

Coates JD, Ellis DJ, Blunt-Harris EL, Gaw CV, Roden EE, and Lovley DR (1998) Recovery of humic-reducing bacteria from a diversity of environments. *Appl Environ Microbiol 64*:1504–1509.

Crozier RH, Agapov P-M, and Pedersen K (1999) Towards complete biodiversity assessment: An evaluation of the subterranean bacterial communities in the Oklo region of the sole surviving natural nuclear reactor. *FEMS Microbiol Ecol 28*:325–334.

Des Marais DJ (1996) Stable light isotope biogeochemistry of hydrothermal systems. In Bock GR and Goode JA (eds): *Evolution of Hydrothermal Ecosystems on Earth (and Mars?)*. Chichester: John Wiley & Sons Ltd., pp 83–98.

Ekendahl S, Arlinger J, Ståhl F, and Pedersen K (1994) Characterization of attached bacterial populations in deep granitic groundwater from the Stripa research mine with 16S-rRNA gene sequencing technique and scanning electron microscopy. *Microbiology 140*:1575–1583.

Ekendahl S and Pedersen K (1994) Carbon transformations by attached bacterial populations in granitic ground water from deep crystalline bed-rock of the Stripa research mine. *Microbiology 140*:1565–1573.

Flodén T and Söderberg P (1994) Shallow gas traps and gas migrations models in crystalline bedrock areas offshore Sweden. *Baltica 8*:50–56.

Fredrickson JK and Onstott TC (1996) Microbes deep inside the earth. *Scien Amer 275*:42–47.

Frieg B, Alexander WR, Dollinger H, Buhler C, Haag P, Mörj A, and Ota K (1998) *In situ* impregnation for investigating radionuclide retardation in fractured repository host rocks. *J Contam Hydrol 35*:115–130.

Fry JC (1990) Direct methods and biomass estimation. In Grigorova R and Norris JR (eds): *Methods in Microbiology*, Vol 22. London: Academic Press, pp 41–85.

Fuhrman JA and Campbell L (1998) Microbial microdiversity. *Nature 393*:410–467.

Ghiorse WC and Balkwill DL (1983) Enumeration and morphological characterization of bacteria indigenous to subsurface sediments. *Develop Indus Microbiol 24*:213–224.

Ghiorse WC and Wilson JT (1988) Microbial ecology of the terrestrial subsurface. *Adv Appl Microbiol 33*:107–172.

Grenthe I, Stumm W, Laaksoharju M, Nilsson A-C, and Wikberg P (1992) Redox potentials and redox reactions in deep ground water systems. *Chem Geol 98*:131–150.

Grigorova R and Norris JR (1990) Techniques in microbial ecology. In Grigorova R and Norris JR (eds): *Methods in Microbiology*, Vol 22. London: Academic Press, p 627.

Gustafsson G, Stanfors R, and Wikberg P (1988) *Swedish Hard Rock Laboratory First Evaluation of Preinvestigations, 1986–1987 and Target Area Characterization*. SKB TR 88-16.

Gàal G and Gorbatschev R (1987) An outline of the Precambrian evolution of the Baltic Shield. *Precam Res 35*:15–52.

Haveman SA, Pedersen K, and Ruotsalainen P (1998) *Geomicrobial Investigations of Groundwater from Olkiluouto, HÄstholmen, Kivetty and Romuvaara, Finland*. Helsinki: POSIVA OY, POSIVA 98-09.

Haveman SH, Pedersen K, and Ruotsalainen P (1999) Distribution and metabolic diversity of microorganims indeep igneous rock aquifers of Finland. *Geomicrobiol J 16*:277–294.

Herbert RA (1990) Methods for enumerating microorganisms and determining biomass in natural environments. In Grigorova R and Norris JR (eds): *Methods in Microbiology*, Vol. 22. London: Academic Press, pp 1–39.

Hindrichs K-U, Hayes JM, Sylva SP, Brewer PG, and DeLong EF (1999) Methane-consuming archaebacteria in marine sediments. *Nature 398*:802–805.

Ishii K, Takki S, Fukunaga S, and Aoki K (2000) Characterization by denaturing gradient gel electrophoresis of bacterial communities in deep groundwater at the Kamaishi Mine. *Japan J Gen Appl Microbiol 46*:85–93.

Jain DK, Stroes-Gascoyne S, Providenti M, Tanner C, and Cord I (1997) Characterization of microbial communities in deep groundwater from granitic rock. *Can J Microbiol 43*:272–283.

Kalyuzhnaya MG, Khmelenina VN, Kotelnikova S, Pedersen K, Holmquist L and Trotsenko YA (1999) *Methylomonas scandinavica*, sp. nov. a new methanotrophic psychrotrophic bacterium isolated from deep igneous rock ground water of Sweden. *System Appl Microbiol 22*:565–572.

Kirchman D, K'nees E, and Hodson R (1985) Leucine incorporation and its potential as a measure of protein synthesis by bacteria in natural aquatic systems. *Appl Environ Microbiol 49*:599–607.

Kotelnikova S, Macario AJL, and Pedersen K (1998) *Methanobacterium subterraneum*, sp. nov., a new alkaliphilic, eurythermic and halotolerant methanogen isolated from deep groundwater. *Internat J System Bacteriol 48*:357–367.

Kotelnikova S and Pedersen K (1997) Evidence for methanogenic *Archaea* and homoaceto-genic *Bacteria* in deep granitic rock aquifers. *FEMS Microbiol Rev 20*:339–349.

Kotelnikova S and Pedersen K (1998) Distribution and activity of methanogens and homo-acetogens in deep granitic aquifers at Äspö Hard Rock Laboratory, Sweden. *FEMS Microbiol Ecol 26*:121–134.

Kotelnikova S and Pedersen K (1999) *The microbe-REX Project: Microbial O_2 Consumption in the Äspö Tunnel*. Stockholm: Swedish Nuclear Fuel and Waste Management Co., SKB-TR 99-17.

Laaksoharju M, Pedersen K, Rhen I, Skårman C, Tullborg E-L, Wallin B, and Wikberg W (1995) *Sulphate Reduction in the Äspö HRL Tunnel*. Stockholm: Swedish Nuclear Fuel and Waste Management Co., SKB-TR 95-25.

Marshall KC (1984) *Microbial Adhesion and Aggregation*. Berlin: Springer-Verlag.

Moreno L, Neretnieks I, and Klockars CE (1985) Analysis of some laboratory tracer runs in natural fissures. *Water Resources Res 21*:951–958.

Motamedi M and Pedersen K (1998) *Desulfovibrio aespoeensis* sp. nov. a mesophilic sulfate-reducing bacterium from deep groundwater at Äspö hard rock laboratory, Sweden. *Internat J System Bacteriol 48*: 311–315.

Murakami Y, Naganuma T, and Iwatuki T (1999) Deep subsurface microbial communities in the Tono area, central Japan. *J Nucl Fuel Cycle Environ 5*:59–66.

Nilsson A-C (1995) *Compilation of Groundwater Chemistry Data from Äspö 1990–1994*. Stockholm: Swedish Nuclear Fuel and Waste Management Co., SKB PR 25-95-02.

Nordstrom DK, Andrews JN, Carlsson L, Fontes J-C, Fritz P, Moser H, and Olsson T (1985) *Hydrogeological and Hydrogeochemical Investigations in Boreholes—Final Report of the Phase 1 Geochemical Investigations of the Stripa Groundwaters*. Stockholm: Swedish Nuclear Fuel and Waste Management Co., SP TR 85-06.

Pace NR (1997) A molecular view of microbial diversity and the biosphere. *Science 276*:734–740.

Pedersen K (1993a) The deep subterranean biosphere. *Earth-Science Rev 34*:243–260.

Pedersen K (1993b) Bacterial processes in nuclear waste disposal. *Microbiol Europe 1*:18–23.

Pedersen K (1997a) Microbial life in granitic rock. *FEMS Microbiol Rev 20*:399–414.

Pedersen K (1997b) Technical Report 97-22, Stockholm, Swedish Nuclear Fuel and Waste Management Co.

Pedersen K (1999) Subterranean microorganisms and radioactive disposal in Sweden. *Eng Geol 52*:163–176.

Pedersen K (2000) Exploration of deep intraterrestrial microbial life: Current perspectives. *FEMS Microbiol Lett 185*:9–16.

Pedersen K and Albinsson Y (1991) Effect of cell number, pH and lanthanide concentration on the sorption of promethium by *Shewanella putrefaciens*. *Radiochimica Acta 54*:91–95.

Pedersen K and Albinsson Y (1992) Possible effects of bacteria on trace element migration in crystalline bed-rock. *Radiochimica Acta 58/59*:365–369.

Pedersen K, Arlinger J, Ekendahl S, and Hallbeck L (1996a) 16S rRNA gene diversity of attached and unattached groundwater bacteria along the access tunnel to the Äspö Hard Rock Laboratory, Sweden. *FEMS Microbiol Ecol 19*:249–262.

Pedersen K, Arlinger J, Erlandson A-C, and Hallbeck L (1997a) Culturability and 16S rRNA gene diversity of microorganisms in the hyperalkaline groundwater of Maqarin, Jordan. In Pedersen K (ed): *Investigations of Subterranean Microorganisms and their Importance for Performance Assessment of Radioactive Waste Disposal. Results and Conclusions Achieved during the Period 1995 to 1997*. Stockholm: Swedish Nuclear Fuel and Waste Management Co., pp 239–262.

Pedersen K, Arlinger J, Hallbeck L, and Pettersson C (1996b) Diversity and distribution of subterranean bacteria in ground water at Oklo in Gabon, Africa, as determined by 16S-rRNA gene sequencing technique. *Molec Ecol 5*:427–436.

Pedersen K and Ekendahl S (1990) Distribution and activity of bacteria in deep granitic groundwaters of southeastern Sweden. *Microbial Ecol 20*:37–52.

Pedersen K and Ekendahl S (1992a) Incorporation of CO_2 and introduced organic compounds by bacterial populations in groundwater from the deep crystalline bedrock of the Stripa mine. *J Gen Microbiol 138*:369–376.

Pedersen K and Ekendahl S (1992b) Assimilation of CO_2 and introduced organic compounds by bacterial communities in ground water from Southeastern Sweden deep crystalline bedrock. *Microbial Ecol 23*:1–14.

Pedersen K, Ekendahl S, Tullborg E-H, Furnes H, Thorseth I-G, and Tumyr O (1997b) Evidence of ancient life at 207 m depth in a granitic aquifer. *Geology 25*:827–830.

Pedersen K, Hallbeck L, Arlinger J, Erlandson A-C, and Jahromi N (1997c) Investigation of the potential for microbial contamination of deep granitic aquifers during drilling using 16S rRNA gene sequencing and culturing methods. *J Microbiol Meth 30*:179–192.

Pedersen K and Karlsson F (1995) *Investigations of Subterranean Microorganisms—Their Importance for Performance Assessment of Radioactive Waste Disposal*. Stockholm: Swedish Nuclear Fuel and Waste Management Co., SKB TR 95-10.

Rhén I, Backblom G, Gustafson G, Stanfors R, and Wikberg P (1997) *Äspö HRL—Geoscientific Evaluation 1997/2. Results from Pre-investigations and Detailed Site Characterization. Summary Report*. Stockholm: Swedish Nuclear Fuel and Waste Management Co., SKB TR 97-03.

Sherwood Lollar B, Frape SK, Fritz P, Macko SA, Welhan JA, Blomqvist R, and Lahermo PW (1993a) Evidence for bacterially generated hydrocarbon gas in Canadian Shield and Fennoscandian Shield rocks. *Geochimica et Cosmochimica Acta 57*:5073–5085.

Sherwood Lollar B, Frape SK, Weise SM, Fritz P, Macko SA, and Welhan JA (1993b) Abiogenic methanogenesis in crystalline rocks. *Geochimica et Cosmochimica Acta 57*:5087–5097.

Smellie J and Wikberg P (1991) Hydrochemical investigations at Finnsjön, Sweden. *J Hydrol 126*:129–158.

Smellie, JAT, Laaksoharju M, and Wikberg P (1995) Äspö, SE Sweden: A natural groundwater flow model derived from hydrogeochemical observations. *J Hydrol 172*:147–169.

Smith DC, Spivack AJ, Fisk MR, Haveman SA, Staudigel H, and ODP Leg 185 Shipboard

Scientific Party (2000) Tracer-based estimates of drilling-induced microbial contamination of deep sea crust. *Geomicrobiol J 17*:207–219.

Stackebrandt E and Goebel BM (1994) Taxonomic note: A place for DNA-DNA reassociation and 16S rRNA sequence analysis in the present species definition in bacteriology. *Internat J System Bacteriol 44*:846–849.

Stanfors R, Erlström M, and Markström I (1997) *Äspö HRL—Geoscientific Evaluation 1997/1. Overview of Site Characterisation 1986–1995.* Stockholm: Swedish Nuclear Fuel and Waste Management Co., SKB TR 97-02.

Stroes-Gascoyne S, Pedersen K, Haveman SA, Dekeyser K, Arlinger J, Daumas S, Ekendahl S, Hallbeck L, Hamon CJ, Jahromi N, and Delaney T-L (1997) Occurrence and identification of microorganisms in compacted clay-based buffer material designed for use in a nuclear fuel waste disposal vault. *Can J Microbiol 43*:1133–1146.

Söderberg P and Flodén T (1991) Pockmark development along a deep crustal structure in the northern Stockholm Archipelago, Baltic Sea. *Beitrage Meereskunde 62*:79–102.

Söderberg P and T Flodéen (1992) Gas seepages, gas eruptions and degassing structures in the seafloor along the Strömma tectonic lineament in the crystalline Stockholm Archipelago, east Sweden. *Continental Shelf Res 12*:1157–1171.

Tabor PS and Neihof RA (1982) Improved microautoradiographic method to determine individual microorganisms active in substrate uptake in natural waters. *Appl Environ Microbiol 44*:945–953.

Tabor PS and Neihof RA (1984) Direct determination of activities for microorganisms of Chesapeake Bay populatiors. *Appl Environ Microbiol 48*:1012–1019.

Torstensson BA (1984) A new system for ground water monitoring. *Groundwater Monitor Rev 3*:131–138.

Torsvik V, Goksoeyr J, and Daae FL (1990) High density in DNA of soil bacteria. *Appl Environ Microbiol 56*:782–787.

Tullborg E-L, Landström O, and Wallin B (1999) Low-temperature trace element mobility influenced by biogenic activity: Indications from ^{18}O, ^{13}C, ^{34}S and trace element analyses of fracture calcite and pyrite in crystalline basement. *Chem Geol 157*:199–218.

Vreeland RH, Piselli Jr AF, McDonnough S, and Meyers SS (1999) Distribution and diversity of halophilitic bacteria in a subsurface salt formation. *Extrmophiles 2*:221–331.

5

VADOSE ZONE
MICROBIOLOGY

THOMAS L. KIEFT

New Mexico Institute of Mining and Technology, Socorro, New Mexico

FRED J. BROCKMAN

Pacific Northwest National Laboratory, Richland, Washington

1 INTRODUCTION

The vadose zone is defined as the portion of the terrestrial subsurface that extends from the land surface downward to the water table. As such, it comprises the surface soil (the rooting zone), the underlying subsoil, and the capillary fringe that directly overlies the water table. The unsaturated zone between the rooting zone and the capillary fringe is termed the "intermediate zone" (Chapelle, 1993). The vadose zone has also been defined as the unsaturated zone, since the sediment pores and/or rock fractures are generally not completely water-filled, but instead contain both water and air. The latter characteristic results in the term "zone of aeration" to describe the vadose zone. The terms "vadose zone," "unsaturated zone," and "zone of aeration" are nearly synonymous, except that the vadose zone may contain regions of perched water that are actually saturated. The term "subsoil" has also been used for studies of shallow areas of the subsurface immediately below the rooting zone. This review focuses almost exclusively on the unsaturated region beneath the soil layer, since there is already extensive literature on surface soil microbial communities and processes, for example, Metting (1993), Richter and

Subsurface Microbiology and Biogeochemistry, Edited by James K. Fredrickson and Madilyn Fletcher. ISBN 0-471-31577-X. Copyright 2001 by Wiley-Liss, Inc.

Markowitz, (1995), Paul and Clark (1996), and Sylvia et al. (1998). The deeper strata of the unsaturated zone have only recently come under scrutiny for their microbiological properties.

The vadose zone has been estimated to contain 0.005% of the earth's water (Bouwer, 1979). This minuscule percentage belies the importance of the vadose zone to the hydrologic cycle. Vadose zone rocks and sediments are of critical importance to groundwater recharge, to bioremediation of subsurface contamination, and to the performance of hazardous waste storage facilities located there. Moreover, the vadose zone can be a habitat for a variety of microorganisms that survive and function under physiologically challenging conditions, and by doing so, participate in the biogeochemical cycling of important elements.

Vadose zones can be relatively thin in regions with high groundwater recharge and shallow water tables. In extreme cases, the water table is contiguous with the soil layer. At the other extreme are the thick vadose zones of arid and semiarid regions, where groundwater recharge can be very slow and the water table may be hundreds of meters below land surface (Chapelle, 1993). These thick vadose zones are of interest as extreme habitats for microorganisms and for their role in hazardous waste management; they are a major focus of this review.

2 PHYSICAL AND CHEMICAL CONDITIONS

The vadose zone is a three-phase system comprising solid, aqueous, and gaseous components. In the uppermost portion of the vadose zone, that is, the surface soil horizons, the soil water content can change dramatically over short time intervals. It can range from saturated conditions to extreme desiccation. The amounts of moisture in the unsaturated zone beneath the surface soil horizons change more slowly and over a shorter range. The water pressure in the vadose zone is less than atmospheric pressure; thus, water will not flow from the vadose zone into a well. The water potential in the unsaturated zone is negative with respect to pure water. It is dominated by the matric water potential, which is controlled by capillary and sorptive forces. These forces restrict the water in the vadose zone to thin layers on solid surfaces and to wedges and pendular rings suspended between solids. The matric water potential in the vadose zone below the soil layer is generally greater than (less negative than) -0.1 MPa (-1 bar), even in the thick, unsaturated zones of arid and semiarid environments. Conditions are thus relatively moist, even in the vadose zones of the driest deserts. At these water potentials, microbes are not subjected to direct desiccation stress. However, the thin, discontinuous water films of the vadose zone can severely retard transport of solutes and microorganisms. This, in turn, limits microbial access to nutrients. The unsaturated hydraulic conductivity in the vadose zone is a function of the texture of the porous medium and the volumetric water content. Together, these determine the thickness of the water films. At matric water potentials that are typical of the vadose zone, finer-textured sediments have greater moisture-holding capacity (and, thus, greater moisture content) and greater hydraulic conductivity at a given matric water potential than coarse-textured sediments. The unsaturated conductivity of small pore throats is greater than in larger pores and in rock fractures. This stands in contrast to saturated or nearly saturated porous media, in which hydraulic conductivity is greater in coarse textured media and in large fractures.

Rates of groundwater recharge through the unsaturated zone are a function of the quantity of precipitation, topography, plant communities, and the hydraulic conductivity of the porous medium. In humid climates, 20–50% of precipitation can reach the water table (Stephens, 1996). At the other extreme, in arid climates where evapotranspiration far exceeds precipitation, ground water recharge rates can be vanishingly low and limited to localized recharge beneath drainage basins, streambeds, playas, etc. Stephens (1996) compiled rates of diffuse recharge (direct recharge of precipitation, exclusive of localized sources like runoff and ponding) estimated by various investigators for a variety of arid and semiarid environments; these ranged from near zero to $100 \, mm \, year^{-1}$. Driscoll (1985) cited rates of $0.03–30 \, mm \, year^{-1}$ for thick alluvial basins in the western United States, $0.3–80 \, mm \, year^{-1}$ for the remaining arid and semiarid western United States, $5–500 \, mm \, year^{-1}$ for the central United States, and $30–500 \, mm \, year^{-1}$ for the eastern United States. Long et al. (1992) reported a recharge rate of $0.001–0.015 \, mm \, year^{-1}$ at the Hanford Site in eastern Washington, site of several microbiological studies (Brockman et al., 1992, 1997b; Kieft et al., 1993; Fredrickson et al., 1994). Nichols (1987) reported a recharge rate of $0.037 \, mm \, year^{-1}$ near Yucca Mountain in Nevada, proposed site of a nuclear waste repository and discussed below with respect to potential microbial activity (Kieft et al., 1997a). Groundwater recharge rates influence nutrient flux and the rates at which microorganisms can be transported through the vadose zone (see the Activities and Transport sections below). The potential for transport of nutrients and of microorganisms is minimal in the thick vadose zones of arid environments. However, groundwater recharge rates can be artificially modified in arid regions by irrigation and wastewater disposal, thereby increasing nutrient fluxes and stimulating microbial activities (Brockman et al., 1992; Kieft et al., 1993; Fredrickson et al., 1994).

A more quantitative presentation of water relations in vadose zones is beyond the scope of this chapter. Treatises on this subject include those by Hillel (1980), Papendick and Campbell (1983), Guymon (1995), and Stephens (1996).

The chemistry of the gaseous phase in vadose zone environments reflects the activities of microorganisms. Gaseous products of microbial metabolism, especially CO_2, occur in higher concentration in soil and in the underlying unsaturated rocks and sediments than in the atmosphere above land surface. Other gaseous metabolites, such as H_2, CH_4, H_2S, NH_3, and volatile fatty acids, may also be present, depending on the availability of energy sources for microbial activity and proximity to locations (such as the water table) that may be a source of microbial metabolites. Oxygen is generally available in all but the most organic-rich microsites of the vadose zone; however, oxygen concentrations may be lower than in the above-ground atmosphere.

Organic carbon may be present as solid material and as dissolved organic carbon (DOC). Particulate organic carbon is generally residual material from the time of burial, for example, in buried soils (paleosols), but in rare cases can be transported in

TABLE 5.1. Microbial Abundance in Various Vadose Zone Environments.

Depth and Recharge,[a] Contaminated?	Vadose Zone Geologic Materials	Heterotrophic Plate Counts (cfu g[-1])	Direct Microscopic Counts (cells g[-1])	Reference
Shallow, high-recharge	Shallow soil, B horizon, Oklahoma (1.2 m)	$3.1 \times 10^1 - 1.9 \times 10^5$	$6.8 \times 10^6 - 9.8 \times 10^6$	Balkwill and Ghiorse (1985)
Shallow, high-recharge	Shallow, sandy coastal plain sediment, Virginia (1–3.5 m)	$3.0 \times 10^2 - 1.0 \times 10^4$	No data	Zhang et al. (1998)
Shallow, high-recharge	Shallow subsoil, agricultural site, Indiana (1–19 m)	$2.0 \times 10^3 - 4.0 \times 10^3$	$2.5 \times 10^7 - 4.2 \times 10^7$	Konopka and Turco (1991)
Shallow, low recharge	Till, Saskatchewan (1–7.45 m)	$<1 \times 10^2 - 7.9 \times 10^6$	No data	Severson et al. (1991)
Shallow, low recharge	Silt loess, Washington state (1–5.54 m)	$2.7 \times 10^3 - 9.0 \times 10^4$	No data	Severson et al. (1991)
Shallow, low recharge	Silt loess paleosols, Washington state (1–10 m)	$<1 \times 10^1 - 1 \times 10^5$	$1 \times 10^7 - 1 \times 10^8$	Kieft et al. (1998)
Deep, low recharge	Basalt and sedimentary interbed, Idaho (70 m)	$2.1 \times 10^1 - 5.0 \times 10^1$	$1.4 \times 10^5 - 4.8 \times 10^5$	Colwell (1989)
Deep, low recharge	Basalt and sedimentary interbeds, Idaho (70–140 m)	$<1 \times 10^2$	$4.0 \times 10^6 - 1.3 \times 10^7$	Kieft et al. (1993)
Deep, low recharge	Volcanic tuff, Nevada Test Site (400 m)	$2.7 \times 10^1 - 4.2 \times 10^4$	$6.9 \times 10^5 - 4.8 \times 10^7$	Haldeman et al. (1993)
Deep, low recharge	Volcanic tuff, Nevada Test Site (50–450 m)	$<1 \times 10^2 - 3.2 \times 10^3$	$4.0 \times 10^5 - 1.3 \times 10^7$	Kieft et al. (1993)
Deep, low recharge	Volcanic tuff, Yucca Mountain, Nevada (300 m)	$1.0 \times 10^1 - 1.0 \times 10^3$	$<3.2 \times 10^4 - 3.2 \times 10^5$	Kieft et al. (1997)
Deep, low recharge	Paleosols, Washington state (54–64 m, plus outcrops[b])	$3.9 \times 10^1 - 6.5 \times 10^2$	$5.2 \times 10^6 - 1.3 \times 10^7$	Brockman et al. (1992)
Deep, low recharge	Paleosols, Washington state (8–11 m)	$<1 \times 10^1$	No data	Brockman et al. (1997)
Deep, low recharge	Hanford Site sediments, Washington state (25–100 m)	$<1 \times 10^2 - 1 \times 10^4$	$2 \times 10^6 - 1 \times 10^7$	Kieft et al. (1993)
Deep, low recharge	Silt loess paleosols, Washington state (10–37 m)	$<1 \times 10^1 - 1 \times 10^2$	$1 \times 10^6 - 1 \times 10^7$	Kieft et al. (1998)
Deep, artificial recharge	Paleosols, Washington state (outcrops[b])	1.9×10^2	5.2×10^6	Brockman et al. (1992)
Deep, artficial recharge and/or contaminated	Hanford Site sediments, Washington state (25–100 m)	$1 \times 10^2 - 2.5 \times 10^7$	$3.2 \times 10^4 - 1.3 \times 10^7$	Kieft et al. (1993)

[a]Shallow: 1–10 m, deep: > 10 m; low recharge: arid and semiarid sites in western United States and Canada with < 80 mm annual recharge, high recharge: midwestern and eastern United States sites with > 80 mm annual recharge.

[b]Outcrops: The estimated depth of these sediments prior to downcutting and erosion of the Columbia River Valley is approximately 125 m; current vertical distances to land surface are 2.7–6.1 m.

colloidal form during storm-associated recharge events. DOC may be transported downward from overlying surface soil horizons; this is generally in low concentration and of low quality as substrates for microbial metabolism. More easily degraded forms of organic C are generally mineralized in the overlying soils before they can be transported to depth. However, where recharge is rapid, for example, in fractured porous material or in storm-associated recharge, easily degraded forms of organic carbon may be transported to relatively deep regions of the vadose zone.

3 MICROBIOLOGY

3.1 Microbial Abundance

Depth is a major determinant of microbial abundance in surface soils, with numbers generally declining with increasing depth (Paul and Clark, 1989; Richter and Markowitz, 1995). Extending this pattern to subsoil layers would suggest that deep vadose zones are devoid of microorganisms. However, studies during the last 10 years have shown that viable microorganisms do exist in a variety of vadose zones, albeit in relatively low numbers (Table 5.1). In general, shallow vadose zone environments have more abundant microorganisms than deeper unsaturated layers (Table 5.1). Rates of moisture recharge also influence microbial abundance in the vadose zone: High-recharge vadose zones usually harbor higher numbers of culturable microorganisms than low-recharge zones (Table 5.1). Compared to surface soils or saturated samples from the same location, the abundance of microorganisms is generally lower in the unsaturated zone samples (Colwell, 1989; Pothuluri et al., 1990; Konopka and Turco, 1991; Severson et al., 1991). For example, Konopka and Turco (1991) quantified microbial populations and activities in samples collected from a 25-m profile encompassing an agricultural surface soil, the underlying vadose zone, and saturated sediments immediately below the water table. Total numbers of microorganisms ranged from 2.5–4.2×10^7 cells g^{-1} and culturable heterotrophs were present from 2–3×10^3 cfu g^{-1} in the vadose zone beneath the soil layer; these populations were lower than in either the surface soil or underlying groundwater. Radiotracer studies of the kinetics of biodegradation of organic amendments showed slower metabolic rates and longer lag times in the unsaturated zone than in either the surface soil or the underlying aquifer (Konopka and Turco, 1991). The lower numbers of microorganisms in the vadose zone were attributed to the low water potential, ~ -0.03 MPa. This is undoubtedly true, in that the lower water potential, or availability, of the vadose zone severely limits microbial access to nutrients (as discussed above). Decreased water availability in the vadose zone, while not causing direct desiccation stress, limits the diffusion and advection of solutes, as well as the movement of microbes (Kieft et al., 1993).

Studies in the arid western United States have confirmed the existence of microorganisms in deep unsaturated zones and have also revealed the extreme hardships of microbial life in these environments. Microorganisms have been detected and quantified in a wide range of unsaturated subsurface habitats, including fractured basalt and clay interbeds within basalt (Colwell, 1989; Kieft et al., 1993), volcanic tuff (Haldeman and Amy, 1993; Haldeman et al., 1993, 1994, 1995; Kieft et al., 1993, 1997a), and in various unconsolidated, unsaturated sediments in Washington state (Severson et al., 1991; Brockman et al., 1992, 1997b; Fredrickson et al., 1994; Kieft et al., 1993, 1998; Stevens and Holbert, 1995; Balkwill et al., 1998). Although these studies represent a diversity of unsaturated environments, the common finding is that the microorganisms are present only in low densities, often near the detection limit for both nonculturing and culturing methods. Comparisons of viable cells, estimated from extracted phospholipid fatty acid (PLFA) and plate counts, to microscopic counts of total cells indicate very low percentages (typically $< 0.01\%$) of culturable cells in deep vadose zones of arid and semiarid environments that are unimpacted by artificial recharge or contaminants (Brockman and Murray, 1997). These percentages are three orders of magnitude lower than equivalent vadose zones in high-recharge environments. Contamination with metabolizable organic substrates can also generate large populations with high percentages being culturable (Table 5.1; Kieft et al., 1993; Fredrickson et al., 1994).

3.2 Microbial Distribution

Although generally low, the numbers and activities of unsaturated zone microorganisms vary spatially with conditions in the subsurface, including texture, water availability, and organic matter content. Residual organic matter in a buried soil (paleosol) can support larger communities of microorganisms (Brockman et al., 1992, 1998a; Kieft et al., 1993, 1998). Paleosols can contain organic carbon that remains from the original period of surface soil development (Retallack, 1990). This organic carbon may be a localized source of energy supporting the growth and survival of vadose zone populations. The microbial communities inhabiting these paleosols may even include populations derived from the original soil inhabitants (Kieft et al., 1998). A number of other conditions can foster locally enhanced vadose zone microbial communities, generally as a result of increased nutrient flux. Artificial recharge, for example, from agricultural irrigation or from disposal of liquid wastes to vadose zone sediments, can stimulate vadose zone communities by increasing advection of dissolved organic matter (Brockman et al., 1992; Fredrickson et al., 1993). Direct addition of organic matter in the form of organic pollutants also results in anomalously large vadose zone microbial communities, as it does in other habitats (Fredrickson et al., 1993; also see the Bioremediation section below). Interfaces between layers often generate localized regions of greater saturation, where microorganisms may exhibit increased numbers or activity due to improved access to sediment-associated nutrients. Finally, any type of physical disturbance of vadose zone sediments can stimulate native microorganisms by bringing them into contact with previously unavailable nutrients, a phenomenon that has been demonstrated by the increase in microbial numbers observed when vadose zone samples are incubated in laboratory experiments (Brockman et al., 1992, 1998a, Haldeman et al., 1994a, 1995b; Fredrickson et al., 1995; Haldeman, 1997). The abundance of microorganisms and their metabolic activities increase during storage of vadose zone samples, even when they are stored intact, that is, without crushing or mixing (Fredrickson et al., 1995; Brockman et al., 1998a; Haldeman, 1997; Haldeman et al., 1994a, 1995b). Stimulation of microbial activity during storage of intact cores may be the result of induced moisture gradients that cause advection of solutes, including microbial nutrients, or increased diffusion of oxygen into the sediment. This stimulation is analogous to what occurs in the field during bioventing (see the Bioremediation section below).

Microbial populations in the vadose zone often exhibit a high degree of spatial discontinuity due to the relatively low numbers of viable microorganisms combined with microscale distributions of physical and chemical properties that exert strong control on water and nutrient availability. At pristine, low-recharge sites, the median percent glucose mineralization in positive samples after 2 months is typically 1–5%, with < 20% of those samples showing >10% mineralization (Brockman et al., 1997). In addition, the great majority of the unutilized glucose was leachable, suggesting that an extremely high fraction of pore spaces is uninhabited (Brockman, unpublished). The impact of water (and by extension nutrient) flux on microbial distribution was studied at sites receiving 15-μm, 1-mm, and 20-cm average annual recharge. At each site, several 50-sample transects (sample spacing of 5–10 cm) were performed, and 1-g aliquots from a 15-g homogenized sample were assayed for ^{14}C-glucose mineralization and ^{3}H-acetate incorporation into membranes in 2- to 3-month unsaturated incubations. Samples were treated with acid at the end of the incubation to recover residual $^{14}CO_2$. ^{3}H-acetate incorporation is highly sensitive and able to detect metabolic activity when cells are limiting expenditure of energy. A large fraction (60–70%) of the samples did not exhibit detectable microbial activity in the low-recharge sites, compared to 0–25% samples with no detectable activity at the 20-cm recharge site (Brockman et al., 1995a, 1996b, 1997b, 1999). Incubation of the low-recharge samples with low levels of inorganic nutrients did not increase the frequency of detection (Brockman et al., 1996b). Lastly, samples at the low-recharge sites were six- to eight-fold more likely to have detectable activity in silt than in the lower water content sand (Brockman et al., 1996a). Because the silt and sand had similarly low organic carbon content, this result is likely due to greater water connectivity at the pore scale in the silt and thus greater survival due to improved diffusion of nutrients to isolated microorganisms.

The large percentage of samples lacking detectable activity at the low-recharge sites suggests that large volumes of sediment (several tens of cubic cm) can be devoid of microorganisms capable of metabolism. More precisely, there were very few, if any, microorganisms able to metabolize the substrate and/or grow under the conditions of the assay (Brockman et al., 1995a, 1996a, 1996b). This pattern may be explained by the gradual death of most microorganisms over time, except in rare microsites where conditions permit long-term survival (Brockman and Murray, 1997; Brockman et al., 1997), and by higher mortality with decreasing recharge (Brockman et al., 1996b). This conclusion was also supported by activity assays of replicate 0.1-, 1-, 10-, and 100-g samples removed (without sample homogenization) from the same core; these assays utilized vessels in which sediment, headspace, and

$^{14}CO_2$ trap volumes and surface areas were kept nearly constant for all sample sizes. In the low-recharge sites, activity was rarely, if ever, detected in the small samples and the frequency of detection increased as sample size increased (Brockman and Murray, 1997; Brockman et al., 1997). In contrast, at the 20-cm recharge site, activity was detected and approximately equal in all eight replicates for all sample sizes (Brockman et al., 1996b). Measurements of viable biomass by PLFA on 75-g samples showed a spatially averaged density of approximately 10^4 viable cells g^{-1} at all three sites. Thus, microbial activity was more evenly distributed at the high-recharge site. This pattern likely results from higher concentrations and fluxes of bioavailable nutrients in pore water (due to higher precipitation and higher plant biomass) and possibly increased transport of microorganisms from higher in the profile (see the Transport section below).

In a separate study, the role of depth, sediment age, pore water age, and pore water organic carbon on fine-scale microbial spatial heterogeneity was examined in a sediment sequence containing multiple ancient buried soils (paleosols). Sediments were deposited by wind from the same source and were of the same general composition over approximately one million years. Cores from eight paleosols with similar degrees of soil development were studied to examine microbial distribution in a setting where geochemical and geohydrologic parameters were as similar as possible. Fifty individual 1-g samples were removed every 0.3 cm in each of the eight cores, and ^{3}H-acetate incorporation and ^{14}C-labeled substrate mineralization were measured. Pore water organic carbon showed the strongest positive correlation to the amount of substrate utilized and the percent of samples in which activity was detected (Brockman et al., 1999). Activity assays conducted on 0.1-, 1-, 10-, and 100-g samples showed the same pattern (as described above) of increasing frequency of detection activity in the replicates with increasing sample size. In addition, as pore water age increased, samples lacking detectable activity became more common as sample size decreased (Brockman et al., 1998b). Thus, expanding upon the results from the low- versus high-recharge sites, the results at this site (recharge of 0.1–0.8 cm) suggest that microsites containing microorganisms capable of activity become increasingly rare with decreasing pore water organic carbon and increasing pore water age. The rare microsite/large extinction volume phenomenon may have important implications for bioremediation in the vadose zone (see the Bioremediation section below).

3.3. Identities of Microorganisms and Community Structure in the Vadose Zone

Relatively few studies have determined the identities of vadose zone microorganisms; however, there is some consensus among the few available studies. Balkwill and Ghiorse (1985) found that 85–90% of cells from shallow, unsaturated sediments at a site in Oklahoma were gram positive, *Arthrobacter*-like bacteria, as determined by transmission electron microscopy. They cultured members of the genera *Arthrobacter* and *Pseudomonas* from vadose zone samples. Colwell (1989) reported a majority (84%) of gram positive bacteria among heterotrophic plate count isolates from deep vadose zone samples. Brown et al. (1994) identified members of the genera *Arthrobacter, Bacillus, Clavibacter, Pseudomonas*, and *Rhodococcus*, as well as three yeasts, among isolates from volcanic tuff samples collected by Hersman et al. (1988) near Los Alamos, New Mexico. Haldeman et al. (1993), Brockman et al. (1997), and Balkwill et al. (1998) identified the genera, or most closely related genera, of bacteria isolated in culture from arid and semiarid sites in the western United States (Table 5.2). Gram positive bacteria appeared to dominate in each of these studies and included endospore-formers (*Bacillus*) and actinomyces, as well as nonspore-formers (e.g., *Arthrobacter, Micrococcus*, and *Nocardioides*). Gram negative bacteria, for example, *Pseudomonas*, were also present.

As in most other natural environments, the total number of microorganisms in vadose zone samples exceeds the number that can be recovered in growth media. Therefore, non-culture-based methods are needed for the full characterization of vadose zone microbial communities. These culture-independent approaches include PLFA analyses and nucleic-acid-based approaches. Balkwill et al. (1998) and Kieft et al. (1998) reported PLFA profiles in samples from various sites in eastern Washington state. These profiles indicated a mix of gram positive and gram negative bacteria, and low, but measurable, proportions of microeukaryotes. The latter is somewhat surprising considering that eukaryotes are rarely detected in subsurface environments (Chapelle, 1993); these are generally protozoa found in aquifers with high nutrient flux or organic contamination and high bacterial counts (Sinclair and Ghiorse, 1987; Madsen et al., 1991). Protozoa in deep vadose zones are unlikely, given the sparse bacterial populations; the eukaryotes are probably fungi that are favored under the low moisture conditions. In a study at the Yucca Mountain

TABLE 5.2 Identities of bacteria cultured from vadose zone samples collected at three sites in arid and semiarid regions of the western United States. Identifications were made using the MIDI system (Microbial ID Incorporated, Newark, DE) (Haldeman et al., 1993; Balkwill et al., 1998) and 16S rRNA sequence analyses (Brockman et al., 1997; Balkwill et al., 1998).

Nevada Test Site (Haldeman et al., 1993)	White Bluffs, Eastern Washington State (Brockman et al., 1997b)	Hanford Site (Balkwill et al., 1998)
Arthrobacter	*Arthrobacter*	Streptomycetes
Micrococcus	*Micrococcus*	*Bacillus*
Bacillus	*Clavibacter*	*Arthrobacter*
Corynebacterium	*Nocardioides*	*Azospirillum*
Gordona	*Planococcus*	*Bradyrhizobium*
Acinetobacter	*Streptomyces*	*Rhizobium*
Acidovorax	*Bacillus*	*Xanthomonas*
Hydrogenophaga	*Blastobacter*	*Pseudomonas*
Pseudomonas	*Paracoccus*	*Telluria*
	Methylobacterium	
	Sphingomonas	

proposed high-level nuclear waste repository in Nevada, PLFA profiles indicated that the ratio of gram negative bacteria to gram positive bacteria was higher in vadose zone volcanic tuff with higher moisture content than in less moist vadose tuff (D. Ringelberg, personal communication). This is consistent with surface soil studies that often show a correlation between the proportion of gram negative cells and moisture content (Kieft, 1991).

Zhou and his colleagues have compared the community structure of soil, vadose zone, and aquifer sediments by amplification, cloning, and sequencing of 16S rRNA genes (Treves et al., 1999). The soil community showed a high species richness, but included a high frequency of a few dominant types (low equitability); the aquifer community had a relatively equal frequency of the various 16S rRNA gene sequences represented (high equitability). Theoretical analysis based on mathematical modeling indicated that this difference could be attributed to the physical environment; in the soil, microbes come into direct contact with each other (at least during wet intervals) and are therefore frequently competing. Microbes in aquifers tend to be physically isolated and not directly competing. Physically, the soil and the aquifer were considered to be high- and low-connectivity environments, respectively. Presumably, the aquifer was considered to have low connectivity due to restrictive pore-throat diameters. Also, relatively sparse microbial distributions in the aquifer would limit competition. High competition in the high-connectivity environment results in some dominant species and, thus, a low species equitability; low competition and low connectivity generate few dominant species and high species equitability. The vadose zones in Treves et al.'s study (1999) had community structures that were intermediate between those of the soil and the aquifer, suggesting an intermediate connectivity of microhabitats. This is consistent with the moisture regimes at these sites. These are mesic sites in the eastern United States, where the vadose zones are shallow and thin, with relatively rapid rates of groundwater recharge. One can hypothesize that in the deep vadose zones of arid and semiarid sites, the connectivity is extremely low and that this fact should be reflected in the community structure. In these relatively static, diffusion-dominated systems, microbes are physically isolated from each other and are not likely to be engaged in competition or other biological interactions. This lack of biological interaction should lead to low species diversity and high equitability, but this remains to be tested.

3.4 Activities of Vadose Zone Microbes

The different types of microbial activity that can occur in vadose zones have not been fully explored. Deep aquifers and other saturated subsurface environments have been shown to harbor representatives of most functional groups of heterotrophic and chemoautotrophic microorganisms (Fredrickson et al., 1989; Chapelle, 1993; Kieft and Phelps, 1997). However, the activities of many of these functional groups of microbes have not yet been studied in vadose zone rocks and sediments. Although numerous studies have shown the potential for vadose zone microorganisms to degrade a variety of organic substrates under aerobic conditions (Severson et al., 1991; Kieft et al., 1993, 1997a, 1998; Kieft and Phelps, 1997), few have tested for facultative and strict anaerobes or for chemolithotrophic bacteria. Balkwill and Ghiorse (1985) determined the population of viable aerobic and anaerobic bacteria using dilution plate count methods on shallow vadose zone samples from sites in Oklahoma; anaerobic populations were two to three orders of magnitude lower than aerobic populations. These may have been predominantly facultative anaerobes. However, one can hypothesize that anaerobic activities occur in anoxic microhabitats of vadose zones, such as in the centers of aggregates, similar to those in soil (Sexstone et al., 1985) or in sites with abundant organic C electron donors. One can also speculate that chemolithotrophs occur in vadose zone habitats. Those that can use gaseous energy sources (e.g., H_2) may be particularly favored.

Estimates of the *in situ* rates of subsurface microbial activities are best made using a geochemical modeling approach; laboratory incubations of subsurface materials generally result in overestimates of activity (Phelps et al., 1994; Kieft and Phelps, 1997). This approach was pioneered for the subsurface by Chapelle and Lovley (1990) and others (Murphy et al., 1992, Phelps et al., 1994) for estimating rates of microbial activities in deep coastal plain aquifers in the southeastern United States. Severson et al. (1991) used a mass balance total organic C (TOC) approach to estimating microbial activity in a deep vadose zone. They estimated that $5.0 \times 10^4 \, g \, TOC \, m^{-2}$ have disappeared from the vadose zone due to leaching $(0.3 \times 10^4 \, g \, TOC \, m^{-2})$ and biodegradation $(4.7 \times 10^4 \, g \, TOC \, m^{-2})$. Assuming a constant rate of biodegradation over distance (7-m depth) and time (11,500 years since the last glaciation), they estimated the rate to be $0.6 \, g \, TOC \, m^{-3} \, year^{-1}$. Wood et al. (1993) formulated a similar estimate for rates of organic mineralization in these sediments, using a geochemical model of CO_2 concentrations. Their model indicated a zone of relatively low activity in the 1- to 6.5-m depth interval below the soil layer and an active region, at least 100 times higher than in the 1- to 6.5-m interval, in the 6.5- to 7.5-m zone immediately above the water table $(2.25 \times 10^{-7} \, g \, CO_2 \, m^{-3} \, sec^{-1}$ equals $7.1 \, g \, CO_2 \, m^{-3} \, year^{-1})$. This active zone includes the capillary fringe, a zone that has often been hypothesized to have greater microbial activity than the overlying vadose zone. This enhanced activity may be the result of fluctuating water levels, that increase nutrient flux by advective transport. There may also be diffusion of electron donors from the underlying saturated zone to the oxygen-rich unsaturated zone.

The rate of microbial activity estimated by Severson et al. (1991) is many orders of magnitude higher than the range of activities estimated by Kieft and Phelps (1997) and Onstott et al. (1998), which were $< 1 \times 10^{-10} \, moles \, CO_2 \, L^{-1}$ porewater year^{-1}. However, if one assumes that rates of biodegradation decline with time as the more easily degraded fractions of the TOC are depleted (leaving more recalcitrant humic compounds), then current *in situ* rates of metabolism are much slower than those estimated by Severson et al. (1991) and Wood et al. (1993). Also, Severson et al. (1991) and Wood et al. (1993) studied relatively shallow sediments beneath a grassland pasture. Deeper vadose zone sediments in areas with very low recharge rates can be expected to have even slower rates of microbial activity. Using the mineralization estimates of Phelps et al. (1994) and Onstott et al. (1998) for vadose zone environments and equal activity among the approximate 10^4 viable cells g^{-1}, and assuming a growth yield of approximately 3.6%, Onstott et al. (1998)

estimated the average generation time for vadose zone microorganisms to be > 3000 years to 300,000 years! These glacially slow rates of microbial metabolism and concomitantly long intervals between cell divisions describe deep vadose zone environments as being among the most extreme on Earth with respect to availability and flux of nutrients. Clearly, the average vadose zone microbe exists in a state of extended starvation survival rather than active cellular metabolism and growth. Subsurface microbial activities have been referred to as "life in the slow lane" (Kieft and Phelps, 1997); a better metaphor for deep vadose zone microbes might be to say that they're not even moving, they're parked with the emergency brake on!

3.5 Long-Term Survival of Microorganisms in Deep Vadose Zones

The generally slow nutrient fluxes in subsurface environments, particularly in diffusion-dominated systems such as deep vadose zone rocks and sediments, cause microorganisms in these habitats to experience extreme nutrient limitation and starvation. Microorganisms in deep unsaturated zones of arid and semiarid sites, where groundwater recharge rates are infinitesimally low, may be capable of surviving for thousands or even millions of years. There have been several credible reports of microorganisms surviving in geologic materials for periods ranging to $> 10^7$ years (Kennedy et al., 1994; Cano and Barucki, 1995; Lambert et al., 1998). In some cases these are spores, in others they are vegetative cells. In the case of unsaturated subsurface environments, microorganisms may persist through millennia by long-term survival of individual cells; however, if nutrient flux is sufficient (even intermittently), it may support occasional DNA replication and cell division as discussed above. In either case, vadose zone microorganisms are nearly always in survival mode (as dormant vegetative cells or spores), an extreme condition that merits further study.

Microbial starvation survival has been studied by two approaches: laboratory incubations and field studies. The traditional laboratory approach has been to cultivate isolates in an organic growth medium, concentrate the cells, resuspend them in a nonnutrient buffer, and monitor them over time for viability, structural changes, and/or metabolic activities (Amy and Morita, 1983; Amy et al., 1993). This laboratory microcosm approach has been adapted for the study of starvation survival in subsurface microorganisms by mixing pure cultures of microorganisms into sterile porous media and monitoring cellular responses (Kieft et al., 1994, 1997b). Laboratory microcosm studies using bacteria of two genera common in the subsurface, *Arthrobacter* and *Pseudomonas*, have shown that these bacteria are capable of starvation survival in porous media for at least 64 weeks, and that their responses are similar to those of bacteria in other environments (Kieft et al., 1994, 1998). In general, viability declined exponentially with time, and cells underwent dwarfing (decline in volume). Dwarfing occurs as cells undergo endogenous metabolism and is a common response in starved bacteria (Kieft, 1999). The *Pseudomonas* strains showed changes in membrane phospholipid fatty acids that typically appear when gram negative bacteria are subjected to nutrient deprivation, that is, increases in the ratios of saturated to unsaturated fatty acids, increases in the *trans*-to *cis* ratio in monoenoic fatty acids, and increases in the ratio of cyclopropyl fatty acids to their monoenoic precursors (Kieft et al., 1994, 1997b). The *Arthrobacter* strains appeared not to change their phospholipid fatty acid composition during starvation. Environmental conditions within the porous medium influenced starvation survival in these studies. In a comparison of starvation survival under saturated conditions versus unsaturated but moist (water potential > -0.1 MPa) conditions, the likelihood of survival was better in the saturated microcosms. The chance of survival was also greater in a silt paleosol than in a similar silt sediment that never underwent soil development. These findings are consistent with the idea that organic carbon and rates of nutrient diffusion are important determinants of microbial survival. Unfortunately, laboratory microcosm studies are limited to relatively short timeframes, nowhere near the thousands to millions of years hypothesized for the survival of some vadose zone microorganisms.

In order to test the hypothesis that microorganisms have survived in deep vadose zone environments for geologic time periods, it is necessary to collect samples that have been geohydrologically isolated for thousands to millions of years, and to test them for the presence of viable microorganisms. In the case of vadose zone rocks and sediments, it is necessary to determine the rate of groundwater recharge (from which the age of the porewater can be calculated) as well as the age (i.e., time since deposition) of the subsurface rock or sediment. These ages can then be used to constrain the age of the microbial community. The upper age limit is the time since geologic deposition; the lower limit is the age of the porewater (Murphy et al., 1992, Kieft et al., 1998). Age of the community here refers to the length of time the microbes have persisted since their introduction to the subsurface environment, by entrainment with the geologic material at or before the time of burial, and/or by transport into the rock or sediment since the time of burial. Persistence can include growth and/or survival; however, survival is likely the dominant mode of persistence in low-nutrient flux environments such as deep vadose zone rocks and sediments (Onstott et al., 1998).

In a study of vadose zone sediments in the Pasco Basin of eastern Washington state, Balkwill et al. (1998) estimated porewater ages of 15,000 years and 30,000 years at depths of about 5 and 8 m, respectively, at two sites with low recharge rates (measured by the chloride mass balance method). Despite the low recharge rates and concomitantly ancient porewater, viable microorganisms were present, as evidenced by viable bacteria determined by dilution plate count methods (approximately 10^1– 10^3 cfu g^{-1}) and by PLFA results. Isolates from plate counts included high proportions of *Bacillus* spp. and actinomycetes. PLFA profiles indicated the presence of gram negative, as well as gram positive, bacteria. These data indicate persistence of a diversity of bacteria in vadose zone sediments for at least 15,000 years. Pore water ages were also determined for a nearby site with a 3-m-thick cemented zone overlying the studied sediments. Viable cells (mostly noncultured) were present at approximately 10^4 cells g^{-1}, as evidenced by phospholipid analysis; a variety of spore-forming and non-spore-forming bacterial isolates were shown to have persisted for a minimum of 12,000 years (perhaps as long as the 4-million-year-old age of the sediment) (Brockman et al., 1997, Table 5.2). Given the low availability of organic carbon in these sediments, the bacteria have likely persisted for much of this time by survival of individual spores or vegetative (morphologically nondifferentiated) cells.

In a similar study at two other sites in eastern Washington state, Kieft et al. (1998) quantified sediment ages, porewater ages, and microbial abundance in unsaturated loess sediments. These sediments underwent continuous soil development while at the surface and were subsequently buried. Sediment ages in the deepest borehole ranged from modern to 1 million years; porewater ages ranged from modern to 1200 years. The numbers of organisms, their biomass, and their activities declined exponentially with depth. Also, the DGFA-to-PLFA ratio [diglyceride fatty acids (indicative of dead cells) to PLFA (indicative of live cells) ratio] increased with depth, indicating that more cells died than survived during long-term sequestration in the unsaturated subsurface. However, even the 1-million-year-old sediment contained microorganisms that were detectable by direct microscopic counts, heterotrophic plate counts, and PLFA analyses. Theoretically, these could have been transported to the buried sediments along with the porewater. However, microorganisms are generally retarded in their transport through vadose zone sediments by sorption processes. The surviving microbial communities are, there-fore, likely older than the 1200-year-old porewater, although not necessarily as old as the million-year-old sediment.

4 TRANSPORT

Laboratory investigations of microbial transport under unsaturated conditions have typically been performed in homogenous sand columns or in simulated porous media such as micromodels. These experiments have used high concentrations of bacteria and/or nutrients. Such conditions poorly represent most unsaturated zones. Bacterial transport in agricultural soil microcosms has been relatively well studied. However, soils are structurally and chemically different from the vadose zone; they possess large microbial populations, and high concentrations of "marked" bacteria are typically added to conduct experiments. Although these laboratory and field studies are poorly representative of most unsaturated zones, several controlling processes are revealed and have relevance to unsaturated zone microbial transport. Transport of microorganisms in the vadose zone is controlled primarily by transient saturated flow, the amount of air- and water-filled porosity, and the concentration of nutrients (natural or as contaminants) available for microbial growth and partitioning of cells into the liquid phase. Cell size, shape, hydrophobicity, and motility are less important factors. Because few results exist on the impact of these factors on microbial transport in the vadose zone, they will not be discussed. Very low permeabilities caused by pore throats smaller than the average size of cells, or occlusion of a high fraction of the pore throats by cementation will prevent or greatly restrict microbial transport. The geochemistry of sediment coatings can also strongly inhibit microbial transport due to selective sorption of cells onto specific minerals.

It is well known that microbial transport in soils occurs primarily by preferential (saturated or near-saturated) flow within soil channels or macropores (Smith et al., 1985; Madsen and Alexander, 1982; Natsch et al., 1996; Breitenbeck et al., 1988; Hagedorn et al., 1978) and matrix flow during transient saturated conditions (Rahe et al., 1978). Although channels and macropores rarely exist in the vadose zone below the rooting and burrowing zone, vadose zones may become transiently or locally saturated and experience preferential flow in pore networks or fracture systems in regions receiving high recharge from snowmelt, storm events, or high annual precipitation. Increasing saturation reduces the adsorption of unattached cells to surfaces. High water velocities in preferential flow paths will transport unattached cells and can cause cells to detach. Most transport of microorganisms in the vadose zone occurs during these events. In the absence of these events, the degree of saturation plays an important role. Water flow under unsaturated conditions occurs in water films by capillary forces. Microorganisms could theoretically be transported or actively move in such water films. However, adsorptive forces and air–water interfaces greatly retard microbial movement in water films. Wong and Griffin (1976) concluded that both active bacterial movement (via cell division or chemotaxis) and passive movement (via advection and diffusion) are unlikely at matric potentials below approximately -0.05 MPa (-0.5 bars) because of discontinuities in water lenses and films (for coarse textures) and very high adsorption and/or filtration (for fine textures that may retain continuous films at much more negative matric potentials).

In laboratory experiments with unsaturated porous media, bacteria preferentially adsorbed to air–water interfaces as opposed to solid–water interfaces (Wan et al., 1994; Powelson and Mills, 1998) and adsorption increased with decreasing saturation (Wan et al., 1994; Schafer et al., 1998). Adsorption to the air–water interface appears to be irreversible and controlled by the degree of cell hydrophobicity (Wan et al., 1994). Thus, microbial transport in an oligotrophic environment at low water saturation and in the absence of preferential flow or transient matrix flow appears close to impossible. Balkwill et al. (1998) studied the distribution of indigenous microorganisms in vertical (4-to-15-m deep), unsaturated flowpaths at five semiarid field sites characterized by low recharge, high recharge accompanied by rare saturated (preferential) flow, and high recharge via solely unsaturated flow. They found several lines of evidence suggesting transport of microbial cells at the high-recharge sites with preferential flow and lack of evidence for microbial transport at the high-recharge unsaturated flow site and low-recharge sites.

Great potential exists for microbial colonization and transport in contaminated vadose zones because contaminants are typically delivered by a transient saturation event or percolation of contaminated water (or free hydrocarbon). Colonization and transport will be especially stimulated if the contaminant supports microbial growth and is present at high but nontoxic concentrations. In saturated columns without water movement, colonization may occur via the displacement of daughter cells into adjacent pore spaces as cells divide (Reynolds et al., 1989; Sharma et al., 1993). Detachment rates from biofilms in liquid bioreactors are growth-rate-dependent (Peyton, 1996) and not significantly affected by shear stress (Peyton and Characklis,

1993). Although similar data are lacking for unsaturated systems, it is reasonable to assume that the same growth-dependent colonization and transport processes can occur in contaminated vadose zones, particularly at high pore saturations and in the presence of relatively high growth rates. Cells are also transported between saturated pores by gas bubbles generated by fermentation (Reynolds et al., 1989). A similar process may occur in the vadose zone during bioventing and in the capillary fringe during biosparging.

Application of municipal and agricultural sludge and septic system and waste-water leachate raises concern about transport of viruses through the vadose zone to aquifers. Results from soil studies provide insight into vertical transport of viruses through deeper layers of the vadose zone. Viral adsorption varied up to four orders of magnitude, depending on the type of virus, type of soil, water content, salt content of the water, and pH (Gerba, 1984; Goyal and Gerba, 1979). Viruses are approximately one hundred times smaller than bacteria so the probability of their encountering a solid or air surface should be much lower. However, viral transport in soil is lower under unsaturated conditions compared to saturated conditions (Lance and Gerba, 1984; Powelson et al., 1990), suggesting adsorption by the air–water interface (Powelson et al., 1990). The majority of viral particles were not recovered in the effluent or on the soil, suggesting rapid inactivation of the virus during unsaturated flow (Powelson et al., 1990). Both natural humic material and sewage-sludge-derived organic matter increased the unsaturated transport of a virus in soil, and the authors speculated that the hydrophobic nature of both viral particles and organic material may result in competition for adsorptive sites at the air–water interface (Powelson et al., 1991). Powelson et al. (1990, 1991) also cited studies that suggest the air–water interface may actually disrupt and inactivate viral particles. Adsorbed viruses can be desorbed when pore water cation concentration decreases, such as in a heavy rainfall (Duboise et al., 1976). These studies indicate that transport of infective viruses will be lowest in vadose zones with low natural organic carbon and low natural recharge, where sludge and leachate percolation rates yield low water saturation, and where sediment texture provides a very high ratio of air–water surface area to water content.

5 PRACTICAL CONSIDERATIONS

5.1 Bioremediation

Vadose zone bioremediation approaches are important for protecting ground water quality because contaminants often reside in the vadose zone and are released, over time, to underlying aquifers. Vadoze zone contaminants can be biotransformed as a result of natural or intrinsic processes, or by accelerating the processes through delivery of nutrients (electron acceptors, electron donors, or inorganic nutrients). In the majority of cases, biomass and/or activity in the vadose zone must be increased many orders of magnitude above the levels that occur via natural processes to have a significant impact on attenuation of contaminant transport. Criteria for proving that bioremediation is occurring or has previously occurred have been summarized elsewhere (National Research Council, 1993).

Extensive literature exists on bioventing, the controlled injection of air to stimulate aerobic biodegradation of petroleum hydrocarbons in the vadose zone. Bioventing promotes biodegradation of both liquid- and vapor-phase petroleum hydrocarbons by volatilizing components with high vapor pressures, distributing vapors through the vadose zone to maximize biodegradation, and overcoming any oxygen limitation that may exist. An excellent review of bioventing is provided by Norris et al. (1994), and compendia of articles exist in Hinchee et al. (1995) and Alleman and Leeson (1997). Bioventing can be supplemented with injection or infiltration of water to increase the bioavailability of sediment-associated nutrients or, if required, water supplemented with inorganic nutrients to further increase biodegrading populations (Dupont et al., 1991). Rates of petroleum hydrocarbon biodegradation during bioventing typically range from 0.2–20 mg/kg/day (Norris et al., 1994). A number of factors control rates, including (1) the type of petroleum hydrocarbon mixture; (2) site physical and geochemical properties such as permeability, redox status, content of reduced minerals that can be oxidized by oxygen, and availability of microbial inorganic nutrients; (3) the length of time since contamination, which impacts the population density of degraders and the chemical form as bioavailability of the contaminant; and (4) toxicity from high-petroleum hydrocarbon concentrations, which can include pure hydrocarbon partially or entirely filling pore spaces at many sites. Bioventing has been successful at depths of 20 m in low-recharge vadose zones (Dupont et al., 1991), demonstrating that at least some of these microbially sparse environments can be transformed to high-population, high-activity "underground bioreactors" in the presence of readily available carbon as an electron donor. Biosparging is the injection of air into the saturated zone. The books cited above for bioventing also have excellent information on biosparging. Movement of air bubbles and gaseous metabolites (potentially volatile fatty acids, hydrogen, and methane) from the saturated zone into the capillary fringe and higher regions of the vadose zone can achieve results similar to those of bioventing.

The genetic potential to biodegrade components of petroleum hydrocarbons is essentially ubiquitous in all natural aerobic environments. In contrast, biotransformation of other classes of contaminants in the vadose zone is more difficult. Effecting detectable rates of biotransformation may require delivery of exogenous nutrients to increase populations and activities by orders of magnitude, and/or injection of biotransforming microorganisms. These refractory contaminant classes include highly chlorinated hydrocarbons, metals and/or radionuclides, synthetic chelating agents, and complex contaminant mixtures containing several contaminant classes (Riley and Zachara, 1992). It is possible that, in some vadose zones, biotransformation of certain contaminants will not occur, will occur so slowly that it cannot be measured, or occur more slowly than desired by policy makers.

Several vadose-zone-relevant issues are to blame for the great difficulty in successfully bioremediating the aforementioned recalcitrant contaminant classes:

1. Prior to contamination, the percentages of the microorganisms in the vadose zone possessing the genetic potential to carry out these biotransformations are orders of magnitude lower than for petroleum hydrocarbons. In addition, reductive dechlorination and reduction of metals and radionuclides require anaerobic metabolisms, and populations of anaerobes (facultative and obligate) are low in pristine vadose zones. If the nutrient and redox conditions imposed during engineered bioremediation are favorable, expression of genetic potential and populations of degraders can increase through microbial growth, selection, and population dynamics.

2. A related issue is the initial distribution of contaminant transformers, and the degree and speed with which they colonize surrounding volumes of "barren" sediment by growth and/or microbial transport. Recall that microorganisms appear to exist in rare microsites in uncontaminated low-recharge vadose zones. For contaminants that are difficult to transform, the distribution of contaminant degraders is also likely to occur in rare microsites even in high-recharge uncontaminated vadose zones. Thus, since microbial colonization and transport in (discontinuous or continuous) water films and partially filled pores are very slow in the presence of contaminants, bioremediation may be unsuccessful or very slow.

3. Contaminant-transforming microorganisms and nutrient amendments may poorly penetrate the regions where contaminants exist. For example, if contaminants exist primarily in clays with low biomass, low permeability, and small pore throats that preclude microbial colonization, then biotransformation will be limited by diffusion of nutrients to the transforming microorganisms and (for some contaminants) by diffusion of contaminant into higher permeability regions where other biotransforming microorganisms may exist. Aqueous delivery of nutrients or microbes may not be feasible for some contaminant classes because it can also spread contaminants in the vadose zone and potentially drive contaminants to the underlying aquifer. If biotransforming microorganisms are injected into the vadose zone, this issue is expanded to include how well the injected microorganisms are delivered to the contaminated regions and microsites. Successful injection is also restricted by very poor transport of microbes in systems of low water saturation, as described above.

4. Diffusion of nutrients and contaminant to microorganisms may be more limiting to the rate of contaminant biotransformation in unsaturated than in saturated systems. For example, the diffusion coefficient for toluene through an unsaturated biofilm (i.e., an air–biofilm interface lacking a significant layer of free water exterior to the biofilm) was found to be approximately two orders of magnitude lower than toluene diffusivity in water (Holden et al., 1997). Thus, mass transfer and biotransformation rates may be very sensitive to whether biotransforming microorganisms exist as isolated cells, microcolonies, or biofilms, and the extent to which the latter morphologies are covered by water films. This concept may explain the observation that TCE and

toluene were not degraded in unsaturated soil at 5% moisture and below, but were degraded at 16–30% moisture (Fan and Scow, 1993).

5. The ability of microorganisms to express and retain biotransformation activity and survive in the presence of toxic contaminants may be more difficult in the vadose zone. Microorganisms and contaminants may become concentrated by physical and chemical processes unique to the dynamics of an air–water–solid system; for example, microorganisms and contaminants may concentrate together at air–water interfaces.

Notwithstanding these potential limitations, significant progress has been made in cometabolic bioventing of chlorinated contaminants in the vadose zone. Gaseous nutrient delivery of electron donor and acceptor (and nitrogen and phosphorus if necessary) has the potential to stimulate microbial metabolism in the vadose zone and to minimize mobilization of contaminants. Although focused on the saturated zone, biosparging of methane, nitrous oxide, and triethyl phosphate (Hazen et al., 1994; Brockman et al., 1995a) stimulated microbial populations and TCE- and PCE-degrading ability in the overlying vadose zone (Hazen et al., 1994; Brockman, 1994), presumably via movement of moist air, unused methane, and possibly gaseous microbial metabolites into the vadose zone. There is ample evidence that anaerobic microsites and/or regions exist in the vadose zone. These are likely due to (1) protection of anaerobic bacteria from oxygen diffusion through high oxygen uptake by aerobic microorganisms (obligate and facultative) and (2) reduced oxygen diffusion into soil aggregates and fine-grained strata of high tortuosity. Anaerobic reductive dechlorination of PCE was demonstrated in a large water-saturated column containing vadose zone sediment (Enzien et al., 1994). This anaerobic process occurred even though the column received O_2, at saturating concentrations, in the influent and no lower than 12% of saturation in the effluent.

Sayles and co-workers have been very active in developing cometabolic bioventing of chlorinated hydrocarbons in the vadose zone (Sayles et al., 1997a, 1997b; Moser et al., 1997). If air is injected, chlorinated compounds can be biodegraded aerobically (e.g., TCE to TCE epoxide) or anaerobically (e.g., PCE to ethylene) in microsites or regions stimulated and/or expanded by the higher levels of substrate and activity or biomass. Successful field-scale aerobic cometabolic bioventing of chloroform, TCE, and 1-1-1-trichloroethane has been demonstrated (Cox et al., 1998). If contaminants require initial dechlorination before aerobic metabolism, an alternative strategy is injection of anaerobic electron donor and acceptor (hydrogen and carbon dioxide) in an anaerobic carrier (nitrogen, plus helium as a tracer) while still allowing aerobic degradation (Sayles et al., 1997a). Aerobic degradation would then occur at the margins of the engineered anaerobic zone or by halting injection and allowing oxygen to diffuse periodically into the injection wells. Thus, with either of these cometabolic bioventing approaches, aerobic and anaerobic processes may occur together, or be separated, in space and/or time. Although these studies have been conducted at high-recharge sites with relatively shallow vadose zones, they suggest that it may also be possible to establish anaerobic microsites and/or larger anaerobic zones when nutrients are injected into low-recharge and deep vadose zones. Other gaseous nutrients, namely, propane, butane, propylene, and ammonia, have been shown to greatly stimulate microbial populations and activity in saturated microcosm studies (Kim et al., 1997; Moser et al., 1997; Palumbo et al., 1995; Tovanabootr et al., 1997) and could also be useful in the vadose zone.

Some metals and radionuclides are transported in the vadose zone because of increased mobility resulting from microbial or geochemical transformations, chemical associations with particulates, or complexation with synthetic or natural organic ligands. Possible scenarios for vadose zone bioremediation of metals and radionuclides include immobilization by microbially assisted mineral formation, mobilization by microbially assisted reductive dissolution by flooding and removing the inorganic contaminants by pumping, and biodegradation of organic ligands. Reductive dissolution of metals/radionuclides would require locally anaerobic conditions in the vadose zone, which may be achievable by the addition of easily metabolized organic substrate, addition of reducing agents, and possibly in-well lifting of anaerobic groundwater and injection into the vadose zone. Although bioremediation concepts and processes for metals and radionuclides have been applied to *ex situ* treatment and, in some cases, evaluated for application to the saturated zone, they have not been tested or evaluated for the vadose zone. This knowledge gap should be addressed in the future.

A final comment is that numerical reactive transport models for the vadose zone are far less developed than for the saturated zone. Although simple reactive transport models have been applied to bioremediation of petroleum hydrocarbons in the vadose zone, the uncertainties are much larger when attempting to apply these models to other contaminant classes. Models that include empirical information on interacting hydrologic and microbiological processes, and account for their spatial variability in unsaturated porous media, do not exist. A major benefit of the development and validation of such models in the future would be the ability to predict outcomes with more acceptable levels of uncertainty and to explore alternative nutrient injection strategies.

5.2 Hazardous Waste Repositories

Deep vadose zone rocks in arid and semiarid environments have been targeted as potential sites for the long-term storage of hazardous materials such as nuclear waste. In the United States, these sites include the Waste Isolation Pilot Plant (WIPP) in southeastern New Mexico and the proposed Yucca Mountain high-level nuclear waste repository in Nevada. These repositories are expected to safely sequester the waste material from the accessible environment for millennia. For Yucca Mountain, this means assuring that radionuclides will not be transported to the underlying groundwater, 150 m below the repository, within the next 10,000 years, even in the event of a spill or failure of the containment vessels. Predicting the fate and transport of radioactive wastes in a repository for 10,000 years is a daunting task; when one considers the myriad of potential microbial influences, the uncertainty becomes even greater.

Hersman (1997) has recently reviewed the potential effects of microorganisms on radionuclide transport in the vadose zone. Briefly, these include (1) sorption and/or bioaccumulation of radionuclides by microbial cells, (2) microbial chelation of radionuclides, (3) microbially mediated oxidation–reduction reactions, (4) biodegradation of organic radionuclide complexes, and (5) microbially induced changes in pH. Depending on which of these processes dominate, microorganisms can accelerate or retard radionuclide transport through the vadose zone. Predicting which of these processes will prevail is next to impossible; however, one can predict with assurance that microbes will exert an influence. Microbiological studies have shown that pristine volcanic tuff at Yucca Mountain contains microorganisms and that the metabolic activities of these microbes are moisture-limited (or limited by low nutrient flux due to low moisture content) (Kieft et al., 1997a). As in other vadose zone environments, the populations of viable microorganisms in the rock were relatively low (10^1–10^3 cfu mL^{-1} measured by heterotrophic plate counts and 5.9×10^3 to 2.2×10^5 cells mL^{-1}, estimated from PLFA concentrations). Although these microbial populations are small and moisture-limited, the effects of their activities will likely be significant when integrated over time and distance (e.g., 10,000 years and 150 m, respectively, for the Yucca Mountain repository). Microbes have also been cultivated from the halite at the WIPP site (Vreeland et al., 1998). Despite the saturated salt solutions in pore waters at this site, microorganisms occur, and their potential activities should not be discounted.

In addition to the naturally occurring microbial populations in the pristine rock surrounding a vadose zone repository (termed the "far field"), the disturbances generated during mining and construction of waste repositories increase microbial abundance and activities in proximity to the waste (the "near field"). David Ringelberg (personal communication) quantified PLFAs in sand and straw within the Yucca Mountain tunnel several months after initial tunnel boring and found a 10^2–10^4-fold increase in microbial biomass compared to the undisturbed rock. This localized stimulation of microbial activity poses the threat of biocorrosion of containment vessels in the near field as well as accelerated transport in the far field.

6 IMPRACTICAL CONSIDERATION: A MARTIAN VADOSE ZONE?

Studies of the surface features of Mars suggest that it is a water-rich planet, which suggests the tantalizing possibility of past or extant Martian life. Apparent stream beds and eroded canyons suggest that water shaped the face of the red planet. Water is thought to exist currently on Mars as polar ice caps, a planet-wide cryosphere extending from near surface to 3–20-km depth, and a deep liquid water-saturated aquifer (Clifford, 1993). In some areas, the cryosphere and the aquifer are thought to be contiguous, whereas in other areas they may be separated by an unsaturated zone that could be as thick as 10 km. Within this unsaturated zone, it has been hypothesized that thermal gradients drive the vertical movement of water. Water that evaporates at the water table would be driven upward toward the underside of the cryosphere. As this water vapor cools, it would condense and drain back downward.

In a closed system, this would result in a dynamic balance of rising water vapor and descending condensate. Clifford (1993) has estimated this water flux to be approximately 1×10^{-4} m year^{-1} (1 m of water is the amount of water that would cover the Martian surface to 1-m depth). An intriguing aspect of this water flux is that it could drive inorganic nutrient fluxes which could support microbial life. Although we know nothing of the chemistry of the Martian subsurface, one can imagine redox gradients containing biologically useful electron donors and acceptors. Thermal cycling of water could conceivably cause advection of electron donors and/or acceptors in sufficient quantity to support life. Of course, this is mere speculation. The search for life on Mars will likely be conducted with a focus on shallow saturated zones (i.e., groundwater environments that may be accessible by drilling in a low-elevation site such as the Valle Marineris). However, the lack of precipitation means Martian aquifers are likely diffusion-dominated systems with little advection of nutrients. The putative unsaturated zone with a thermally driven moisture flux may have a nutrient flux that could support life, although populations would likely be very low. Microorganisms may persist in a predominantly dormant form, and activity may be near or below the limit of detection. Knowledge about microbial life in the deep, low-recharge, unsaturated zones of Earth is likely to be of high value in searching for and detecting Martian vadose zone microorganisms, should they exist.

7 SUMMARY AND CONCLUSIONS

Although much has been learned about the microbiology of unsaturated subsurface environments during the last decade, we still lack a comprehensive understanding of microbial patterns and processes in the vadose zone. We can generalize about the low abundance and patchy distributions of microorganisms, and we have some information relating physical/chemical conditions to their population sizes. Nonetheless, more data are needed before we can accurately predict the numbers, biotransformation capabilities, and activities of microorganisms based on those chemical and physical conditions. Similarly, we have only a limited knowledge of the identities of microorganisms in the vadose zone. Fortunately, the growing use of molecular approaches, especially 16S rRNA-based techniques, should rapidly expand our knowledge of microbial phylogeny in the vadose zone. The subsurface vadose zone has characteristics in common with both surface soils and saturated subsurface environments, so it is not surprising that similar microorganisms are present in both of these environments. However, further investigation may reveal microorganisms that are unique to the vadose zone. The deep vadose zones of arid environments typically have extremely low nutrient fluxes, and thus indigenous microorganisms may be uniquely adapted to this extreme environment. Increased knowledge of the basic physiology and ecology of vadose zone microorganisms, including the spatial and temporal aspects of their colonization responses to contaminants, will likely also be of practical benefit to bioremediation efforts and to the successful implementation of hazardous waste repositories in the vadose zone.

REFERENCES

Alleman BC and Leeson A (eds) (1997) In Situ *and On-Site Bioremediation*, Vol 1. Columbus, OH: Battelle Press.

Amy PS (1997) Microbial dormancy and survival in the subsurface. In Amy PS and Haldeman DL (eds): *The Microbiology of the Terrestrial Subsurface*. Boca Raton, FL: CRC Press, pp 185–203.

Amy PS, Durham C, Hall D and Haldeman DC (1993) Starvation survival of deep subsurface isolates. *Curr Microbiol 26*:345–352.

Amy P S and Morita RY (1983) Starvation-survival patterns of sixteen freshly isolated open-ocean bacteria. *Appl Environ Microbiol 45*:1109–1115.

Balkwill DL and Ghiorse WC (1985) Characterization of subsurface bacteria associated with two shallow aquifers in Oklahoma. *Appl Environ Microbiol 50*:580–588.

Balkwill DL, Murphy EM, Fair DM, Ringelberg DB, and White DC (1998) Microbial communities in high and low recharge environments: Implications for microbial transport in the vadose zone. *Microbial Ecol 35*:156–171.

Breitenbeck GA, Yang H, and Dunigan EP (1988) Water-facilitated dispersal of inoculant *Bradyrhizobium japonicum* in soils. *Biol Fertil Soils 7*:58–62.

Bouwer H (1979) *Groundwater Hydrology*. New York: McGraw-Hill.

Brockman FJ (1994) *Biomolecular Probe Analysis of Bioremediation Organisms: Non-arid VOC Integrated Demonstration*. Pacific Northwest National Laboratory, Richland, WA, final rept. PNNL-10162.

Brockman FJ, Griffiths RP, Murray CJ, Li SW, Spadoni CM, and Braby J (1995a) Spatial heterogeneity of microbial activity in subsurface sediments. In: *Abstracts of the American Society for Microbiology Annual Meeting*, p 348.

Brockman FJ, Kieft TL, Fredrickson JK, Bjornstad BN, Li SW, Spangenburg W, and Long PE (1992) Microbiology of vadose zone paleosols in south-central Washington State. *Microbial Ecol 23*:279–301.

Brockman FJ, Li SW, Fredrickson JK, Ringelberg DB, Kieft TL, Spadoni CM, White DC, and McKinley JP (1998a) Post-sampling changes in microbial community composition and activity in a subsurface paleosol. *Microbial Ecol 36*:152–164.

Brockman FJ, Li SW, Spadoni CM, and Braby J (1996a) Spatial distribution and continuity of microbial activity in the unsaturated subsurface at an arid site. In: *Abstracts of the American Society for Microbiology Annual Meeting*, p 331.

Brockman FJ and Murray CJ (1997) Microbiological heterogeneity in the terrestrial subsurface and approaches for its description. p. 75–102. In Amy PS and Haldeman DL (eds): *The Microbiology of the Terrestrial Subsurface*. Boca Raton, FL: CRC Press, pp 75–102.

Brockman FJ, Murray CJ, Griffiths RP, Palumbo TV, Pfiffner SM, Li SW, Spadoni CM, Braby J, and Mazurkiewicz Y (1996b) Effect of recharge on the spatial heterogeneity of microbial activity in subsurface sediments. In: *Abstracts of the 1996 International Symposium on Subsurface Microbiology*, p 77.

Brockman FJ, Murray C, Murphy E, Bjornstad B, Balkwill D, Ringelberg D, Pfiffner S, and Griffiths R (1997) Microbial life in the unsaturated subsurface under conditions of extremely low recharge: An extreme environment. In Hoover RB (ed): *Proceedings of SPIE, Instruments, Methods, and Missions for the Investigation of Extraterrestrial Microorganisms, San Diego, CA, 27 July–1 August, 1997*, Vol. 3111. Washington, DC: SPIE, pp 388–394.

Brockman FJ, Payne W, Workman DJ, Soong A, Manley S, and Hazen TC (1995b) Effect of gaseous nitrogen and phosphorus injection on *in situ* bioremediation of a trichloroethylene-contaminated site. *J Hazard Mat 41*:287–298.

Brockman FJ, Stults J, Li S, Spadoni C, Pfiffner S, Murray C, and Murphy E (1999) Patchy distribution of microbial activity in low recharge vadose zone sediments. In: *Abstracts of the 1999 International Symposium on Subsurface Microbiology*, p 50 American Society for Microbiology.

Brockman FJ, Stults J, Miick R, Li SW, Spadoni CS, and Murphy EM (1998b) Pore water age affects the frequency, extent, and distribution of microbial growth in deep vadose zone sediments. In: *Abstracts of the American Society for Microbiology Annual Meeting*, p 369.

Brown RL, Bowman RW, and Kieft TL (1994) Microbial effects on nickel and cadmium sorption and transport in volcanic tuff. *J Environ Qual 23*:723–729.

Cano RJ and Barucki MK (1995) Revival and identification of bacterial spores in 25- to 40-million-year-old Dominican amber. *Science 268*:1060–1064.

Chapelle FH (1993) *Ground-water Microbiology and Geochemistry.* New York: John Wiley & Sons.

Chapelle FH and Lovley DR (1990) Rates of microbial activity in deep coastal plain aquifers. *Appl Environ Microbiol 56*:1865–1874.

Clifford SM (1993) A model for the hydrologic and climatic behavior of water on Mars. *J Geophys Res 98*:10973–11016.

Colwell FS (1989) Microbial comparison of surface soil and unsaturated subsurface soil from a semiarid high desert. *Appl Environ Microbiol 55*:2420–2423.

Cox EE, McAlary TA, Major DW, Allan J, Lehmicke L, and Neville SL (1998) Cometabolic bioventing of chlorinated solvents at a former waste lagoon. In Wickramanayake GB and Hinchee RE (eds): *Bioremediation and Phytoremediation: Chlorinated and Recalcitrant Compounds.* Columbus, OH: Battelle Press, pp 227–232.

Driscoll FG (1985) *Groundwater and Wells*, 2nd ed. St. Paul, MN: Johnson Division.

Duboise SM, Moore BE, and Sagik BP (1996) Poliovirus survival and movement in a sandy forest soil. *Appl Environ Microbiol 31*:536–543.

Dupont RR, Doucette WJ, and Hinchee RE (1991) Assessment of *in situ* bioremediation potential and the application of bioventing at a fuel-contaminated site. In Hinchee RE and Olfenbuttel RF (eds): In Situ *and On Site Bioreclamation.* Stoneham, MA: Butterworth-Heinemann, pp 262–282.

Fan S and Scow KM (1993) Biodegradation of trichloroethylene and toluene by indigenous microbial populations in soil. *Appl Environ Microbiol 59*:1911–1918.

Fredrickson JK, Brockman FJ, Bjornstad BN, Long PE, Li SW, McKinley JP, Conca JL, Kieft TL and Balkwill DL (1993) Microbiological characteristics of pristine and contaminated deep vadose sediments from an arid region. *Geomicrobiol J 11*:95–107.

Fredrickson JK, Garland TR, Hicks RJ, Thomas JM, Li SW, McFadden KM (1989) Lithotrophic and heterotrophic bacteria in deep subsurface sediments and their relation to sediment properties. *Geomicrobiol J 7*:53–66.

Fredrickson JK, Li SW, Brockman FJ, Haldeman DL, Amy PS, and Balkwill DL (1995) Time-dependent changes in viable numbers and activities of aerobic heterotrophic bacteria in subsurface samples. *J Microbiol Meth 21*:253–265.

Enzien MV, Picardal F, Hazen TC, Arnold RG, and Fliermans CB (1994) Reductive dechlorination of trichloroethylene and tetrachloroethylene under aerobic conditions in a sediment column. *Appl Environ Microbiol 60*(6):2200–2204.

Fuller ME, Mu DY, and Scow KM (1995) Biodegradation of trichloroethylene and toluene by indigenous microbial populations in vadose sediments. *Microbial Ecol 29*:311–325.

Gerba CP (1984) Applied and theoretical aspects of virus adsorption to surfaces. *Adv Appl Microbiol 30*:133–168.

Goyal SM and Gerba CP (1979) Comparative adsorption of human enteroviruses, simian rotovirus, and selected bacteriophages to soil. *Appl Environ Microbiol 38*:241–247.

Guymon GL (1995) *Unsaturated Zone Hydrology.* Englewood Cliffs, NJ: Prentice Hall.

Hagedorn C, Hansen DT, and Simonson GH (1978) Survival and movement of fecal indicator bacteria in soil under conditions of saturated flow. *J Environ Qual 7*:55–59.

Haldeman DL (1997) The storage-related phenomenon: implications for handling and analysis of subsurface samples. In Amy PS and Haldeman DL (eds): *The Microbiology of the Terrestrial Subsurface.* Boca Raton, FL: CRC Press, pp 61–73.

Haldeman DL and Amy PS (1993) Bacterial heterogeneity in deep subsurface tunnels in Rainier Mesa, Nevada Test Site. *Microbial Ecol 25*:183–194.

Haldeman DL, Amy PS, Ringelberg D, and White DC (1993) Characterization of the microbiology within a $21\,m^3$ section of rock from the deep subsurface. *Microbial Ecol 26*:145–159.

Haldeman DL, Amy PS, Ringelberg D, White DC, Gharen DC, and Ghiorse WC (1995a) Microbial growth and resuscitation alter community structure after perturbation. *FEMS Microbiol Ecol 17*:27–38.

Haldeman DL, Amy PS, Russell CE, and Jacobson R (1995b) Comparison of drilling and mining as methods for obtaining microbiological sampling from the deep subsurface. *J Microbiol Meth 21*:305–316.

Haldeman DL, Amy PS, White DC, and Ringelberg DB (1994a) Changes in bacteria recoverable from subsurface volcanic rock samples during storage at 4°C. *Appl Environ Microbiol 60*:2697–2703.

Haldeman PL, Pitonzo B, Story SP, and Amy PS (1994b) Comparison of the microbiota recovered from surface and deep subsurface rock, water, and soil samples along an elevational gradient. *Geomicrobiol J 12*:99–111.

Hazen TC, Lombard KH, Looney BB, Enzien MV, Dougherty JM, Fliermans CB, Wear J, and Eddy-Dilek CA (1994) Summary of *in-situ* bioremediation demonstration (methane biostimulation) via horizontal wells at the Savannah River site integrated demonstration project. In Gee FW and Wing NR (eds): In-Situ *Remediation: Scientific Basis for Current and Future Technologies.* Columbus, OH: Battelle Press, pp 137–150.

Hersman LE (1997) Subsurface microbiology: effects on the transport of radionuclides in the vadose zone. In Amy PS and Haldeman DL (eds): *The Microbiology of the Terrestrial Subsurface.* Boca Raton, FL: CRC Press, pp 299–323.

Hersman LE, Purtymun W, and Sinclair J (1988) Preliminary microbial analysis of the vadose zone, Pajarito Plateau, NM. In: *Abstracts of the American Socity for Microbiology Annual Meeting,* p 252.

Hillel D (1980) *Fundamentals of Soil Physics.* New York: Academic Press.

Hinchee RE, Miller RN, and Johnson PC (1995) *In situ Aeration: Air Sparging, Bioventing, and Related Remediation Processes.* Columbus, OH: Battelle Press.

Holden PA, Hunt JR, and Firestone MK (1997) Toluene diffusion and reaction in unsaturated *Pseudomonas putida* biofilms. *Biotechnol Bioeng 56*:656–670.

Kennedy MJ, Reader SL, and Swierczynski LM (1994) Preservation records of microorganisms: Evidence of the tenacity of life. *Microbiology 140*:2513–2529.

Kieft TL (1991) Soil microbiology in reclamation of arid and semi-arid lands. In Skugins J (ed): *Semiarid Lands and Deserts: Soil Resource and Reclamation.* New York: Marcel Dekker. pp 209–256.

Kieft TL (2000) Dwarf cells in soil and subsurface terrestrial environments. In Colwell RR and Grimes DJ (eds): *Non-culturable Microorganisms in the Environment.* New York: Chapman and Hall pp 19–46.

Kieft TL, Amy PS, Bjornstad BN, Brockman FJ, Fredrickson JK, and Rosacker LL (1993) Microbial abundance and activities in relation to water potential in the vadose zones of arid and semiarid sites. *Microbial Ecol 26*:59–78.

Kieft TL, Kovacik WP Jr, Ringelberg DB, White DC, Haldeman DL, Amy PS, and Hersman LE (1997a) Factors limiting to microbial growth and activity at a proposed high-level nuclear repository, Yucca Mountain, Nevada. *Appl Environ Microbiol 63*:3128–3133.

Kieft TL, Murphy EM, Haldeman DL, Amy PS, Bjornstadt BN, McDonald EV, Ringelberg DB, White DC, Stair JO, Griffiths RP, Gsell TC, Holben WE, and Boone DR (1998) Microbial transport, survival, and succession in a sequence of buried sediments. *Microbial Ecol 36*:336–348.

Kieft TL and Phelps TJ (1997) Life in the slow lane: Activities of microorganisms in the subsurface. In Amy PS and Haldeman DL (eds): *The Microbiology of the Terrestrial Subsurface.* Boca Raton, FL: CRC Press, pp 137–163.

Kieft TL, Ringelberg DB and White DC (1994) Changes in ester-linked phospholipid fatty acid profiles of subsurface bacteria during starvation and desiccation in a porous medium. *Appl Environ Microbiol 60*:3292–3299.

Kieft TL, Wilch E, O'Connor K, Ringelberg DB, and White DC (1997b) Survival and phospholipid fatty acid profiles of surface and subsurface bacteria in natural sediment microcosms. *Appl Environ Microbiol 63*:1531–1542.

Kim Y, Semprini L, and Arp DJ (1997) Aerobic cometabolism of chloroform, 1,1,1-trichloroethane, 1,1-dichloroethylene, and the other chlorinated aliphatic hydrocarbons by butane-utilizing microorganisms. In: In Situ *and On-Site Bioremediation*, Alloman BC and Leeson A, eds, Vol 3. Columbus, OH: Battelle Press, pp. 107–112.

Konopka A and Turco R (1991) Biodegradation of organic compounds in vadose zone and aquifer sediments. *Appl Environ Microbiol 57*:2260–2268.

Lambert LH, Cox T, Mitchell K, Rosello-Mora RA, Del Cuet C, Dodge DE, Orkand P, and Cano RJ (1998) *Staphylococcus succinus* sp. nov., isolated from Dominican amber. *Internat J System Bacteriol 48*:511–518.

Lance JC and Gerba CP (1984) Virus movement in soil during saturated and unsaturated flow. *Appl Environ Microbiol 47*:335–337.

Long PE, Rawson SA, Murphy E, and Bjornstad B (1992) Hydrologic and geochemical controls on microorganisms in subsurface formations. In: *Pacific Northwest Laboratory Annual Report for 1991 to the DOE Office of Energy Research, Part 2, Environmental Sciences* (PNL-8000 pt. 2), Richland, WA, pp 49–71.

Madsen EL and Alexander M (1982) Transport of *Rhizobium* and *Pseudomonas* through soil. *Soil Sci Soc Amer J 46*:557–560.

Madsen EL, Sinclair JL, and Ghiorse WC (1991) *In situ* biodegradation: Microbiological pattern in a contaminated aquifer. *Science 252*:830–833.

Metting FB Jr. (ed) (1993) *Soil Microbial Ecology: Applications in Agricultural and Environmental Management*. New York: Marcel Dekker.

Moser LE, Sayles GD, Gannon DJ, Lee MD, Kampbell DJ, and Vogel CM (1997) Comparison of methodologies for cometabolic bioventing treatability studies. In: In Situ *and On-Site Bioremediation*, Vol 3, Columbus, OH: Battelle Press.

Murphy EM, Shramke JA, Fredrickson JK, Bledsoe HW, Francis AJ, Sklarew DS, and Linehan JC (1992) The influence of microbial activity and sediment organic carbon on the isotope geochemistry of the Middendorf aquifer. *Water Resources Res 28*:723–740.

National Research Council (1993) In Situ *Bioremediation: When Does It Work?* Washington DC: National Academy Press.

Natsch A, Keel C, Troxler J, Zala M, von Albertini N, and Defago G (1996) Importance of preferential flow and soil management in vertical transport of biocontrol strain of *Pseudomonas fluorescens* in structured field soil. *Appl Environ Microbiol 62*:33–40.

Nichols WD (1987) Geohydrology of the unsaturated zone at the burial site for low level radioactive waste near Beatty, Nye County Nevada. Denver, CO: US Government Printing Office, USGS Water Supply Paper 2312.

Norris RD, Hinchee RE, Brown R, McCarty PL, Semprini L, Wilson JT, Kampbell DH, Reinhard M, Bouwer EJ, Borden RC, Vogel TM, Thmas JM, Ward CJ, and Wilson JT (1994) *Handbook of Bioremediation*. Boca Raton, FL: Lewis Publishers.

Onstott TC, Phelps TJ, Kieft TL, Colwell FS, Balkwill DL, Fredrickson JK, and Brockman FJ (1998) A global perspective on the microbial abundance and activity in the deep subsurface. In Seckbach J (ed): *Enigmatic Microorganisms and Life in Extreme Environments: Cellular Origin and Life in Extreme Habitats*. Kluwer Publications, Norwell, MA, pp 1–14.

Palumbo AV, Scarborough SP, Pfiffner SM, and Phelps TJ (1995) Influence of nitrogen and phosphorous on the in-situ bioremediation of trichloroethylene. *Appl Biochem Biotechnol 55/56*:635–647.

Papendick RI and Campbell GS (1981) Theory and measurement of water potential. In Parr JF, Gardner WR, and Elliott LF (eds): *Water Potential Relations in Soil Microbiology*. Madison, WI: Soil Science Society of America, pp 1–22.

Paul EA and Clark FE (1996) *Soil Microbiology and Biochemistry*, 2nd edn. San Diego: CA: Academic Press.

Peyton BM (1996) Improved biomass distributions using pulsed injections of electron donor and acceptor. *Water Resources Res 30*:756–758.

Peyton BM and Characklis WG (1993) A statistical analysis of the effects of substrate utilization and shear stress on the kinetics of biofilm detachment. *Biotechnol Bioeng 41*: 728–735.

Phelps TJ, Murphy EM, Pfiffner SM, and White DC (1994) Comparison of geochemical and biological estimates of subsurface microbial activities. *Microbial Ecol 28*:335–350.

Pothuluri JV, Moorman TB, Obenhuber DC, and Wauchope RD (1990) Aerobic and anaerobic degradation of alachlor in samples from a surface-to-groundwater profile. *J Environ Qual 19*:525–530.

Powelson DK, Simpson JR, and Gerba CP (1990) Virus transport and survival in saturated and unsaturated flow through soil columns. *J Environ Qual 19*:396–401.

Powelson DK, Simpson JR, and Gerba CP (1991) Effects of organic matter on virus transport in unsaturated flow. *Appl Environ Microbiol 57*:2192–2196.

Powelson DK and Mills AL (1998) Water saturation and surfactant effects on bacterial transport in sand columns. *Soil Sci 163*(9):694–704.

Rahe TM, Hagedorn C, McCoy EL, and Kling GF (1978) Transport of antibiotic-resistant *Escherichia coli* through western Oregon hillslope soils under conditions of saturated flow. *J Environ Qual 7*:487–494.

Retallack GJ (1990) Soils of the Past. Winchester, MA: Unwin Hyman.

Reynolds PJ, Sharma P, Jenneman GE, and McInerney MJ (1989) Mechanisms of microbial movement in subsurface materials. *Appl Environ Microbiol 55*:2280–2286.

Richter DD and Markewitz D (1995) How deep is soil? *BioScience 45*:600–609.

Riley RG and Zachara JM (1992) Chemical contaminants on DOE lands and selection of contaminant mixtures for subsurface science research. DOE/ER-0547T. U.S. Department of Energy, Washington, DC.

Sayles GD, Hihopoulos P, and Suidan MT (1997a) Anaerobic bioventing of PCE. In: In Situ *and On-Site Bioremediation*, Vol 1, Alleman BC and Leeson A (eds). Columbus, OH: Battelle Press, pp 353–359.

Sayles GD, Moser LE, Gannon DJ, Campbell DH, and Vogel CM (1997b) Development of cometabolic bioventing for the *in-situ* bioremediation of chlorinated solvents. In: In Situ *and On-Site Bioremediation*, Vol 3. Columbus, OH: Battelle Press, p 285.

Schafer A, Ustohal P, Harma H, Stauffer F, Dracos T, and Zehnder AJB (1998) Transport of bacteria in unsaturated porous media. *J Contam Hydrol 33*:149–169.

Severson KJ, Johnstone DL, Keller CK, and Wood DB (1991) Hydrologic parameters affecting vadose-zone microbial distributions. *Geomicrobiol J 9*:197–216.

Sexstone AJ, Revsbeck NP, Parkin TB, and Tiedje JM (1985) Direct measurement of oxygen profiles and denitrification rates in soil aggregates. *Soil Sci Soc Amer J 49*:645–651.

Sharma PK, McInerney MJ, and Knapp RM (1993) *In situ* growth and activity and modes of penetration of *Eschericia coli* in unconsolidated porous materials. *Appl Environ Microbiol 59*:3386–3694.

Sinclair JL and Ghiorse WC (1987) Distribution of protozoa in subsurface sediments of a pristine groundwater study site in Oklahoma. *Appl Environ Microbiol 53*:1157–1163.

Smith MS, Thomas GW, White RE, and Ritonga D (1985) Transport of *Escherichia coli* through intact and disturbed soil columns. *J Environ Qual 14*:87–91.

Stephens DB (1996) *Vadose Zone Hydrology.* Boca Raton, FL: Lewis.

Stevens TO and Holbert BS (1995) Variability and density dependence of bacteria in terrestrial subsurface samples: Implications for enumeration. *J Microbiol Meth 21*:283–292.

Sylvia DM, Fuhrmann JJ, Hartel PG, and Zuberer DA (eds) (1998) *Principles and Applications of Soil Microbiology.* Upper Saddle River, NJ: Prentice-Hall.

Tovanabootr A, Russel S, Stoffers NH, Arp DJ, and Semprini L (1997) An evaluation of five aerobic cometabolic substrates for trichloroethylene treatment by microbes stimulated from the subsurface of McClellan Air Force Base. In: In Situ *and On-Site Bioremediation*, Vol 3. Columbus, OH: Battelle Press, pp 93–99.

Treves DS, Xia B, Zhou J, Tiedje JM (1999) What drives the level of microbial diversity in soils? In: *Abstract of the American Society for Microbiology Annual Meeting*, p 493.

Vreeland RH, Piselli AF Jr, McDonnough S, Meyers SS (1998) Distribution and diversity of halophilic bacteria in a subsurface salt formation. *Extremophiles 2*:321–331.

Wan J, Wilson JL and Kieft TL (1994) Influence of the gas-water interface on transport of microorganisms through unsaturated porous media. *Appl Environ Microbiol 60*:509–516.

Wong PTW, and Griffin DM (1976) Bacterial movement at high matric potentials. I. In artificial and natural soils. *Soil Biol Biochem 8*:215–218.

Wood BD, Keller CK, and Johnstone DL (1993) *In situ* measurement of microbial activity and controls on microbial CO_2 production in the unsaturated zone. *Water Resources Res 29*: 647–659.

Zhang C, Palumbo AV, Phelps TJ, Beauchamp JJ, Brockman FJ, Murray CJ, Parsons BS, Swift DJP (1998) Grain size and depth constraints on microbial variability in coastal plain subsurface sediments. *Geomicrobiol J 15*:171–185.

BIOGEOCHEMICAL
PROCESSES

6

THE USE OF GEOCHEMISTRY AND THE IMPORTANCE OF SAMPLE SCALE IN INVESTIGATIONS OF LITHOLOGICALLY HETEROGENEOUS MICROBIAL ECOSYSTEMS

JAMES P. MCKINLEY

Pacific Northwest National Laboratory Interfacial Geochemistry Group, Richland, Washington

1 INTRODUCTION

Studies of subsurface microbial ecology have become increasingly linked to contextual studies of geology and geochemistry. Even though the point of such studies is to investigate the microbial population, the interplay between the three components—living, physical, and chemical—needs to be understood to make sensible use of information about the kinds and numbers of organisms present. Fortunately, practice in the geologic sciences is readily adaptable to cross-disciplinary investigations. The system of geologic nomenclature and classification is a tool

Subsurface Microbiology and Biogeochemistry, Edited by James K. Fredrickson and Madilyn Fletcher.
ISBN 0-471-31577-X. Copyright 2001 by Wiley-Liss, Inc.

for characterizing the physicochemical heterogeneity of the subsurface and provides a basic description of each environment through geologic maps and publications. Because stratigraphic nomenclature is concerned with depositional environment, mineralogy, and texture, these geologic variables can be viewed as a description of the physical constraints on microbial movement and nutrient flux. Since subsurface lithology is diversely heterogeneous, rocks that are sharply dissimilar can be juxtaposed and have significant intercommunity effects across lithologic boundaries. Geochemistry provides the currency for exchange between microbial communities across these boundaries. The correspondence of chemistry to geologic context can be observed by examining the groundwater chemistry of disparate, juxtaposed strata (e.g., Pucci and Owens, 1989). Within a given lithologic regime, geochemical conditions can determine microbial community function by setting the overall supply of nutrients and the broad values of intensive variables that control how the microbial community expresses itself. At the same time, the microbial community disposes the provided nutrients to transform the chemical environment by transferring electrons from organic carbon to a variety of electron acceptors. In aquifers where the water has traveled away from the source at recharge and where dissolved oxygen cannot be replenished by the dissolution of atmospheric O_2, the anaerobic metabolism of organic carbon can create a variety of reactive compounds and transform the general geochemistry of the aquifer. This chapter describes the dependence of microbial communities on contrasts in geochemistry and geology, with a particular focus on fine-scale geochemistry as an investigative tool.

2 FIELD MICROBIOLOGY AND CHEMISTRY

In the terrigenous subsurface, microbes derive metabolic energy from remnant, clastic organic carbon, or from electron-donating inorganic compounds. The vastly predominant energy source is organic carbon derived from photosynthesis. Photosynthesis proceeds by the disproportionation of energy into localized centers of highly negative $p\epsilon$ and a reservoir of O_2 (Stumm and Morgan, 1996), where $p\epsilon$ is the relative tendency of a solution to accept or transfer electrons. Photosynthesis thus imposes a state of disequilibrium on the chemical components that comprise the microbial ecosystem. Nonphotosynthetic organisms exploit this disequilibrium, and act to restore it, by using catalysts to decompose the unstable organic products of photosynthesis. They may do this by respiration, using an external acceptor (e.g., oxygen or sulfate) for electrons liberated from electron donors, or by fermentation, in which an organic compound is oxidized (electron removed) inside a cell and the liberated electron is transferred to another, partially oxidized, compound to reduce it (Ehrlich, 1990). In any case, the microbes facilitate thermodynamically feasible chemical reactions, realizing a fraction of the available free energy (ΔG) for metabolic purposes.

As summarized by Stumm and Morgan (1996), organisms act as redox catalysts, mediating the electron transfer within a substrate oxidation half-reaction and a coupled electron acceptor half-reaction. It is convenient to combine the ideas of $p\epsilon$

and half-reactions to construct a relatively simple thermodynamic method for evaluating the feasibility of hypothetical chemical reactions for driving microbial metabolism.

The quantity $p\epsilon$ can be defined in a way that is analogous to pH:

$$p\epsilon = -\log\{e^-\} \tag{1}$$

giving the hypothetical electron activity at equilibrium and indicating the tendency of a solution to accept or transfer electrons. In a highly reducing solution, the electron activity is relatively high, and in an oxidizing environment, the electron activity is low. By convention, the free-energy change for the oxidation of H_2 is fixed at zero. The $p\epsilon$ can be related to the standard electrode potential (E_H°) and free energy of reaction by the relations

$$E_H^{\circ} = -\Delta G^{\circ}/nF = (RT/nF)\ln K = (2.3RT/nF)\log K = (2.3RT/F)p\epsilon^{\circ} \tag{2}$$

where ΔG° is the standard Gibbs free-energy change in the oxidation–reduction reaction, K is the thermodynamic equilibrium constant, n is the number of electrons in the reaction as written, R is the gas constant, and F is the Faraday constant.

Note the basic definition of free energy of reaction:

$$\Delta G = \Delta G^{\circ} + RT\ln(\Pi_i\{\text{products}\}^{ni}/\Pi_j\{\text{reactants}\}^{nj}) \tag{3}$$

where the quantity in parentheses is defined as the equilibrium constant K. At equilibrium, the net ΔG is zero, so the relationship becomes

$$-\Delta G^{\circ} = RT\ln K \tag{4}$$

and specifies that disequilibrium must exist to provide energy for metabolism. Permissible reactions have a free energy that is negative. The energetics of half-reactions may be characterized by the value of K for the reaction, by the standard potential (E_H°), or by the $p\epsilon^{\circ}$. By adding appropriate half-reactions, the net potential for any given oxidation–reduction reaction can be readily derived.

A further convenient modification of $p\epsilon$ as a tool for understanding energy-yielding reactions in natural waters is to define a subsidiary constant $p\epsilon^{\circ}(W)$. This symbol is analogous to $p\epsilon^{\circ}$, except that $\{H^+\}$ and $\{OH^-\}$ are assigned their activities in neutral water to unambiguously negate the effects of variable pH on comparisons based on $p\epsilon^{\circ}$. Standard conditions for $p\epsilon^{\circ}(W)$ are thus defined as 25°C, unit activities of oxidant and reductant, and pH= 7.00. $p\epsilon^{\circ}(W)$ is defined by

$$p\epsilon^{\circ}(W) = p\epsilon^{\circ} + (n_H/2)\log K_W \tag{5}$$

where n_H is the number of moles of protons exchanged per mole of electrons. If we use this convention, half-reactions can be evaluated according to $p\epsilon^{\circ}(W)$, so that a ranked list will contain reactions ordered so that any higher (more positive) system will oxidize any lower (more negative) system. For example, given a table of half-reactions with values for $pe°(W)$, one can readily establish that NO_3^- can oxidize HS^- to SO_4^{2-} [details of this approach along with tables and examples are given in Stumm and Morgan (1996)]. The construction of ranked lists of half-reactions has become widespread and has proved useful in understanding redox reactions that support life and the competition and interplay between microbial metabolic processes, and in exploring the possibilities of life elsewhere (e.g., Klass, 1984; Lovley and Goodwin, 1988; Stumm and Morgan, 1996; Madigan et al., 1997; Gaidos et al., 1999).

In geochemical studies of microbial ecology (biogeochemistry), one is attempting to detect and trace the effects of the microbial community on groundwater or sediment chemistry. Measurements with respect to microbial community function are therefore based on chemical species that incorporate the products or reactants within microbially catalyzed chemical reactions. For example, in the examination of a subsurface environment where sulfate-reducing bacteria are active through the use of acetate as an electron donor, the overall reaction can be represented by

$$2CH_2O = SO_4^{2-} \rightarrow H_2S + 2HCO_3^- \tag{6}$$

One could attempt to observe changes in any of these constituents for evidence of sulfate reduction. Of all the reactants and products, however, the obvious target is dissolved sulfide, since acetate is not particular to sulfate reducers as an electron donor, bicarbonate is ubiquitous, and the presence of sulfate alone does not indicate the presence of an active sulfate-reducing community. Sulfide is not a common abiotic component of natural groundwaters. A useful tool in geochemical investigations of microbial systems is the measurement of multiple analytical components (e.g., sulfide, ferrous iron, methane, etc.) to determine their variability in space as an indicator of microbial activity and the transition of microbial activity from the predominance of one microbial guild to another. In that context, it is worthwhile to track changes in as many components as logistics allow.

3 THE CHEMICAL INVESTIGATION OF MICROBIAL COMMUNITIES

Subsurface microbiology has been more widely studied in recent years than previously, largely due to the importance of bioremediation at contaminated industrial sites. Since bioremediation occurs as a byproduct of the functions of a viable interactive microbial community, the community structure would seem to be an important component of most bioremediation efforts. However, the utility of characterizing microbial communities to assess bioremediation potential has been questioned. For example, early bioremediation efforts focused on petroleum hydrocarbons in surface soils, and the industry found that hydrocarbon biodegradation could be stimulated in most soils, so characterization of the microbial population was found to provide little useful information. In fact, a basic tenet of near-surface bioremediation is the assumption that microbes needed to effect chemical transformations of hazardous waste are present and can be stimulated to remediate a given site. This is because microbes are often assumed to be distributed so that essentially any functional guild (e.g., sulfate reducers) could be present and culturable in any given sediment. In this view, the dispersal of soil microorganisms is considered to be complete, whereas their functional distribution is controlled by the temporal selective properties of particular environments. Consequently, if conditions arose (such as introduction of an exogenous substrate), such that a particular oxidative pathway were possible and favored, an extant, *in situ* group of microorganisms would become rapidly enriched to take advantage of this potential energy.

In contrast, in the deep subsurface—below the zone where soil processes can directly affect microbial ecology—the dispersal of disparate functional guilds is much more restricted than in near-surface environments. The hypothesis that representatives of all guilds are present and able to be enriched may be false; extensive exploration has shown that, underground, everything is *not* everywhere. In these environments, a useful approach has been to characterize microbial communities by enumerating numbers of microorganisms in different functional groups (guilds), defined by the ability to utilize particular terminal electron acceptors. This is because the specific phylogeny of organisms able to carry out a particular bioremediation step is usually not known, and, many times, these traits are not monophyletic. For instance, dissimilatory metal reduction occurs among many unrelated groups of microorganisms. With this approach, a single organism may be counted in more than one functional group, due to facultative metabolic capabilities. In addition, the enumeration methods require growth or at least active metabolism, and so may be somewhat selective, for example, in choice of electron donor, or biased against very slow metabolizers. Nevertheless, this approach allows evaluation of the ability of the microbial community to carry out major biogeochemical processes relevant to bioremediation. If a microbial activity cannot be stimulated in a properly preserved sample in the laboratory, it is likely to be stimulated *in situ* by neither the systematic ecological perturbation represented by the introduction of a contaminant, nor by the subsequent introduction of supposedly stimulating nutrients.

Microbial guilds able to transform metals are of particular interest in subsurface ecosystems. These include dissimilatory iron-reducing bacteria (DIRB) and sulfate-reducing bacteria (SRB), among others. Research has shown that metal-reducing bacteria may be absent from a sediment profile, or may be narrowly restricted.

A number of research efforts have focused on the evolution of aqueous geochemistry along a groundwater flowpath (e.g., Murphy et al., 1992) and on the abrupt or cross-stratigraphic heterogeneity that exists at formation boundaries (e.g., McMahon and Chapelle, 1991). These studies have demonstrated some characteristic relationships between geochemistry and microbiology. In general, the microbial community is dependent on geochemical variables, but the geochemical environment is simultaneously modified by bacteria contained within it. Geochemical conditions may thus change under the influence of cumulative microbial processes. Much research has focused on the succession of dominant metabolic pathways as an aqueous biogeochemical system undergoes a transition from oxic to anoxic and subsequent progressive changes as groundwater moves and is modified within an anaerobic system. Redox zonation, the segregation of different terminal electron-accepting processes (TEAP) during the degradation of sedimentary organic matter, takes place along groundwater flowpaths or between distinct sedimentary environments. A sedimentary system containing organic matter and admixed terminal electron acceptors may exhibit zonation in a sequence that is defined by the relative energy yield of each pathway (e.g., Claypool and Kaplan, 1974; Chapelle, 1993; Postma and Jakobsen, 1996; Stumm and Morgan, 1996; reactions from Berner, 1981):

Reaction	$^{\circ}G_r$ (kJ/mol)
$CH_2O + O_2 \rightarrow CO_2 + H_2O$	-475
$5CH_2O + 4NO_3^- \rightarrow 2N_2 + 4HCO_3^- + CO_2 + 3H_2O$	-448
$CH_2O + 3CO_2 + H_2O + 2MnO_2 \rightarrow Mn_2^+ + 4HCO_3^-$	-349
$CH_2O + 7CO_2 + 4Fe(OH)_3 \rightarrow 4Fe_2^+ + 8HCO_3^- + 3H_2O$	-114
$2CH_2O + SO_4^{2-} \rightarrow H_2S + 2HCO_3^-$	-77
$2CH_2O \rightarrow CH_4 + CO_2$	-58

If one assumes that H_2 is an important electron donor and is therefore a valuable characterization tool, then an equivalent set of successive TEAPs can be produced by reformulating these reactions in terms of H_2 as an electron donor (Lovley and Goodwin, 1988; Postma and Jakobsen, 1996). As each electron acceptor in this series is depleted, the succeeding TEAP becomes favored, leading to the biogeochemical succession of predominant TEAPs and zonation. Zonation is also influenced by physiological controls on microbial metabolism and competition between functional groups within a given community (Lovley and Chapelle, 1995). In some systems where more favored and less favored guilds compete, the favored guild has been theorized to drive a necessary substrate to concentrations where the competing organism cannot utilize it. The inhibition of methanogenesis by SRB, for example, is thought to be a result of SRB lowering H_2-concentrations below the threshold at which methanogens can utilize them (Lovley and Goodwin, 1988). For a CO_2-reducing methanogen, for example,

$$CO_2 + 4H_2 = CH_4 + 2H_2O \qquad (7)$$

the reaction must proceed under conditions that provide sufficient minimum free energy for ATP synthesis. The smallest quantum of energy for this process, the energy required to transport an ion across the cytoplasmic membrane, is about $-20 \, kJ/mol$ (Schink, 1997; Conrad, 1999), or a bit lower (9–$15 \, kJ/mol$; Hoehler et al., 1998).

Redox zonation via a succession of energetically favored TEAPs is a manifestation of the free-energy disequilibrium systematics that govern microbial ecology generally. However, the simplified thermodynamic explanation of microbial behavior may be complicated by microbial community dynamics and physical limitations on nutrient availability. The succession is dependent on fermentation to maintain an adequate supply of appropriate electron donors (Jakobsen and Postma, 1999) and maintain disequilibrium conditions. Other complicating factors can arise, including such circumstances as elevated reaction-product concentrations, and (in the case of iron reducers) the microscale unavailability of the reactants acting as electron acceptors. Regardless of the standard-state free-energy value for a given reaction, if the operative reaction free energy is positive on the relevant spatial scale, the reaction will not proceed as an overall source of metabolic energy. This relationship is explicitly or implicitly recognized in most studies involving redox zoning (Lovley and Goodwin, 1988; Chapelle et al., 1995).

The reactions listed above along with their standard-state free energies are components in reaction chains, with fermentation of complex organic matter initiating the process of organic carbon oxidation by producing simple organic substrates, for example, acetate and formate, and H_2, which are consumed by the listed TEAPs (Lovley and Chapelle, 1995; Postma and Jakobsen, 1996). Fermentation is rate-limiting, and the products and reactants within each TEAP can approach equilibrium values. Under natural conditions, therefore, competing TEAPs may function under conditions yielding similar net free energies (Hoehler et al., 1998), suggesting that the segregation of TEAPs need not be strictly adhered to in nature. The role of fermentation aside, consideration of the hypothetical boundary between a sulfate-reducing and methanogenic zone in an anaerobic aquifer may illustrate the point. As sulfate is depleted so that methanogenesis can become dominant, the sulfate concentration must approach a value at which sulfate reduction becomes less negative in free-energy terms, and consequently, SRB must become less able to maintain H_2 below a hypothetical threshold required by methanogens. It is not likely that sulfate reduction will cease in such a zone, and methanogens will not be inhibited; the free energy of both pathways will be negative and both processes will proceed. Postma and Jakobsen (1996) present a similar argument, along with example calculations from field data, with respect to zonation of sulfate and iron reduction, and show subsequently (Jakobsen and Postma, 1999) the coexistence of iron reduction, sulfate reduction, and methanogenesis. In a study of organic-rich sediment, Holmer and Kristensen (1994) also observed unchecked simultaneous sulfate reduction and methanogenesis. Explanations for these observations differed markedly, however. Jakobsen and Postma (1999) advanced a "partial equilibrium" model, in which the rate-limiting, slow fermentation step is coupled to fast oxidation–reduction steps (TEAPs) that approach chemical equilibrium. Whereas the partial equilibrium model (Postma and Jakobsen, 1996; Jakobsen and Postma, 1999) suggested that electron donor-starved systems may have permitted different TEAPs to coexist productively when their free-energy yields were approximately equal, Holmer and Kristensen (1994) concluded that the concentration of substrates had exceeded the level at which competition was effective (one organism could not indirectly limit the other through restraints on free energy). Microbial dominance in either case was not exclusive of other TEAPs and was dependent on the concentrations of all candidate products and reactants. As recognized by Lovley and Goodwin (1988), the zonation of TEAPs into zones of dominance occurs under steady-state conditions; under conditions of shifting geochemistry, zonation is diffuse and dominance may occur counter to expectations (Postma and Jakobsen, 1996).

Uncontaminated samples from the subsurface are difficult to obtain, and laboratory incubations may not mimic subsurface processes; microbial activity in the subsurface must often be inferred from nonmicrobiological analyses of groundwater (Lovley and Chapelle, 1995). These inferences are not always precise. The analysis of key electron donors, acceptors, and "indicator" species such as dissolved H_2 within aquifers may yield good estimates of dominant metabolic processes, or may leave ambiguities (Chapelle et al., 1995).

The idea of H_2-concentrations as indicators of the predominant metabolic pathway (TEAP) is widely supported by field data. The underlying cause of this relationship is the free-energy limitation of competing microbial metabolic processes. With fermentation acting as a throttle on microbial activity, bacteria operate at the lower limit of ΔG_r, so that H_2 is driven to a value that maintains a constant ΔG_r (Hoehler et al., 1998). The use of H_2 alone as an indicator of the dominant TEAP has been successful (e.g., Lovley and Goodwin, 1988), but may be better combined with companion measurements of electron acceptors. In a comparative study of E_h and H_2 as indicators of the dominant TEAP (Chapelle et al., 1995), H_2 was found to be superior to E_h in discriminating between dominant iron reduction, sulfate reduction, and methanogenesis. The combination of H_2-measurements with electron acceptor concentrations and their products was found more reliable than H_2 alone. Spatial analysis of TEAP products, however, demonstrated that methanogenesis, sulfate reduction, and iron reduction overlapped, regardless of microbial dominance. These results could be due to the coexistence of competing TEAPs, or because of the spatial coarseness of sampling, could reflect the heterogeneous distribution of organisms at a fine scale.

4   THE IMPORTANCE OF SCALE

The scale of geochemical heterogeneity and the practical limitations on sampling density are important factors in setting the sampling scale. Subsurface geochemical systems are heterogeneous, often at a centimeter scale or smaller (Brockman and Murray, 1997a,b), as is obvious at a glance when observing strata in the field. The effects of this heterogeneity on groundwater and solute movement, on the function of enveloped microbial communities, and on the fate of entrained contaminants are the subjects of intense investigation in hydrology, geochemistry, and microbial ecology.

Sampling scale in groundwater studies has importance whenever comparisons are drawn between samples: whenever more than a single sample is collected. In viewing the results from a single sample from a single well, one accepts that the information represents some process or processes within the volume sampled. When results from two wells are compared, one must exert some real or conceptual control over the origin of the samples relative to parameters controlling microbial processes at depth. It is not valid to conclude, for example, that chemical changes *a priori* represent changes in biogeochemical processes between sampling points, because the subsurface is massively and intricately heterogeneous and differing chemical environments are readily sampled. In examining regional changes in biogeochemistry, single samples from multiple wells can represent a "fine" scale in defining broad changes in microbial communities; alternatively, within single wells spanning significant stratigraphic changes, the scale of sampling must incorporate the scale of stratigraphic variability to adequately reflect significant biogeochemical changes. The scale of microbial heterogeneity may be smaller than any achievable sampling scale; the finest scale of sampling that logistics allow will provide the best information on variations in microbiology and geochemistry.

The potential advantage of fine-scale over coarse-scale estimates of microbial function in the field may be illustrated by an example (Conrad et al., 1986). In a system containing the fermentation products, H_2, propionate, and butyrate, H_2 concentrations were calculated, on thermodynamic grounds, to be too high to permit syntrophic degradation of propionate and butyrate. These compounds were known to be degraded nonetheless. The conclusion was drawn that an environmental niche—resulting from fine-scale heterogeneity—must have existed when the processes were favored. Specifically, the anaerobic oxidation of butyrate and propionate must have taken place where zones of active methanogenesis (drawing down H_2-concentration) drove the pertinent oxidation reactions through mass action on the overall system. Sampling at a fine-enough scale may have revealed direct evidence for such zones.

A similar analysis on a large scale along a regional flowpath was undertaken for the Middendorf Aquifer in South Carolina (Murphy et al., 1992; Schramke et al., 1996; Murphy and Schramke, 1998). The aquifer geochemistry was controlled by a complex interplay of chemical and microbial factors, as illustrated and summarized in Fig. 6.1 (Murphy and Schramke, 1998). The changes in aquifer chemistry along the aquifer flowpath show microbial influence. In this aquifer, high-lignite zones were noted, along with organic carbon-stained sands, and inorganic carbon increased along the flowpath due to the oxidation of organic carbon. Aerobes were most numerous, overall, in zones without lignite, and vice versa. Where organic acids were produced by fermentation, sulfate- and iron-reducing bacteria were present. Along the flowpath, a modified system of redox zonation was observed: The aquifer underwent transition from aerobic waters to anaerobic waters containing Fe^{2+} and Mn^{2+} to waters with increasing SO_4^{2-}. However, ferrous iron (Fe^{2+}) increased along the flowpath in an oxygenated portion of the aquifer; this should not occur, since iron reduction is an obligately anaerobic process, where oxygen should reach undetectable levels before Fe^{2+} is observed. The occurrence of Fe^{2+} was coupled to the occurrence of high-lignite zones. Careful observation, modeling, and deduction indicated the existence, as illustrated, of a complex fine-scale relationship between anaerobic microsites, where DIRB could use fermentation-produced organic acids and contribute ferrous iron as a solute, observable in oxygenated groundwaters.

In the cited examples, conclusions were necessarily deductive, in part, because relatively coarse sampling produced an integrated geochemical signature for coexistent fine-scale processes. Deductive processes may not always be successful: The underlying heterogeneity may be effectively invisible. Although fine-scale sampling multiplies the labor required for sample collection and analysis, it may provide the

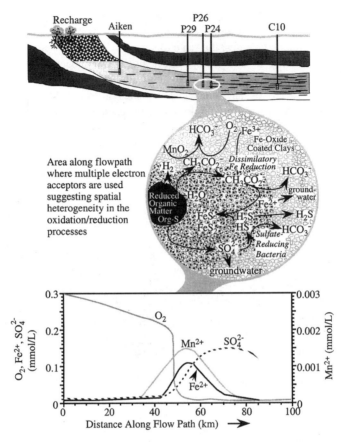

FIGURE 6.1 The complex relationship between aerobic and anaerobic bacteria in the southeastern coastal plain. Geochemical measurements obtained from drilling sites (e.g., P29, P26) along the groundwater flowpath indicate the existence of fine zones harboring anaerobic bacteria in an otherwise aerobic aquifer. (From Murphy et al., 1992; Murphy and Schramke, 1998.)

best means to answering specific questions, and its benefits, on balance, may be justified as most efficient. For example, in bioremediation applications, sampling at a fine scale may be the best way to show the local interplay between microbial communities, which could be useful in estimating the rate of remediation.

5 SAMPLING AT FINER SCALE

5.1 Cross-Lithologic Sampling

Over the past decade or so, it has become more common to investigate the variation of microbiology and geochemistry in different lithologies, either within a single borehole or within multiple boreholes as a set in an aquifer flowpath. Pucci and Owens (1989) examined the chemical variations across numerous beds (seven aquifers and eight confining beds) within a single borehole. They established that large geochemical gradients were established between lithologies, with greater total dissolved solid concentrations within confining beds (where pore water residence was relatively long) than within aquifers. Water–rock interaction was concluded to have significant impact on solute quality and to be minimal at depth, where equilibrium between solute and solid phases was approached.

Other efforts were driven by microbiology (e.g., Chapelle et al., 1987; Jones et al., 1989; McMahon et al., 1990; Chapelle and Lovley, 1992) to investigate diverse aspects of microbial ecology. These included, for example, the substrate source and microbial release of CO_2 to groundwater, the identity and distribution of different functional groups, and the exclusion or coexistence of competing TEAPs across lithologic boundaries or within single strata. When multiple zones within multiple boreholes were studied, some conclusions could be reached concerning the overall evolution of system biogeochemistry along flowpaths, even within alternating sequences of aquifers and aquitards. Among other observations in these studies were the following: DIRB can compete successfully with SRB (Chapelle and Lovley, 1992); the mineralization of organic carbon can cause a general rise in dissolved inorganic carbon concentrations (Chapelle et al., 1987; McMahon et al., 1990); and bacteria tend to be more abundant in coarse-grained sediment than in juxtaposed fine-grained sediment (Jones et al., 1989). By analyzing sediments and pore waters from an alternating sequence of aquifer and aquitard sediments, McMahon and Chapelle (1991) demonstrated a cross-stratigraphic relationship between companion communities. In aquitard sediments, fermentation rates exceeded respiration, and organic acids accumulated in sediment pores. In adjacent aquifer sediments, respiration exceeded fermentation. Aquitard fermenters thus acted as producers of organic acids available to respiring organisms in adjacent sediments through diffusion across the lithologic boundary. Sample intervals for the cited studies, over vertical depths of perhaps 100 m, were in general 5–10 m, targeting contrasting lithologies in boreholes through the stratigraphic column.

5.2 Intensive Sampling Across Lithologic Boundaries

An effort to sample deeply buried, isolated, anaerobic strata at depth, at relatively fine scale (ca. 1-m vertical spacing at a depth of 173–197 m) produced unusual results, relative to shallower, coarser sampling efforts (Fredrickson et al., 1995; Kieft et al., 1995). The borehole location was in a semiarid environment with a deep (100-m) water table, in south central Washington. The sampled sequence (Fig. 6.2) was an upper, 12-m thick, lacustrine deposit, over an 8-m-thick paleosol, grading downward into fluvial river sands. The overall sequence was anaerobic and was sandwiched within an aquifer of very coarse, aerobic river gravels. These sediments were all Miocene in age (6–8 My). We hypothesized before drilling and sampling that the fine-grained, carbon-rich (ca. 1% organic carbon by weight) lacustrine interval would provide nutrients to microbes in the coarser paleosol, consistent with the results of

McMahon and Chapelle (1991) and Brockman et al. (1992). The expected low number of fermenting bacteria, limited by fermentation products, would explain the persistence over millions of years of relatively high organic carbon concentrations in lacustrine sediments. Results were not as expected. Microbial numbers were highest in the fine-grained sediments (2×10^5 cells/gdw) and were lower than those of the thick vadose zone sampled earlier in the same well (Brockman et al., 1992; Fredrickson et al., 1939). Lipid analysis and 16S RNA-directed probes indicated a community within the lacustrine sediments that included active DIRB and SRB, with relatively low sulfate concentrations (ca. 20 mM or less) and the HCl-extractable Fe predominantly as Fe(II). In this unusual subsurface environment, electron acceptor limitations apparently prevented coupled fermentative and respiring bacteria from consuming the abundant remnant organic carbon that was laid down along with its extant microbial population in Miocene time.

An examination of the spatial distribution of samples, selected chemical parameters, and culturable microbial numbers for fermenters, SRB, and DIRB (Fig. 6.2) provides some information on the biogeochemistry of this system (McKinley et al., 1997). Organic carbon exceeded 1 wt.% in lacustrine sediments and was, with one exception, below 0.2 wt.% in other sediments. HCl-extractable Fe was highest also in lacustrine sediments. This measure of bioavailable Fe was adapted from Lovley and Phillips (1987), targeting poorly crystalline Fe(III) oxyhydroxides. [Poorly crystalline Fe oxyhydroxides are appropriate sources for DIRB coexisting with SRB under the geochemical conditions encountered: abundant Fe^{2+}, $-\log[SO_4]$ of ca. 2, and pH$=$ 8–9 (Postma and Jakobsen, 1996)]. Sulfate concentrations described an apparent gradient from a high of $>$ 2 mM (200 mg/L) in paleosols to $<$ 0.1 mM (10 mg/L) at the center of the lacustrine sediments, suggesting that sulfate was supplied to the lacustrine interval by diffusion from the paleosol. Decimal dilution enumerations of the microbial population suggested a spatial relationship with nutrient availability. DIRB predominated in lacustrine sediments and at the lacustrine–paleosol boundary and showed a one-decimal (log) increase in numbers in a single sample at 189 m of depth. These zones did not consistently coincide with highs in bioavailable Fe (e.g., at 185 m), but did coincide with zones of high numbers of fermenters; notable coincidences occurred between the elevated DIRB and fermenter numbers at 185 and 189 m. High values for the number of fermenters, in turn, corresponded to high values of organic carbon, as might be expected, again notably at 185 and 189 m. DIRB apparently predominated where organic carbon supported abundant fermentative bacteria capable of providing them with sufficient organic-acid substrate for metabolism. SRB occurred where conditions favored them (and did not favor DIRB). The notable occurrence of SRB was between 181 and 184 m. Sulfate was more abundant there than elsewhere in the lacustrine sediment. Fermentative bacteria and DIRB were low in numbers also; the pH in this interval approached 9 (data not shown), and this may also have disfavored DIRB. SRB may have competed well with DIRB due to the mobility and abundance of sulfate relative to Fe(III), and other environmental factors may have acted against DIRB. The other occurrence of SRB was at 193 m; again, SO_4 was elevated, and fermenters were common also.

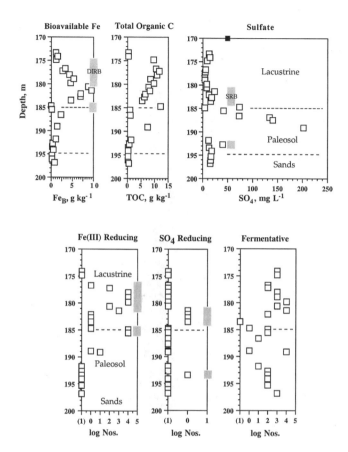

FIGURE 6.2 Geochemistry and microbiology in an anaerobic sediment sequence in an aerobic aquifer. Zones of significant SRB or DIRB activity are indicated with shading; (1) indicates nondetection. [From (McKinley et al. 1997) and unpublished data.]

The relatively closely spaced sample intervals in this sedimentary sequence permitted the derivation of information about the spatial and biogeochemical relationships that could not have been observed with the more common practice of collecting one or several samples from each of several lithologies in a sequence. The microbial community within the lacustrine sequence was apparently diverse, in a compact, fine-grained system limited overall by its poor hydraulic conductivity (small pore size) and its concomitant limited access to electron-accepting ions.

The papers by Postma and Jakobsen (1996, 1999) are good examples of the merits of sampling at a fine, submeter scale and of the use of free-energy calculations to understand biogeochemical processes. In the first of these (Postma and Jakobsen, 1996), an elegant argument (outlined above) was advanced against the zoning of TEAPs as determined by ranking TEAP reactions according to their standard-state free-energy changes. Essentially, if fermentation products occurred at low concentrations, then slow fermentation would be rate-limiting and the kinetics of the overall reaction with different TEAPs could not be predicted from their energy yields. Thermodynamic equilibrium should then be approached by the terminal inorganic electron acceptors and their reduction products, and chemical equilibrium could be used as an argument in establishing the sequence of TEAP succession. Two field examples illustrated the use of this partial equilibrium model. In the first, data at a vertical scale of 1 cm, in lake-bottom sediment, was used to demonstrate the sequential predominance, in a narrow zone, of sulfate reduction over iron reduction due to the free-energy advantage provided to SRB by the relatively stable (i.e, crystalline) form of iron oxides available to DIRB; after the depletion of sulfate as an electron acceptor, iron reduction proceeded and predominated. In the second example, data at a scale of 20-cm vertical spacing, in an aquifer, showed that iron reduction gave way, over a few meters' depth, to sulfate reduction, not as a result of the exhaustion of reducible iron (which remained plentiful), but due to a shift in pH to conditions in which the reduction of sulfate was energetically favored. Both examples demonstrated the control of overall, as opposed to standard-state, free-energy considerations over favored metabolic processes in the subsurface. In a more recent report (Jakobsen and Postma, 1999), the same approach was extended in a study of aquifer sediments to show similar *in situ* free-energy yields for sulfate reduction, iron reduction, and methanogenesis. Sulfate reduction and iron reduction were therefore predicted and demonstrated to coexist. For methane, however, the relationship was not clear and was inferred to take place in micro-niches occurring at a scale below the ca. 20-cm sampling scale. The field examples used in both reports originated in shallow environments where fine-scale sampling was readily accomplished.

5.3   Multidisciplinary Sampling at Cerro Negro, New Mexico

Investigating deeper biogeochemical systems at a fine (centimeter) scale to define the role of different microbial guilds and microbial processes operating in variable lithologies poses significant challenges. The removal of closely spaced aseptic samples from depth and their characterization at an appropriate scale for microbes, solid-phase chemistry, and pore water are difficult due to constraints on sample volume, drilling logistics, and sample preservation. Fine-scale sampling of an alternating sandstone–shale sequence taking several approaches to the definition of biogeochemical variations was undertaken at a site near the land-grant village of Seboyeta, New Mexico. Site location, drilling details and strategy, and results from cored samples were presented in Fredrickson et al. (1997). Stratigraphically, the site consisted of the alternating transgressive and regressive sequences of the Late Cretaceous (90–93 Myo) Mancos shale and Dakota sandstone (Fig. 6.3.) Sampling was focused on the ca. 150 to 250-m-deep sequence Paguate Sandstone—Clay Mesa shale—Cubero sandstone—Oak Canyon shale, with termination of drilling in the upper (Jurassic) Morrison formation. Sample intervals for coring over this sequence were usually ca. 1 m apart, but were greater where logistical problems arose, and groundwater samples were collected by packing and pumping selected transmissive zones. A parallel investigation showed predominant eastward groundwater flow

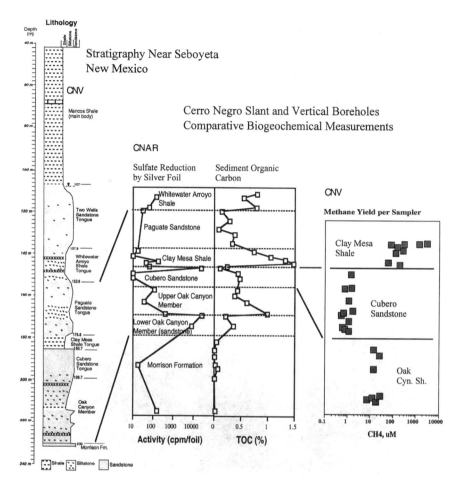

FIGURE 6.3 Lithology and microbiological activity at a site near Seboyeta, New Mexico. Results are from vertical (CNV) and slanted (CNAR) boreholes, from sediments (CNAR borehole) and formation waters (CNV borehole); actual depths varied, but stratigraphic intervals and the depths between them were equivalent. [From (Fredrickson et al. 1997; Krumholz et al. 1997) and unpublished data.]

originating in uplands to the west, with a local upward vertical hydrologic gradient (Walvoord et al., 1999). Results indicated highly variable *in situ* biomass as estimated by phospholipid fatty acid (PLFA) analysis, ranging from below detection to 31.9 pmol PLFA g^{-1} (of rock). No metabolic activities were detected via mineralization of [^{14}C]glucose or [^{14}C]acetate or by reduction of $^{35}SO_4$ in samples with pore throats of diameter less than 0.2 μm, but such activity was present in samples with larger pores. Enrichments for SRB showed their presence in some small-pore lithologies. Organic carbon was not degraded in small-pore lithologies (primarily shales). These results suggested that interconnected pores were required for sustained activity, but that viable bacteria could be maintained on poorly permeable rocks, nonetheless.

Subsequent detailed analysis of $^{35}SO_4$-reduction (using a technique utilizing a silver foil as a trap for radioactive sulfide) compared to the concentration of sediment organic carbon (Krumholz et al., 1997) indicated a relationship between microbial communities across sandstone–shale boundaries (Fig. 6.3). The silver foil technique detected isolated colonies less than 1 cm in diameter, 10 cm in cores. The experimental interaction of microbes from the alternating carbon-rich and carbon-poor lithologies was used to confirm this relationship. Shales showed little sulfate-reducing activity, and sandstones illustrated variable activity. Marked highs in sulfate reduction were found in sandstones at the sandstone–shale boundary, indicating that organic substrates (e.g., organic acids from fermentation), derived from shales perhaps by fermentation, were acting as substrates for microbes in adjacent sandstones. These results were consistent with those of McMahon and Chapelle (1991) for aquifer–aquitard sediments in South Carolina. Also, an upward hydrologic gradient (Walvoord et al., 1999) might tend to compress zones of concentrated organic acids diffusing downward from shales, while dispersing organic acids at the complementary upward shale–sandstone boundary, perhaps explaining the lack of increased sulfate reduction there. These results were obtained through the integration of detected sulfate reduction at the finest of scales, even though the overall sample interval was at meter scale.

A subsequent study of biogeochemical processes across a short sequence encompassing the Clay Mesa shale—Cubero sandstone—Oak Canyon shale lithologies was done following core sampling (McKinley et al., unpublished data) and using a passive multilevel sampler. This sampling method has been used in a wide variety of shallow groundwater investigations, providing discrete samples of limited volume at centimeter scale without necessary pumping or purging of screened wells (Ronen et al., 1987a, 1987b; Magaritz et al., 1989; Kaplan et al., 1991). Preliminary results are shown in Fig. 6.4. Sulfide production was high within sandstones at the upper shale–sandstone boundary and was negligible in the overlying shale. This geochemical indicator of sulfate reduction conformed well to previous results (Fredrickson et al., 1997; Krumholz et al., 1997), indicating no sulfate reduction in fine-pore shales and active reduction in just-adjacent sandstones. In addition, sulfide production was indicated in a narrow, more porous zone within the central portion of the Cubero sandstone, where pore throats were larger than in adjacent sediments and where acetate and glucose mineralization was observed, but sulfate reduction was not observed at coarser sampling intervals (Fredrickson et al., 1997). Minor sulfide production was also evident in the gradational zone between the Cubero sandstone and the underlying shale. Other dissolved components suggested a relationship of microbial activity to aquifer geochemistry.

Dissolved organic carbon (DOC) and dissolved inorganic carbon (DIC) were related to sulfate reduction (Fig. 6.4). The DOC concentration was elevated near the upper sandstone–shale boundary, consistent with diffusion from the overlying shale, and was elevated also at the sandstone-central depth where sulfate-reducing activity was indicated by sulfide generation. The *patterns* of sulfide concentrations and DIC

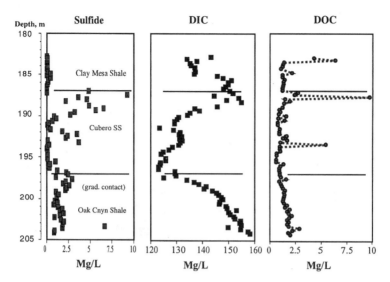

FIGURE 6.4 Formation water geochemistry sampled passively at 15-cm intervals, from the Cerro Negro site near Seboyeta, New Mexico.

in the Cubero sandstone are strikingly similar. Where sulfide production was most evident, the concentration of DIC was relatively high also; SRB have apparently mineralized organic carbon, adding to the DIC pool within the Cubero sandstone. The pattern of increasing DIC concentrations moving downward within the Oak Canyon shale, where some sulfide was detected, is apparently the result also of organic carbon mineralization. Finally, methanogenesis has apparently occurred in the presence of predominant sulfate reduction within these sulfate-rich sediments (Fig. 6.3). Methane was sampled using purpose-built gas samplers (McKinley and Stevens, unpublished data); results in Fig. 6.3 are for molar quantities of methane removed from pore waters into gas-filled samplers (not dissolved concentrations). The data, from which below-detection values have been removed, show a consistent pattern of methane generation within each lithologic unit. No methanogens were detected or cultured from the core samples at this site (Fredrickson et al., 1997; Krumholz et al., 1997). Methanogens evidently existed within the sediments, however. They may have inhabited microniches where free-energy conditions permitted them to function despite surrounding SRB populations [e.g., in juxtaposition to hydrogen or organic acid producers (Schink, 1997)], or they may have coexisted with SRB directly under geochemical conditions permitting simultaneous competitive TEAPs (Postma and Jakobsen, 1996; Jakobsen and Postma, 1999).

6 CONCLUSION

Geochemistry provides an excellent tool in the exploration of complex subsurface microbial ecosystems. It often provides convincing evidence for processes and functional groups that may be missed by more direct sampling and culturing methods. The complexities of the microbial community should be matched, however, by chemical measurements that can reveal underlying microbial effects along with the thermodynamic constraints on microbial function. The use of free-energy calculations, for example, as an indicator of limitations on microbial processes, can explain the transitions observed from one TEAP to another, or the coexistence of TEAPs presumed to be incompatible due to competitive exclusion and zoning. Sample scale is an important factor in the success of sample-dependent investigations. Although fine-scale sampling is labor-intensive, it may be a necessary component of investigations where marked changes in microbial communities, lithologies, and geochemistry are observed.

REFERENCES

Berner RA (1981) A new geochemical classification of sedimentary environments. *J Sedimentary Petrol 51*:359–365.

Brockman FJ, Kieft TL, Fredrickson JK, Bjornstad BN, Li SW, Spangenburg W, and Long PE (1992) Microbiology of vadose zone paleosols in south-central Washington State. *Microbial Ecol 23*:279–301.

Brockman FJ and Murray CJ (1997a) Microbiological heterogeneity in the terrestrial subsurface and approaches for its description. In Amy P and Haldeman D (eds): *Microbiology of the Terrestrial Deep Subsurface*. Boca Raton, FL: CRC Lewis Press. 75–102.

Brockman FJ and Murray CJ (1997b) "Subsurface microbiological heterogeneity: current knowledge, descriptive approaches and applications." *FEMS Microbiology Reviews 20*:231–247.

Chapelle FH, Zelibor J, Grimes DJ, and Knobel LL (1987) Bacteria in deep coastal plain sediments of Maryland: A possible source of CO_2 to groundwater. *Water Resources Res 23*:1625–1632.

Chapelle FH and Lovley DR (1992) Competitive exclusion of sulfate reduction by Fe(III)-reducing bacteria: A mechanism for producing discrete zones of high-iron ground water. *Ground Water 30*:29–36.

Chapelle FH, McMahon PB, Dubrovsky NM, Fujii RF, Oaksford ET, and Vroblesky DA (1995) Deducing the distribution of terminal electron-accepting processes in hydrologically diverse groundwater systems. *Water Resources Res 31*:359–371.

Claypool GE and Kaplan IR (1974) The origin and distribution of methane in marine sediments. In Kaplan IR (ed): *Natural Gases in Marine Sediments*. New York: Plenum, pp 99–139.

Conrad R (1999) Contribution of hydrogen to methane production and control of hydrogen concentrations in methanogenic soils and sediments. *FEMS Microbial Ecol 28*:193–202.

Conrad R, Schink B, and Phelps TJ (1986) Thermodynamics of H_2-consuming and H_2-producing metabolic reactions in divers methanogenic environments under *in situ* conditions. *FEMS Microbiol Ecol 38*:353–360.

Ehrlich HL (1990) *Geomicrobiology*. New York: Marcel Dekker.

Fredrickson JK, Brockman FJ, Bjornstad BN, Long PE, Li SW, McKinley JP, Wright JV, Conka

JL, Kieft TL, and Balkwill DL (1993) Microbiological characteristics of pristine and contaminated deep vadose sediments from an arid region. *Geomicrobiol J 11*: 95–107.

Fredrickson JK, McKinley JP, Bjornstad BN, Long PE, Ringelberg DB, White DC, Krumholtz LR, Suflita JM, Colwell FS, Lehman RM, Phelps TJ, and Onstott TC (1997) Pore-size constraints on the activity and survival of subsurface bacteria in a Late Cretaceous shale-sandstone sequence, northwestern New Mexico. *Geomicrobiol J 14*:183–202.

Fredrickson JK, McKinley JP, Nierzwicki-Bauer SA, White DC, Ringelberg DB, Rawson SA, Li S-M, Brockman FJ, and Bjornstad BN (1995) Microbial community structure and biogeochemistry of Miocene subsurface sediments: Implications for long-term microbial survival. *Mol Ecol 4*:619–626.

Gaidos EJ, Nealson KH, and Kirschvink JL (1999) Life in ice-covered oceans. *Science 284*:1631–1633.

Hoehler TM, Alperin MJ, Albert DB, and Martens CS (1998) Thermodynamic control on hydrogen concentrations in anoxic sediments. *Geochimica et Cosmochimica Acta 62*:1745–1756.

Holmer M and Kristensen E (1994) Coexistence of sulfate reduction and methane production in an organic-rich sediment. *Marine Ecol Progress Series 107*:177–184.

Jakobsen R and Postma D (1999) Redox zoning, rates of sulfate reduction and interactions with Fe-reduction and methanogenesis in a shallow sandy aquifer, Romo, Denmark. *Geochimica et Cosmochimica Acta 63*:137–151.

Jones RE, Beeman RE, and Suflita JM (1989) Anaerobic metabolic processes in the deep terrestrial subsurface. *Geomicrobiol J 7*:117–130.

Kaplan E, Banerjee S, Ronen D, Magaritz M, Machlin A, Sosnow M, and Koglin E (1991) Multilayer sampling in the water-table region of a sandy aquifer. *Ground Water 29*:191–198.

Kieft TL, Fredrickson JK, McKinley JP, Bjornstad BN, Rawson SA, Phelps TJ, Brockman FJ, and Pfiffner SM (1995) Microbiological comparisons within and across contiguous lacustrine, paleosol, and fluvial subsurface sediments. *Appl Environ Microbiol 61*:749–757.

Klass DL (1984) Methane from anaerobic fermentation. *Science 223*:1021–1028.

Krumholz LR, McKinley JP, Ulrich GA, and Suflita JM (1997) Confined subsurface microbial communities in Cretaceous rock. *Nature 386*:64–66.

Lovley DR and Chapelle FH (1995) Deep subsurface microbial processes. *Rev Geophys 33*:365–381.

Lovley DR and Goodwin S (1988) Hydrogen concentrations as an indicator of the predominant terminal electron-accepting reactions in aquatic sediments. *Geochimica et Cosmochimica Acta 52*:2993–3003.

Lovley DR and Phillips EJP (1987) Rapid assay for microbially reducible ferric iron in aquatic sediments. *Appl Environ Microbiol 53*:1536–1540.

Madigan MT, Martinko JM, and Parker J (1997) *Brock Biology of Microorganisms*. Upper Saddle River, NJ: Prentice-Hall.

Magaritz M, Wells M, Amiel AJ, and Ronen D (1989) Application of a multi-layer sampler based on the dialysis cell technique for the study of trace metals in groundwater. *Appl Geochem 4*:617–624.

McKinley JP, Stevens TO, Fredrickson JK, Zachara JM, Colwell FS, Wagnon KB, Smith SC, Rawson SA, and Bjornstad BN (1997) Biogeochemistry of anaerobic lacustrine and paleosol sediments within an aerobic unconfined aquifer. *Geomicrobiol J 14*:23–29.

McMahon PB and Chapelle FH (1991) Microbial production of organic acids in aquitard sediments and its role in aquifer geochemistry. *Nature 349*:233–235.

McMahon PB, Williams DF, and Morris JT (1990) Production and carbon isotopic composition of bacterial CO_2 in deep coastal plain sediments of South Carolina. *Ground Water 28*:693–702.

Murphy EM and Schramke JA (1998) Estimation of microbial respiration rates in groundwater by geochemical modeling constrained with stable isotopes. *Geochimica et Cosmochimica Acta 62*:3395–3406.

Murphy EM, Schramke JA, Fredrickson JK, Bledsoe HW, Francis AJ, Sklarew DS, and Linehan JC (1992) The influence of microbial activity and sedimentary organic carbon on the isotope geochemistry of the Middendorf aquifer. *Water Resources Res 28*:723–740.

Postma D and Jakobsen R (1996) Redox zonation: Equilibrium constraints on the Fe(III)/SO_4-reduction interface. *Geochimica et Cosmochimica Acta 60*:3169–3175.

Pucci Jr AA and Owens JP (1989) Geochemical variations in a core of hydrogeologic units near Freehold, New Jersey. *Ground Water 27*:802–812.

Ronen D, Magaritz M, Gvirtzman H, and Garner W (1987a) Microscale chemical heterogeneity in groundwater. *J Hydrol 92*:173–178.

Ronen D, Magaritz M, and Levy I (1987b) An *in situ* multilevel sampler for preventive monitoring and study of hydrochemical profiles. *Ground Water Mon Rev 7*:69–74.

Schink B (1997) Energetics of syntrophic cooperation in methanogenic degradation. *Microbiol Mol Biol Rev 61*:262–280.

Schramke JA, Murphy EM, and Wood BD (1996) The use of geochemical mass-balance and mixing models to determine groundwater sources. *App Geochem 11*:523–539.

Stumm W and Morgan JW (1996) *Aquatic Chemistry*. New York: John Wiley & Sons.

Walvoord MA, Pegram P, Phillips FM, Person M, Kieft TL, Fredrickson JK, McKinley JP, and Swenson J (1999) Groundwater flow and geochemistry in the southeastern San Juan Basin: Implications for microbial transport and activity. *Water Resources Res 35*:1409–1424.

7

REDUCTION OF IRON AND HUMICS IN SUBSURFACE ENVIRONMENTS

DEREK R. LOVLEY

Department of Microbiology, University of Massachusetts, Massachusetts

1 INTRODUCTION

It has long been recognized that microorganisms influence the geochemistry and water quality of subsurface environments (Pedersen, 1993; Lovley and Chapelle, 1995; Fredrickson and Onstott, 1996). However, until recently, abiotic reactions were generally thought to control the redox speciation of many metals in the subsurface, especially Fe(III) (Lovley, 1991). The findings that microorganisms could conserve energy for growth from oxidizing organic compounds coupled to the reduction of Fe(III) (Lovley et al., 1987; Lovley and Phillips, 1988b), and that abiological processes were ineffective in Fe(III)-reduction under environmental conditions (Lovley et al., 1991b), led to the realization that enzymatic processes may, in large part, control the redox speciation of metals in sedimentary environments.

This chapter summarizes the current understanding of the microbiology of Fe(III)-reduction in subsurface environments and illustrates the ways in which microbial Fe(III)-reduction can influence both the inorganic and organic geochemistry of the subsurface. To date, all microorganisms that have the ability to use Fe(III) as a terminal electron acceptor also have the ability to use humic substances (humics) as an electron acceptor (Lovley et al., 1996a, 1998, 1999a). As detailed

Subsurface Microbiology and Biogeochemistry, Edited by James K. Fredrickson and Madilyn Fletcher.
ISBN 0-471-31577-X. Copyright 2001 by Wiley-Liss, Inc.

below, microbial reduction of humics may be an important intermediate step in the reduction of Fe(III) in the subsurface. Thus, this review will also summarize the current understanding of microbial humics reduction.

2 DISSIMILATORY FE(III)-REDUCTION

Dissimilatory Fe(III)-reduction is the process in which microorganisms transfer electrons to external Fe(III), reducing it to Fe(II) for purposes other than assimilation of iron. Dissimilatory Fe(III)-reduction is not necessarily involved in conservation of cellular energy in all organisms (Lovley, 1987, 1991; Coleman et al., 1993; Lovley et al., 1993b). However, there is a large number of microorganisms that can conserve energy for growth via the oxidation of organic compounds, hydrogen, or S ° with the reduction of Fe(III) (Lovley et al., 1997).

2.1 Dissimilatory Fe(III)-Reducing Microorganisms Recovered from the Subsurface

A wide phylogenetic diversity of microorganisms that can grow via Fe(III)-reduction have been recovered from subsurface environments (Fig. 7.1). Characteristics of these organisms and their distribution in the subsurface are briefly summarized below.

2.1.1 The Geobacteraceae Much of the current thinking about the mechanisms for Fe(III)-reduction in subsurface environments is derived from studies that have been conducted with microorganisms recovered from aquatic sediments. The first organism found to conserve energy to support growth from the oxidation of organic compounds coupled to Fe(III)-reduction was *Geobacter metallireducens* (formerly strain GS-15) (Lovley et al., 1987, 1993a; Lovley and Phillips, 1988b). *G. metallireducens* was recovered from freshwater sediments of the Potomac River. It oxidizes acetate and other short-chain fatty acids to carbon dioxide, with Fe(III), Mn(IV), or nitrate serving as the sole electron acceptor. It can also completely oxidize a number of monoaromatic compounds, including the hydrocarbon toluene, with Fe(III) as the electron acceptor. It is a strict anaerobe. Since the discovery of *G. metallireducens*, a number of other closely related organisms have been isolated. This phylogenetically coherent group of Fe(III) reducers has been designated the *Geobacteraceae* (Lonergan et al., 1996). A number of *Geobacteraceae* have been recovered from subsurface environments. These include *Geobacter chapellei* (formerly strain 172), which was isolated from Fe(III)-reducing deep subsurface sediments from an Atlantic Coastal Plain aquifer (Lovley et al., 1990; Coates et al., 1996) and *Geobacter hydrogenophilus*, which was isolated from the Fe(III)-reduction zone of a petroleum-contaminated aquifer (Coates et al., 1996). Numerous other, but not yet fully characterized, subsurface isolates, which are also in the *Geobacteraceae*, have been recently recovered.

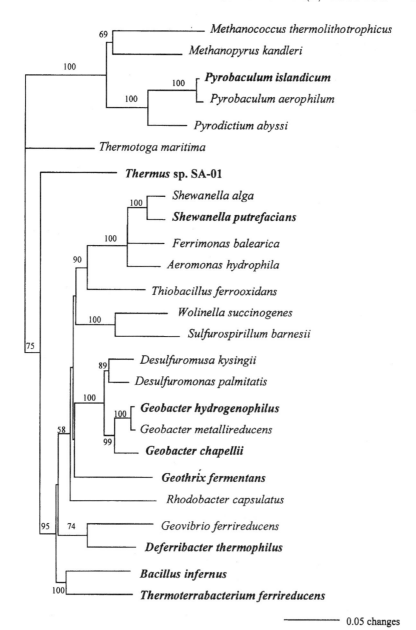

FIGURE 7.1 Phylogenetic tree based on 16S rDNA sequences of Fe(III)-reducing microorganisms recovered from subsurface environments. Organisms recovered from subsurface environments are in bold. Other Fe(III)-reducing microorganisms have been included to aid in defining the tree.

Molecular studies have suggested that strains of *Geobacter* closely related to *Geobacter* strains that have been cultured are the dominant Fe(III)-reducing microorganisms in several sandy aquifers. This was first indicated in studies of a petroleum-contaminated aquifer in Bemidji, Minnesota (Anderson et al., 1998; Rooney-Varga et al., 1999). Analysis of the 16S rDNA sequences in the aquifer sediments demonstrated that organisms with sequences closely related to known cultures of *Geobacter* species were among the most prominent organisms recovered in the zone of the most intense Fe(III)-reduction (Rooney-Varga et al., 1999). Enumeration of *Geobacteraceae* sequences demonstrated that these sediments had several orders of magnitude more *Geobacteraceae* than sediments in which Fe(III) was not important (Anderson et al., 1998).

There was a similar enrichment of *Geobacters* in field experiments in which electron donors were added to sandy aquifers (Synoeyenbos-West et al., 1999). Benzoate was added to one aquifer in order to promote reductive dechlorination. In another study, petroleum hydrocarbons were released into an aquifer in order to evaluate the potential for natural attenuation of aromatic hydrocarbons in the subsurface. In both instances, addition of the organic electron donors led to anaerobic conditions and the stimulation of microbial Fe(III)-reduction. Molecular analysis demonstrated that enhanced Fe(III)-reduction was associated with significant growth in *Geobacter* species.

Geobacter strains were also the predominant Fe(III)-reducing microorganisms in laboratory studies, in which various electron donors were added to stimulate Fe(III)-reduction (Synoeyenbos-West et al., 1999). Adding glucose, lactate, benzoate, or acetate resulted in a significant enrichment of *Geobacter* strains. Regardless of the electron donor added, strains of other intensively studied Fe(III) reducers, such as *Shewanella* species, were not detected, even when PCR primers specific for these organisms were employed.

These field and laboratory studies suggest that in order to understand Fe(III)-reduction in sandy aquifers, it will be important to understand the physiology of *Geobacter* strains that live in the subsurface. Preliminary studies have demonstrated that it may be possible to recover the relevant subsurface *Geobacter* strains in culture. When sediments from both field (Rooney-Varga et al., 1999) and laboratory studies (Synoeyenbos-West et al., 1999) were diluted into culture media, the highest dilutions that were positive for Fe(III)-reduction contained *Geobacter* sequences that were closely related to the predominant *Geobacter* sequences found in the aquifer sediments. Thus, at least in these instances, the general finding that it is difficult to culture the most environmentally significant organisms may not hold. This is fortuitous, because it means that controlled physiological studies on these organisms may provide useful information on the factors controlling the rate and extent of Fe(III)-reduction in these sandy aquifers.

2.1.2 Shewanella putrefaciens *and Closely Related Organisms*

Shewanella species and closely related organisms in the gamma subdivision of the *Proteobacteria* have also been intensively studied as model Fe(III)-reducing microorganisms (Nealson and Saffarini, 1994; Lovley et al., 1997). *Shewanella putrefaciens*, which has been recovered from a variety of environmental sources (including aquatic sediments), was found to reduce Fe(III) with lactate, formate, or hydrogen as the electron donor (Lovley et al., 1989b). A closely related organism, *Shewanella alga*, was recovered from estuarine sediments (Caccavo et al., 1992). *Shewanella* species have also recently been recovered from subsurface environments (Pedersen et al., 1996; Fredrickson et al., 1998).

Although *Shewanella* species can be cultured from various subsurface environments, the significance of these organisms in Fe(III)-reduction in the subsurface has yet to be determined. As noted above, studies on Fe(III)-reduction in several sandy aquifers failed to detect any *Shewanella* sequences among the dominant organisms in the Fe(III)-reduction zone. *Shewanella* species could not even be detected when Fe(III)-reduction was stimulated with lactate and formate (which are preferred electron donors for *Shewanella* species) (Synoeyenbos-West et al., 1999).

2.1.3 Geothrix *Geothrix fermentans* was initially isolated from aquifer sediments of a petroleum-contaminated aquifer in which Fe(III)-reduction was stimulated by adding the chelator, nitrilotriacetic acid (Coates et al., 1999b). Closely related strains were subsequently recovered from freshly collected sediments in the Fe(III)-reduction zone of another petroleum-contaminated aquifer. Like *Geobacter* species, *G. fermentans* can oxidize short-chain fatty acids to carbon dioxide, with Fe(III) serving as the sole electron acceptor. It cannot use aromatic compounds as electron donors.

G. fermentans forms a deeply branching, novel line of descent in the *Bacteria*, both with its closest relatives *Holophaga foetida* and *Acidobacterium capsulatum*, and with a diversity of other *Bacteria* that have not been isolated, but whose sequences are abundant in soils (Ludwig et al., 1997). Studies in which Fe(III)-reducing microorganisms were recovered in culture media suggested that organisms closely related to *G. fermentans* might be as numerous as *Geobacter* sp. in the Fe(III)-reduction zone of a petroleum-contaminated aquifer (Anderson et al., 1998). However, subsequent molecular studies, which do not suffer from potential biases of culture media composition, indicated that *Geobacter* species, not *Geothrix*, were the predominant organisms in the Fe(III)-reduction zone (Rooney-Varga et al., 1999; Synoeyenbos-West et al., 1999).

2.1.4 Thermophilic and Hyperthermophilic Fe(III) Reducers The increasing temperature associated with greater depths in the subsurface and hydrothermal regions may select for Fe(III)-reducing communities that grow at high temperatures. For example, the thermophile, *Bacillus infernus*, was isolated from subsurface sediments collected at a depth of 2.7 km below land surface in the Taylorsville Triassic Basin in Virginia (Boone et al., 1995). This organism can grow at 60°C with formate or lactate as the electron donor and Fe(III) or Mn(IV) as the electron acceptor. Enrichment cultures of other Fe(III)-reducing microorganisms that could grow at temperatures up to 75°C were also recovered from the Taylorsville site as well as from depths of up to 2.9 km from the Cretaceous Piceance Basin in

Colorado (Liu et al., 1997). These enrichments could use various organic acids, and at the Colorado site, hydrogen, as the electron donor for Fe(III)-reduction.

A thermophilic organism, *Deferribacter thermophilius*, was isolated from the production waters of a petroleum reservoir (Greene et al., 1997). The closest known relative of this organism is the mesophile, *Geovibrio ferrireducens*, which was recovered from hydrocarbon-contaminated surface sediments (Caccavo et al., 1996). The ability of both organisms to oxidize a variety of organic compounds and hydrogen with the reduction of Fe(III) is similar, but *D. thermophilius* has a temperature optimum of 65°C compared with 35°C for *G. ferrireducens*.

Terrestrial hot springs may provide a window for examining the life in deep, hot subsurface environments. Several Fe(III)-reducing microorganisms have been recovered from such environments. These include *Thermoterrabacterium ferrireducens*, which can oxidize hydrogen with the reduction of Fe(III), and use Fe(III) to incompletely oxidize glycerol, lactate, and several other organics to acetate (Slobodkin et al., 1997). Hydrogen-oxidizing, Fe(III)-reducing enrichment cultures were also established with a variety of hot spring samples (Slobodkin and Wiegel, 1997). *Thermus* species recovered from a South African gold mine and a thermal spring were capable of Fe(III)-reduction at 65°C, with lactate as the electron donor (Kieft et al., 1999).

Geothermal springs were also the source of the Fe(III)-reducing hyperthermophile, *Pyrobaculum islandicum*. Although *P. islandicum* was initially isolated as an S°-reducing microorganism (Huber et al., 1987), subsequent studies demonstrated that *P. islandicum* can grow as a hydrogen-oxidizing, Fe(III)-reducing microorganism at 100°C (Vargas et al., 1998; Kashefi and Lovley, 1999). In fact, a wide diversity of hyperthermophilic *Archaea* and *Bacteria* are capable of Fe(III)-reduction (Vargas et al., 1998). The facts that all the hyperthermophilic microorganisms evaluated to date can reduce Fe(III) and that hyperthermophiles are the extant organisms considered to be most closely related to the last common ancestor of modern microorganisms suggest that Fe(III)-reduction was a characteristic of the last common ancestor (Vargas et al., 1998). If so, this is consistent with geochemical models which suggest that Fe(III) may have been abundant on prebiotic Earth and that Fe(III)-reduction may have been the first globally significant form of microbial respiration (Walker, 1984, 1987; Cairns-Smith et al., 1992; de Duve, 1995).

2.2 Types of Subsurface Environments in Which Fe(III)-Reduction is Important

Anaerobic conditions are required in order for Fe(III)-reduction to be the predominant terminal electron-accepting process (TEAP). Therefore, subsurface environments in which Fe(III)-reduction is likely to be important are those with limited oxygen input [such as the deep subsurface (Lovley and Chapelle, 1995)] and shallower environments in which high concentrations of organic matter generate a significant oxygen demand (Lovley, 1997b). For example, Fe(III)-reduction is a significant process in deep pristine aquifers of the Atlantic Coastal Plain (Lovley et al., 1990; Chapelle and Lovley, 1992). In these and other deep pristine aquifers, the zone of Fe(III)-reduction is often extensive and is found downgradient of the zones of oxygen, nitrate, and Mn(IV)-reduction (Fig. 7.2*a*) and upgradient from the zones of sulfate reduction and methanogenesis (Lovley and Chapelle, 1995). Shallow aquifers contaminated with organic compounds from sources such as landfill leachate and underground petroleum pollution also typically have large zones of Fe(III)-reduction (Lovley et al., 1989a, 1994; Christensen et al., 1994; Anderson et

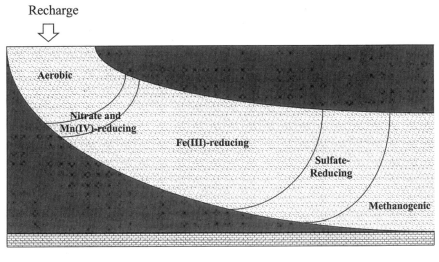

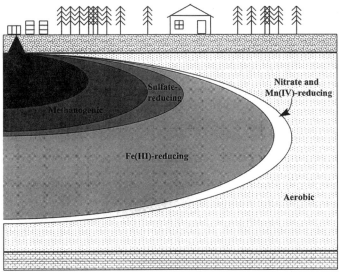

FIGURE 7.2 Idealized distribution of anaerobic terminal electron-accepting processes in (*a*) deep pristine aquifers and (*b*) shallow aquifers contaminated with petroleum or landfill leachate.

al., 1998; Anderson and Lovley, 1999). In such aquifers, Fe(III)-reduction typically predominates downgradient of a methanogenic zone, near the source of the contamination (Fig. 7.2*b*). In instances when there is sufficient sulfate, a sulfate reduction zone also precedes Fe(III)-reduction (Lovley, 1991, 1997a). Another instance in which Fe(III)-reduction may become important is in engineered bioremediation strategies for the removal of chlorinated solvents or the immobilization of metals, in which organic compounds are added to the subsurface in order to generate anaerobic conditions to promote reductive dechlorination or metal reduction (Nevin and Lovley, 1999; Synoeyenbos-West et al., 1999).

2.3   Electron Donors for Fe(III)-Reduction in the Subsurface

In pristine subsurface environments in which complex particulate or dissolved organic matter is the primary source of electron donors for microbial metabolism, acetate is likely to be the major electron donor for Fe(III)-reduction (Fig. 7.3) (Lovley, 1991; Lovley et al., 1997). This is because both culturing studies and studies of the metabolism of microbial communities in sedimentary environments, including the subsurface, have indicated that fermentable substrates, such as sugars and amino acids, are primarily metabolized to fermentation acids in environments in which Fe(III)-reduction is the predominant TEAP. For example, most Fe(III)-reducing microorganisms available in culture that can utilize glucose transfer only a minor portion of the electron equivalents in glucose to Fe(III), with most of the electron equivalents recovered in fermentation products (Lovley, 1987). Many of these fermentative Fe(III)-reducing microorganisms are unlikely to conserve significant energy from Fe(III)-reduction. One conserves energy from the oxidation of glucose coupled to Fe(III)-reduction; it has been recently described (Coates et al., 1999b). However, this organism incompletely oxidizes the glucose to acetate (Coates et al., 1999b). Studies in which the metabolism of glucose in Fe(III)-reducing

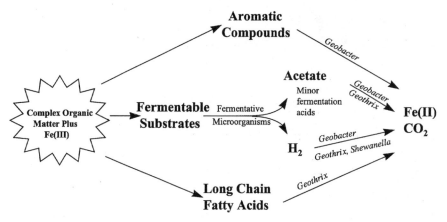

FIGURE 7.3   Model for pathways for organic matter degradation in sedimentary environments in which Fe(III)-reduction is the predominant terminal electron-accepting process.

sediments was evaluated with [^{14}C]-glucose demonstrated that microbial communities in these environments first ferment glucose, primarily to acetate, which is then oxidized to carbon dioxide (Lovley and Phillips, 1989). Although one Fe(III)-reducing microorganism can completely oxidize one amino acid to carbon dioxide (Caccavo et al., 1996), complete oxidation of readily fermentable compounds does not appear to be a common phenomenon. Thus, carbon flow from such substrates can be expected to proceed through fermentation acid intermediates, with acetate being the most common acid produced (Lovley and Chapelle, 1995). As summarized above, subsurface microorganisms such as *Geobacter* and *Geothrix* species can oxidize acetate and other fermentation acids with the reduction of Fe(III), and many of these organisms, as well as *Shewanella* species, can oxidize the hydrogen that may be produced from fermentation.

Some Fe(III)-reducing microorganisms can oxidize a variety of monoaromatic compounds to carbon dioxide, including the hydrocarbons toluene and benzene (Lovley et al., 1989a, 1994; Lovley and Lonergan, 1990; Lonergan and Lovley, 1991; Rooney-Varga et al., 1999). There are also those that can oxidize long-chain fatty acids directly to carbon dioxide (Coates et al., 1995, 1999b).

These metabolic capabilities of Fe(III)-reducing isolates and sedimentary communities of Fe(III) reducers suggest that a consortia of microorganisms are involved in the oxidation of complex organic matter coupled to Fe(III)-reduction (Fig. 7.3). This is analogous to the pathways for degradation of complex organic matter found in sediments in which sulfate reduction is the terminal electron acceptor (Lovley and Chapelle, 1995).

The pathways for organic matter oxidation coupled to Fe(III)-reduction may be much simpler in environments in which organic contaminants are the primary electron donors. For example, in aquifers contaminated with petroleum hydrocarbons, soluble aromatic hydrocarbons, such as benzene and toluene, may be the primary electron donors for Fe(III)-reduction. The composition of the microbial community in subsurface environments may be relatively simple (Rooney-Varga et al., 1999), as these substrates can be directly oxidized to carbon dioxide within single Fe(III)-reducing microorganisms.

2.4 Geochemical Impact of Dissimilatory Fe(III) Reducers in the Subsurface

2.4.1 *Organic Matter Oxidation with the Production of High-Iron Groundwaters*   Worldwide, the most prevalent groundwater quality problem is undesirably high concentrations of dissolved Fe(II) (Anderson and Lovley, 1997). Dissolved Fe(II) in groundwater is a concern because when Fe(II)-containing waters are pumped to the surface and contact oxygen, Fe(III) oxides are formed. These Fe(III) oxides can clog plumbing systems and/or cause discoloration of surfaces they contact.

Prior to the discovery that microorganisms could effectively couple the oxidation of organic matter to the reduction of Fe(III), high-iron groundwaters were primarily discussed in an abiological context in terms of redox potential and pH. However, studies within high-iron zones in deep pristine aquifers of the Atlantic Coastal Plain suggested that the distribution of high-iron groundwaters is better explained as a microbiological phenomenon (Lovley et al., 1990; Chapelle and Lovley, 1992). In these studies, Fe(III)-reducing microorganisms were recovered specifically from sediments in which geochemical evidence indicated that Fe(III)-reduction was taking place. There was a direct correlation between the accumulation of dissolved inorganic carbon and the concentration of Fe(II) in the high-iron groundwaters. Carbon isotope evidence suggested that the dissolved inorganic carbon released into the groundwater was derived from the oxidation of organic matter in the aquifer. Organic matter was being oxidized to carbon dioxide with the reduction of Fe(III) in the zones of high dissolved iron. As there is no known abiological mechanism for oxidizing organic matter to carbon dioxide with Fe(III) at the temperatures found in these aquifers, this indicated that the high-iron groundwaters resulted from the activity of the Fe(III)-reducing microorganisms.

2.4.2 Production of Fe(II) Minerals Although Fe(III)-reduction in sedimentary environments releases dissolved Fe(II) into the surrounding water, in most systems the vast majority of the Fe(II) remains in solid forms. This is apparent from studies in which the production of dissolved and total Fe(II) is monitored over time (Lovley and Phillips, 1988b). A variety of Fe(II) minerals are produced in cultures of Fe(III)-reducing microorganisms (Lovley et al., 1987; Lovley and Phillips, 1988b; Lovley, 1990; Fredrickson et al., 1998). Commonly observed Fe(II) minerals in culture include magnetite, siderite, vivianite, and less defined mixed Fe(II)-Fe(III) forms, such as green rust. Which minerals predominate depend on the culture conditions.

Information on the formation of Fe(III) minerals during microbial Fe(III)-reduction is somewhat circumstantial. For example, the magnetic mineral, magnetite, is often the predominant end-product of the microbial reduction of poorly crystalline Fe(III) oxide in cultures (Lovley et al., 1987; Lovley, 1990; Fredrickson et al., 1998; Vargas et al., 1998). High concentrations of ultrafine-grained magnetite in aquatic sediments known to have been artificially enriched with Fe(III) have suggested that magnetite may be an important product of microbial Fe(III)-reduction in these sediments (Gibbs-Eggar et al., 1999). Magnetite observed in a petroleum-contaminated aquifer in which Fe(III)-reduction is known to be important (Baedecker et al., 1992) may also be the product of microbial Fe(III)-reduction. Magnetite that often accumulates around subsurface hydrocarbon deposits could be the result of microbial Fe(III)-reduction (Lovley et al., 1987). Large accumulations of ultrafine-grained magnetite recovered from over 6 km below the Earth's surface were attributed to the activity of Fe(III)-reducing microorganisms (Gold, 1992). No direct evidence for Fe(III)-reducing microorganism in these deep, hot sediments was provided, but thermophilic and hyperthermophilic microorganisms from other subsurface environments have been found to produce magnetite at temperatures of up to 100°C (Liu et al., 1997; Slobodkin and Wiegel, 1997; Vargas et al., 1998; Kashefi and Lovley, 1999).

Siderite, an Fe(II) carbonate, often forms in cultures of Fe(III)-reducing microorganisms in which the media has a significant bicarbonate content (Lovley and Phillips, 1988b; Mortimer and Coleman, 1997; Fredrickson et al., 1998). Microbial Fe(III)-reduction is responsible for the formation of siderite in some aquatic sediments (Coleman et al., 1993) and might account for the accumulation of siderite and Fe(II)-bearing calcites observed in contaminated subsurface environments (Baedecker et al., 1992).

2.4.3 Removal of Organic Contaminants from the Subsurface Fe(III)-reducing microorganisms can affect the fate of both organic and inorganic contaminants in the subsurface. In terms of organic contaminants, the most intensive investigations to date have dealt with degradation of aromatic petroleum hydrocarbons. Monoaromatic hydrocarbons, such as benzene, toluene, ethylbenzene, and xylenes, are common groundwater contaminants resulting from the introduction of petroleum into the subsurface. In instances in which petroleum contamination is substantial, dissolved oxygen is removed from the groundwater as the result of aerobic degradation. Extensive anaerobic zones develop (Anderson and Lovley, 1997; Lovley, 1997a,b). As the anaerobic conditions develop, Fe(III) is generally the most abundant electron acceptor for organic matter oxidation. Direct measurement of the TEAP in aquifer sediments and/or estimation of the TEAP via measurements of dissolved hydrogen in groundwater have indicated extensive zones of Fe(III)-reduction in contaminated aquifers (Lovley et al., 1994a,b; Anderson et al., 1998; Anderson and Lovley, 1999).

The role of Fe(III)-reduction in removing aromatic hydrocarbons from contaminated groundwater was first documented, in detail, in a petroleum-contaminated aquifer in Bemidji, Minnesota. Selective removal of aromatic hydrocarbons from the groundwater was associated with an accumulation of isotopically light carbon dioxide and Fe(II) in the groundwater, and a depletion of Fe(III) from the sediments (Lovley et al., 1989; Baedecker et al., 1993).

Subsequent studies have revealed that toluene is readily oxidized to carbon dioxide in Fe(III)-reducing sediments from all petroleum-contaminated aquifers that have been examined (Anderson et al., 1998; Anderson and Lovley, 1999). Toluene oxidation coupled to Fe(III)-reduction has been described in pure culture with *G. metallireducens* (Lovley et al., 1989a; Lovley and Lonergan, 1990). *G. metallireducens* can oxidize other monoaromatic compounds, including important contaminants such as phenol, and *p*-cresol, which presumably will be degraded under Fe(III)-reducing conditions in aquifer sediments. A variety of other aromatic contaminants have been degraded in enrichment cultures in which Fe(III) was the sole electron acceptor (Lonergan and Lovley, 1991).

Degradation of unsubstituted aromatic hydrocarbons, such as benzene and naphthalene, has also been observed in aquifer sediments in which Fe(III)-reduction is the TEAP, but not at all sites. In screening several petroleum-contaminated aquifers for the capacity for anaerobic benzene oxidation, only sediments from a narrow zone within the aquifer in Bemidji, Minnesota, oxidized benzene when sediments were incubated under *in situ* conditions (Anderson et al., 1998).

Naphthalene was also oxidized to carbon dioxide in these sediments (Anderson and Lovley, 1999). Both benzene and naphthalene were oxidized without a lag period, suggesting that the microorganisms were adapted for oxidation of the aromatic hydrocarbons *in situ*. This conclusion was further supported by the finding that benzene was depleted from the groundwater as it moved into the portion of the Fe(III)-reduction zone that exhibited the greatest potential for benzene degradation in laboratory incubations (Lovley and Anderson, 2000).

Although benzene persisted under Fe(III)-reducing conditions in other petroleum-contaminated aquifer sediments, benzene degradation could be stimulated in these sediments when compounds that chelate Fe(III) were added (Lovley et al., 1994b, 1996b). The chelators stimulate Fe(III)-reduction by solubilizing Fe(III), which makes it more accessible for microbial reduction. This alleviates the need for the Fe(III) reducers to establish contact with insoluble Fe(III) oxides (Lovley and Woodward, 1996). Addition of humic acids also stimulated anaerobic benzene degradation in these sediments (Lovley ct al., 1996b). As detailed below, this has been attributed to humics acting as a soluble electron shuttle between Fe(III)-reducing microorganisms and Fe(III) oxides, which also eliminates the necessity for direct contact between them.

Further investigation of benzene degradation at Bemidji demonstrated that the capacity for it was located within the portion of the Fe(III)-reduction zone furthest downgradient from the contaminant plume, where the sediments contained the highest concentrations of microbially reducible Fe(III) (Anderson et al., 1998; Anderson and Lovley, 1999). This was also the zone of the most rapid oxidation of added toluene and naphthalene. As discussed above, molecular analysis of the microbial community demonstrated that the sediments which had the capacity to oxidize benzene were highly enriched in microorganisms closely related to known *Geobacter* species (Rooney-Varga et al., 1999). Furthermore, molecular analysis of a benzene-oxidizing, Fe(III)-reducing enrichment culture revealed that a *Geobacter* species was a dominant member of the culture. These results suggest that the *Geobacter* species are responsible for the aromatic hydrocarbon degradation in the Fe(III)-reduction zone of the Bemidji sediments. This is consistent with the ability of some *Geobacter* species available in culture to oxidize monoaromatic compounds, with Fe(III) serving as the sole electron acceptor.

Fe(III)-reducing microorganisms might also be involved in the reductive dechlorination of organics in the subsurface. A microorganism in the *Geobacteraceae*, the same family that contains *Geobacter* species, uses tetrachlorethylene or trichloroethylene as the sole electron acceptor (Krumholz et al., 1996). *Desulfitobacterium dehalogenans* reductively dechlorinates chlorophenolic compounds (Utkin et al., 1994). Both these organisms also have the ability to reduce Fe(III) (Krumholz et al., 1996; Lovley et al., 1998). However, the actual role of Fe(III)-reducing microorganisms in the dechlorination of contaminants in the subsurface is yet to be evaluated.

Fe(II) minerals that accumulate as the result of microbial Fe(III)-reduction can have an impact on the fate of some organic contaminants in the subsurface. For example, reduction of Fe(III) by *Geobacter* inoculated into columns of subsurface materials, as well as Fe(III)-reduction by indigenous organisms, produced Fe(II) minerals that could reduce nitroaromatic compounds (Heijman et al., 1993, 1995; Hofstetter et al., 1999). Studies in an aquifer contaminated with landfill leachate demonstrated that Fe(II) minerals generated in this manner could be important reductants in contaminated aquifers (Rugge et al., 1998). Reduction of nitroaromatics by microbially produced Fe(II) minerals could be the first step in a sequential anaerobic/aerobic treatment scheme for nitroaromatics (Hofstetter et al., 1999). These minerals may also serve to dechlorinate some chlorinated solvents (Fredrickson and Gorby, 1996).

2.4.4 Immobilization of Contaminant Metals Dissimilatory Fe(III)-reducing microorganisms have the capacity to reduce various contaminant metals and metalloids, which can lead to immobilization of these contaminants in the subsurface. For example, most Fe(III) reducers that have been evaluated have the capacity to reduce U(VI) to U(IV) (Lovley et al., 1997). As previously reviewed (Lovley et al., 1991a; Lovley 1995), U(VI) is soluble in most groundwaters, whereas U(IV) generally precipitates as the insoluble mineral uraninite. This can seen, for example, in the subsurface in the formation of roll-front uranium deposits in which oxidized groundwater containing U(VI) enters an anaerobic zone in which U(IV) is precipitated. The formation of such deposits in the subsurface has not been directly linked to the activity of U(VI)-reducing microorganisms, but microbial reduction seems like the most likely mechanism for uranium reduction since potential abiotic mechanisms for U(VI) reduction have been shown to be ineffective (Lovley et al., 1991a; Lovley and Phillips, 1992a). In laboratory studies, microorganisms reduced U(VI) in uranium-contaminated groundwater, resulting in precipitation of the uranium from solution (Lovley and Phillips, 1992).

Thus, stimulation of U(VI)-reducing microorganisms in order to precipitate uranium from contaminated waters may be a useful strategy for immobilizing uranium *in situ*. In order to effectively remove uranium, it is also necessary to reduce Fe(III) that may oxidize U(IV) back to U(VI). Studies with organic-poor subsurface materials have indicated that the addition of electron donors for Fe(III) reducers and/or extracellular quinones can greatly accelerate the rate of Fe(III)-reduction (Nevin and Lovley, 1999; Synoeyenbos-West et al., 1999). As noted above, in both field and laboratory studies, stimulation of Fe(III)-reduction in several sandy aquifers has invariably led to an enrichment of *Geobacter* species. This suggests that strategies for *in situ* remediation of uranium contamination should consider the physiological characteristics of organisms of this genus in the treatment design.

Stimulation of metal reduction *in situ* is also likely to help prevent the mobility of other contaminant metals in the subsurface. For example, technetium is another radioactive metal contaminant that is found in the subsurface. Microorganisms such as *G. metallireducens* and *S. putrefaciens* can reduce soluble Tc(VII) to less soluble forms (Lloyd and Macaskie, 1996). Some Fe(III)-reducing microorganisms can reduce radioactive Co(III), which is typically complexed with EDTA in contaminated subsurface environments, to Co(II), which tends to disassociate from EDTA and adsorb onto aquifer surfaces (Caccavo et al., 1994; Gorby et al., 1998). Many microorganisms can reduce soluble, highly toxic Cr(VI) to less soluble, less toxic Cr(III) (Lovley, 1993; Chen and Hao, 1998). Furthermore, Fe(II) produced as the result of Fe(III)-reduction may abiotically reduce Cr(VI) (Buerge and Hug, 1997). Some Fe(III)-reducing microorganisms, as well as organisms that do not reduce Fe(III), can reduce the soluble metalloid contaminant selenate to insoluble, elemental selenium, but the microbial reduction of selenate in subsurface environments has yet to be studied in detail (Oremland, 1994). Microbial reduction of arsenate to arsenite can also affect the mobility of arsenic but, like selenate reduction, has not been studied in the subsurface (Ahmann et al., 1994, 1997; Laverman et al., 1995; Newman et al., 1998).

3 HUMICS REDUCTION

The finding that Fe(III)-reducing microorganisms can also use humics as electron acceptors resulted from studies on potential mechanisms for stimulating Fe(III)-reduction in petroleum-contaminated aquifers (Lovley et al., 1996a). As discussed above, a variety of synthetic Fe(III) chelating agents enhanced anaerobic benzene degradation in aquifer sediments in which Fe(III)-reduction was the TEAP. In order to determine if naturally occurring organics that chelate Fe(III) might also stimulate this process, the addition of humics was also evaluated. Humics stimulated anaerobic benzene degradation better than any of the synthetic Fe(III) chelators, even though the ability of humics to chelate Fe(III) is much less than that of the synthetic chelators.

Subsequent studies demonstrated that the primary mechanisms by which humics stimulate reduction of insoluble Fe(III) oxides is not by solubilizing Fe(III), but rather by acting as an electron acceptor for Fe(III)-reducing microorganisms (Lovley et al., 1996a, 1998). All the microorganisms that have been evaluated to date have the capacity to also use humics as an electron acceptor (Lovley et al., 1996, 1998, 1999a,b; Coates et al., 1998). Microbially reduced humics can abiotically transfer electrons received from humics-reducing microorganisms to Fe(III) oxides. Thus, humics can serve as an electron shuttle (Fig. 7.4), eliminating the need for contact between Fe(III) reducers and Fe(III) oxide. The transfer of electrons from humics to Fe(III) oxide oxidizes the humics, which may then accept electrons from humics-reducing microorganisms. In this manner, even a low concentration of humics may function as an important electron acceptor because each humics molecule may be reduced multiple times.

Electron shuttling between Fe(III) reducers and insoluble Fe(III) oxides via humics can stimulate the reduction of a variety of defined Fe(III) oxides, such as poorly crystalline Fe(III) oxide, goethite, and hematite (Lovley et al., 1996a, 1998). Humics also stimulated the reduction of structural Fe(III) in clay (Lovley et al., 1998), as well as the Fe(III) oxides found in sandy aquifer sediments (Nevin and Lovley, 1999). Not only do humics alleviate the need for Fe(III) reducers to contact Fe(III) oxides, they may also permit reduction of Fe(III) that would otherwise not be accessible to Fe(III) reducers. For example, *G. metallireducens* could not reduce

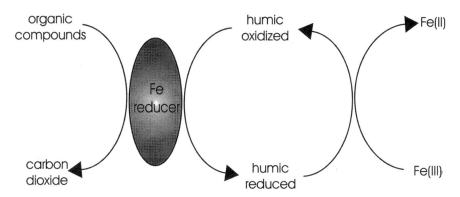

FIGURE 7.4 Model for electron shuttling serving as an electron shuttle between Fe(III)-reducing microorganisms and Fe(III).

Fe(III) oxides that were occluded within microporous beads with pore sizes too small to permit the entry of the organism (Nevin and Lovley, 1999). However, when the humics analog, anthraquinone-2,6-disulfonate (AQDS), was added to the cultures, the Fe(III) oxide in the beads was readily reduced. This was because the AQDS that *G. metallireducens* reduced outside the beads could enter the beads and react with the Fe(III). This finding has implications for subsurface habitats in which tight pore throats might prevent microorganisms from accessing Fe(III) oxides.

Quinones appear to be the primary electron-accepting moieties for microbial reduction in humics. Electron spin resonance studies demonstrated a direct correlation between the electron-accepting capacity of a wide diversity of humics and the quinone free-radical content (Scott et al., 1998). Furthermore, when humics were microbially reduced, there was an increase in quinone free radicals in direct proportion to the electron-accepting capacity of the humics. All microorganisms that are known to transfer electrons to humics also have the ability to transfer electrons to extracellular quinones, whereas organisms that do not reduce extra-cellular quinones do not reduce humics (Lovley et al., 1996a, 1998; Coates et al., 1998). AQDS and other extracellular quinones stimulate Fe(III)-oxide reduction in a manner similar to that observed with humics (Lovley et al., 1998) and can enhance anaerobic benzene oxidation in the same aquifer sediments in which the addition of humics stimulated this process (Anderson and Lovley, 1999).

Studies with AQDS have demonstrated that extracellular quinones are excellent electron shuttles between Fe(III) reducers and Fe(III) oxides because the hydro-quinone produced from microbial reduction of the quinone can readily donate electrons to Fe(III) with the regeneration of the quinone (Lovley et al., 1996a, 1998). Without this ability to undergo repeated reduction–oxidation cycles, humics could not be major electron acceptors in most environments. This is because, on a per gram basis, the electron-accepting capacity of humics is much less than that of other soluble electron acceptors, such as sulfate or nitrate. When coupled with the relatively low concentration of humics in many groundwaters, it is clear that the impact of humics on electron flow would be rather limited if humics could not undergo multiple reduction/oxidation cycles.

Iron bound in humics might be another redox active species involved in electron transfer (Benz et al., 1998). Evaluation of this hypothesis demonstrated that *G. metallireducens* could reduce Fe(III) bound in humics to Fe(II) (Lovley and Blunt-Harris, 1999). However, most of the humics tested contained little, if any, Fe(III). The Fe(III) typically accounted for ca. 10% or less of the total electron-accepting capacity of the humics. Furthermore, once electrons were transferred to the Fe(III) in humics [reducing it to Fe(II)], the Fe(II) could not transfer electrons to Fe(III) oxides. Thus, transfer of electrons to Fe(III) in humics can only take place once. This contrasts with quinone moieties, which can be recycled numerous times in the presence of Fe(III). The combination of low Fe(III)-concentrations in humics and the inability of iron to act as an electron shuttling agent suggest that iron bound in humics is not an important electron transfer agent.

The environmental significance of microbial humics reduction in subsurface environments has yet to be quantified. When Fe(III) is available to continuously recycle humics into an oxidized form, a significant electron flow may proceed through humics even if the concentration of humics is quite low. For example, concentrations as low as $10\,\mu M$ AQDS (the lowest concentration evaluated) stimulated reduction of Fe(III) oxides in aquifer sediments (Nevin and Lovley, 1999) and as little as 200 nM AQDS stimulated Fe(III)-oxide reduction by washed cell suspensions of *Geobacter sulfurreducens* (Lloyd et al., 1999). It is clear from these studies that even groundwaters that are low in dissolved organic matter might contain enough electron-shuttling organics for electron flow through humics or other extracellular quinones to be important.

3.1 Microbial Oxidation of Reduced Humics and Interspecies Electron Transfer

Fe(III) is not the only potential oxidant for microbially reduced humics in anaerobic sedimentary environments. Once microbially reduced, humics can serve as electron donors for microbial reduction of electron acceptors such as nitrate, fumarate, arsenate, and selenate (Lovley et al., 1999a). Depending on the microorganisms involved, nitrate is reduced to nitrogen gas or to ammonia. Electron spin resonance studies demonstrated that microbial oxidation of humics is associated with the oxidation of reduced quinone moieties in humics. Furthermore, studies with the reduced humics analog, anthrahydroquinone-2,6-disulfonate (AHQDS), demonstrated that humics-oxidizing microorganisms oxidize AHQDS to AQDS.

To date, most of the microorganisms with the capacity to oxidize microbially reduced humics are organisms that also have the ability to reduce humics. However, this may be because initial studies focused on such organisms. Screening a greater diversity of organisms, as well as attempting to recover humics-oxidizing microorganisms from the environment, with reduced humics or humic analogs as the electron donor may identify a wider diversity of such organisms.

The findings that some microorganisms can use humics and other extracellular quinones as electron acceptors and that other organisms may oxidize the reduced extracellular quinones indicate that it is possible for extracellular quinones to mediate an interspecies electron transfer between microorganisms. Such interspecies electron transfer has been demonstrated in laboratory cultures (Lovley et al., 1999a). However, the environmental significance of this process has yet to be evaluated. One instance in which this might be important would be within organic-rich aggregates that have an anoxic interior and oxic exterior. Such aggregates would not be expected to have high concentrations of Fe(III) that could abiotically oxidize humics. Thus, humics reduced within the center of the particle may diffuse out and serve as an electron donor for environmentally significant processes, such as denitrification, nearer to the particle's surface.

4 MECHANISMS FOR Fe(III)- AND HUMICS REDUCTION IN THE SUBSURFACE

The fact that reduction of Fe(III) is actually an abiological process when humics or other extracellular quinones serve as an electron shuttle raises the question of whether Fe(III) is primarily reduced via this abiotic mechanism or whether direct enzymatic reduction of Fe(III) is most important in sedimentary environments. The answer to this question will depend, in part, on the concentration of soluble extracellular quinones in a given environment and on the accessibility of the Fe(III) oxides for microbial contact. The relative importance of electron shuttling versus direct reduction has yet to be quantified in any sedimentary environment. However, such measurements will be important in order to fully understand the factors controlling the rate of Fe(III)-reduction in subsurface environments. In addition to Fe(III), microbially reduced humics can transfer electrons to other oxidized metals, such as Mn(IV), U(VI), Cr(VI), and Tc(VII) (Lovley et al., 1998 and unpublished data), which suggests that such considerations are important for understanding the fate of other metals as well.

The fact that numerous Fe(III)-reducing microorganisms have the capacity to reduce Fe(III) oxide in defined cultures without electron-shuttling compounds suggests that Fe(III)-reducing microorganisms can reduce Fe(III) directly, without the need for an electron shuttle. However, an alternative model in which *G. sulfurreducens* releases its own electron shuttle has been proposed (Seeliger et al., 1998). Culture filtrates of *G. sulfurreducens* were found to have an absorbance spectra characteristic of *c*-type cytochromes. A 9.6 kDa *c*-type cytochrome purified from the periplasm of *G. sulfurreducens* was oxidized by Fe(III). It was hypothesized that *G. sulfurreducens* releases the 9.6 kDa cytochrome into its environment to serve as an electron shuttle between the organism and Fe(III) oxide (Seeliger et al., 1998). However, direct investigation of the potential of the cytochrome to act as an electron shuttle revealed that it is not effective (Lloyd et al., 1999). Furthermore, when the cytochromes in the filtrates of *G. sulfurreducens* cultures were directly examined, the cytochrome responsible for the absorbance spectra of the filtrates was found to be a 41 kDa cytochrome that is associated with the outer membrane rather than the 9.6 kDa cytochrome (Lloyd et al., 1999). Thus, the available data do not support the concept of the 9.6 kDa cytochrome serving as an electron shuttle for Fe(III)-oxide reduction.

Energy considerations suggest that using an extracellular cytochrome as an electron shuttle primarily for Fe(III)-oxide reduction in sediments would not be cost-effective. The high energetic cost of producing the electron-shuttling protein must be balanced against the strong likelihood that a cytochrome released into sediments will be rapidly degraded by other microorganisms. This is exacerbated by the fact that each Fe(III) can only accept one electron. There has to be extensive shuttling by each shuttle molecule just to meet the energetic cost of producing the shuttle. However, these considerations do not rule out the possibility that Fe(III) reducers might produce an electron shuttle that is less subject to degradation than cytochromes in order to promote Fe(III)-reduction.

Most investigations into the reduction of Fe(III) oxides have been based on the concept that the microorganisms directly transfer electrons to the insoluble Fe(III) oxides, presumably through an Fe(III) reductase that is located in the outer membrane (Gorby and Lovley, 1991; Myers and Myers, 1992, 1993). In both *Shewanella* and *Geobacter* species, it is clear that c-type cytochromes are involved in electron transport to Fe(III) (Myers and Myers, 1992, 1997; Lovley et al., 1993). However, these studies did not determine whether the c-type cytochromes were intermediary electron carriers in the electron transport chain to Fe(III) or the terminal Fe(III) reductase. However, a membrane-bound NADH-dependent Fe(III) reductase has been purified from *G. sulfurreducens* (Magnuson et al., 1999). This complex, which is comprised of five major subunits, contains a 90 kDa c-type cytochrome capable of transferring electrons to Fe(III). When the 90 kDa cytochrome is removed from the Fe(III) reductase complex, the capacity for Fe(III)-reduction is lost. These results demonstrate that the 90 kDa cytochrome is essential for Fe(III)-reduction and suggest that the cytochrome may be the Fe(III) reductase.

The periplasmic c_3 cytochrome was found to function as a U(VI) and Cr(VI) reductase in *Desulfovibrio vulgaris* (Lovley et al., 1993c; Lovley and Phillips 1994). Unlike *G. metallireducens* and *S. putrefaciens*, which can grow via U(VI) reduction, *D. vulgaris* does not grow in medium with U(VI) as the sole electron acceptor (Lovley et al., 1993b). Unlike the Fe(III) reductase activity in the *Geobacter* and *Shewanella* species, which, as discussed above, is primarily membrane-bound, the hydrogen-dependent U(VI) reductase activity in *D. vulgaris* is localized in the soluble fraction (Lovley et al., 1993c).

As with Fe(III)-reduction, the mechanisms by which microorganisms reduce, or for that matter oxidize, humics are not understood. The large size of humics precludes their entry into the cell. Therefore, as with Fe(III), reduction of humics is likely to take place near the outer cell surface. Preliminary studies have suggested that the humics reductase is distinct from the Fe(III) reductase.

5 SUMMARY

Studies to date have indicated that a wide phylogenetic diversity of Fe(III)- and humics-reducing microorganisms can be recovered from a variety of subsurface environments, and that these microorganisms can have an important impact on both the organic and metal chemistry of the subsurface. Based on the known physiological characteristics of Fe(III)-reducing microorganisms and studies of carbon metabolism in environments in which Fe(III)-reduction is the predominant TEAP, naturally occurring complex organic matter in subsurface environments appears to be degraded by a consortia of fermentative microorganisms and Fe(III)-reducing microorganisms that can oxidize organic acids and hydrogen. In environments contaminated with aromatic hydrocarbons, Fe(III)-reducing microorganisms can directly oxidize the aromatic compounds to carbon dioxide, with Fe(III) serving as the sole electron acceptor. Molecular studies have indicated that *Geobacter* species play an important role in aromatic hydrocarbon degradation. *Geobacters* are also the primary organisms that respond to the stimulation of Fe(III)-reduction with the addition of other organic electron donors to sandy aquifers. Fe(III)-reducing microorganisms also have the capacity to transfer electrons to a variety of radioactive and toxic metals. This can aid in the remediation of metal-contaminated subsurface environments because the reduction of contaminant metals typically converts them to less mobile forms.

Fe(III)-reducing microorganisms have the ability to transfer electrons to humics and other extracellular quinones and, once reduced, these organics can transfer electrons to Fe(III) oxide. Laboratory studies have suggested that this electron shuttling between Fe(III) reducers and Fe(III) oxides by extracellular quinones may greatly increase the rate and extent of Fe(III)-reduction in the subsurface, but the amount of Fe(III)-reduction that actually proceeds via electron shuttling in subsurface environments has not been quantified. Once Fe(III)-reducing microorganisms reduce extracellular quinones, the hydroquinone moieties that are generated may serve as electron donors for other respiratory processes, such as denitrification.

The biochemical mechanisms for microbial Fe(III)-reduction and humics reduction have not been fully elucidated, but studies to date suggest that Fe(III) reducers can reduce Fe(III) oxides via a membrane-bound Fe(III) reductase. Further study of the physiology and biochemistry of not only Fe(III)- and humics reduction, but also other aspects of the metabolism of Fe(III)- and humics-reducing microorganisms is needed in order to better predict the activity of these organisms in the subsurface, and to help design strategies for employing them in the remediation of contaminated subsurface environments.

Acknowledgment The research from the author's laboratory summarized in this chapter was supported by grants from the NABIR program of the Department of Energy, the National Science Foundation, the Office of Naval Research, and the American Petroleum Institute.

REFERENCES

Ahmann D, Krumholz LR, Hemond HF, Lovley DR, and Morel FMM (1997) Microbial mobilization of arsenic from sediments of the Aberjona watershed. *Environ Sci Technol* *31*:2923–2930.

Ahmann D, Roberts AL, Krumholz LR, and Morel FMM (1994) Microbe grows by reducing arsenic. *Nature 371*:750.

Anderson RT and Lovley DR (1997) Ecology and biogeochemistry of *in situ* groundwater bioremediation. *Adv Microbial Ecol 15*:289–350.

Anderson RT and Lovley DR (1999) Naphthalene and benzene degradation under Fe(III)-reducing conditions in petroleum-contaminated aquifers. *Bioremediation J 3*:121–134.

Anderson RT, Rooney-Varga J, Gaw CV, and Lovley DR (1998) Anaerobic benzene oxidation in the Fe(III)-reduction zone of petroleum-contaminated aquifers. *Environ Sci Technol 32*:1222–1229.

Baedecker MJ, Cozzarelli IM, Evans JR, and Hearn PP (1992) Authigenic mineral formation in aquifers rich in organic material. 7th International Symposium on Water–Rock Interaction, Park City, UT.

Baedecker MJ, Cozzarelli IM, Siegel DI, Bennett PC, and Eganhouse RP (1993) Crude oil in a shallow sand and gravel aquifer: 3. Biogeochemical reactions and mass balance modeling in anoxic ground water. *Appl Geochem 8*:569–586.

Benz M, Schink B, and Brune A (1998) Humic acid reduction by *Propionibacterium freudenreichii* and other fermenting bacteria. *Appl Environ Microbiol 64*:4507–4512.

Boone DR, Liu Y, Zhao Z-J, Balkwill DL, Drake GT, Stevens TO, and Aldrich HC (1995) *Bacillus infernus* sp. nov., an Fe(III)- and Mn(IV)-reducing anaerobe from the deep terrestrial subsurface. *Internat J Syst Bacteriol 45*:441–448.

Buerge IJ and Hug SJ (1997) Kinetics and pH dependence of chromium(VI) reduction by iron(II). *Environ Sci Technol 31*:1426–1432.

Caccavo F Jr, Blakemore RP, and Lovley DR (1992) A hydrogen-oxidizing, Fe(III)-reducing microorganism from the Great Bay Estuary, New Hampshire. *Appl Environ Microbiol 58*:3211–3216.

Caccavo F Jr, Coates JD, Rossello-Mora RA, Ludwig W, Schleifer KH, Lovley DR, and McInerney MJ (1996) *Geovibrio ferrireducens*, a phylogenetically distinct dissimilatory Fe(III)-reducing bacterium. *Arch Microbiol 165*:370–376.

Caccavo F, Lonergan DJ, Lovley DR, Davis M, Stolz JF, and McInerney MJ (1994) *Geobacter sulfurreducens* sp. nov., a hydrogen- and acetate-oxidizing dissimilatory metal-reducing microorganism. *Appl Environ Microbiol 60*:3752–3759.

Cairns-Smith AG, Hall AJ, and Russell MJ (1992) Mineral theories of the origin of life and an iron sulfide example. *Orig Life Evol Biosphere 22*:161–180.

Chapelle FH and Lovley DR (1992) Competitive exclusion of sulfate reduction by Fe(III)-reducing bacteria: A mechanism for producing discrete zones of high-iron ground water. *Ground Water 30*:29–36.

Chen JM and Hao OJ (1998) Microbial chromium (VI) reduction. *Crit Rev Environ Sci Technol 28*:219–251.

Christensen TH, Kjeldsen P, Albrechtsen H-J, Heron G, Nielsen PH, Bjerg PL, and Holm PE (1994) Attenuation of landfill leachate pollutants in aquifers. *Crit Rev Environ Sci Technol 24*:109–202.

Coates JD, Councell TB, Ellis DJ, and Lovley DR (1998a) Carbohydrate-oxidation coupled to Fe(III) reduction, a novel form of anaerobic metabolism. *Anaerobe 4*277–282.

Coates JD, Ellis DJ, and Lovley DR (1999) *Geothrix fermentans* gen. nov. sp. nov., an acetate-oxidizing Fe(III) reducer capable of growth via fermentation. *Internat J Syst Bacteriol 49*:1615–1622.

Coates JD, Ellis DJ, Roden E, Gaw K, Blunt-Harris EL, and Lovley DR (1998) Recovery of humics-reducing bacteria from a diversity of sedimentary environments. *Appl Environ Microbiol 64*:1504–1509.

Coates JD, Lonergan DJ, Jenter H, and Lovley DR (1996) Isolation of *Geobacter* species from diverse sedimentary environments. *Appl Environ Microbiol 62*:1531–1536.

Coates JD, Lonergan DJ, and Lovley DR (1995) *Desulfuromonas palmitatis* sp. nov., a long-chain fatty acid oxidizing Fe(III) reducer from marine sediments. *Arch Microbiol 164*:406–413.

Coleman ML, Hedrick DB, Lovley DR, White DC, and Pye K (1993) Reduction of Fe(III) in sediments by sulphate-reducing bacteria. *Nature 361*:436–438.

de Duve C (1995) *Vital Dust*. New York: Basic Books.

Fredrickson JK and Gorby YA (1996) Environmental processes mediated by iron-reducing bacteria. *Curr Opin Biotech 7*:287–294.

Fredrickson JK and Onstott TC (1996) Microbes deep inside the earth. *Scien Amer 275*:68–73.

Fredrickson JK, Zachara JM, Kennedy DW, Dong H, Onstott TC, Hinman NW, and Li S-M (1998) Biogenic iron mineralization accompanying the dissimilatory reduction of hydrous ferric oxide by a groundwater bacterium. *Geochim Cosmochim Acta 62*:3239–3257.

Gibbs-Eggar Z, Jude B, Dominik J, Loizeau J-L, and Oldfield F (1999) Possible evidence for dissimilatory bacterial magnetite dominating the magnetic properties of recent lake sediments. *Earth Planet Sci Lett 168*:1–6.

Gold T (1992) The deep, hot biosphere. *Proc Natl Acad Sci USA 89*:6045–6049.

Gorby Y and Lovley DR (1991) Electron transport in the dissimilatory iron-reducer, GS-15. *Appl Environ Microbiol 57*:867–870.

Gorby YA, Caccavo F Jr, and Bolton H Jr (1998) Microbial reduction of Cobalt^{III}EDTA$^-$ in the presence and absence of manganese(IV) oxide. *Environ Sci Technol 32*:244–250.

Greene AC, Patel BKC, and Sheehy AJ (1997) *Deferribacter thermophilus* gen. nov., sp. nov., a novel thermophilic manganese- and iron-reducing bacterium isolated from a petroleum reservoir. *Internat J Syst Bacteriol 47*:505–509.

Heijman CG, Grieder E, Holliger C, and Schwarzenbach RP (1995) Reduction of nitroaromatic compounds coupled to microbial iron reduction in laboratory aquifer columns. *Environ Sci Technol 29*:775–783.

Heijman CG, Holliger C, Glaus MA, Schwarzenbach RP, and Zeyer J (1993) Abiotic reduction of 4-chloronitrobenzene to 4-chloroaniline in a dissimilatory iron-reducing enrichment culture. *Appl Environ Microbiol 59*:4350–4353.

Hofstetter TB, Heijman CG, Haderlein SB, Holliger C, and Schwarzenbach RP (1999) Complete reduction of TNT and other (poly)nitroaromatic compounds under iron-reducing subsurface conditions. *Environ Sci Technol 33*:1479–1487.

Huber R, Kristjansson JK, and Stetter KO (1987) *Pyrobaculum* gen. nov., a new genus of neutrophilic, rod-shaped archaebacteria form continental solfataras growing optimally at 100°C. *Arch Microbiol 149*:95–101.

Kashefi K and Lovley DR (1999) Reduction of Fe(III), Mn(IV), and toxic metals at 100°C by *Pyrobaculum islandicum*. *Appl Environ Microbiol 66*::1050–1056.

Kieft TL, Fredrickson JK, Onstott TC, Gorby YA, Kostandarithes HM, Bailey TJ, Kennedy DW, Li W, Plymale AE, Spadoni CM, et al. (1999) Dissimilatory reduction of Fe(III) and other electron acceptors by a *Thermus* isolate. *Appl Environ Microbiol 65*:1214–1221.

Krumholz LR, Sharp R, and Fishbain SS (1996) A freshwater anaerobe coupling acetate oxidation to tetrachloroethylene dehalogenation. *Appl Environ Microbiol 62*:4108–4113.

Laverman AM, Switzer Blum J, Schaefer JK, Phillips EJP, Lovley DR, and Oremland RS (1995) Growth of strain SES-3 with arsenate and other diverse electron acceptors. *Appl Environ Microbiol 61*:3556–3561.

Liu SV, Zhou J, Zhang C, Cole DR, Gajdarziska-Josifovska M, and Phelps TJ (1997) Thermophilic Fe(III)-reducing bacteria from the deep subsurface: The evolutionary implications. *Science 277*:1106–1109.

Lloyd JR, Blunt-Harris EL, and Lovley DR (1999) The periplasmic 9.6 kDa *c*-type cytochrome is not an electron shuttle to Fe(III). *J Bacteriol 181*:7647–7649.

Lloyd JR and Macaskie LE (1996) A novel phosporimager-based technique for monitoring the microbial reduction of technetium. *Appl Environ Microbiol 62*:578–582.

Lonergan DJ, Jenter H, Coates JD, Phillips EJP, Schmidt T, and Lovley DR (1996) Phylogenetic analysis of dissimilatory Fe(III)-reducing bacteria. *J Bacteriol 178*:2402–2408.

Lonergan DJ and Lovley DR (1991) Microbial oxidation of natural and anthropogenic aromatic compounds coupled to Fe(III) reduction. In Baker RA (ed) *Organic Substances and Sediments in Water*. Chelsea, MI: Lewis Publishers, pp 327–338.

Lovley DR (1987) Organic matter mineralization with the reduction of ferric iron: A review. *Geomicrobiol J 5*:375–399.

Lovley DR (1990) Magnetite formation during microbial dissimilatory iron reduction. In: Frankel RB and Blakemore RP (eds): *Iron Biominerals*. New York: Plenum Press, pp 151–166.

Lovley DR (1991) Dissimilatory Fe(III) and Mn(IV) reduction. *Microbiol Rev 55*:259–287.

Lovley DR (1993) Dissimilatory metal reduction. *Ann Rev Microbiol 47*:263–290.

Lovley DR (1995) Bioremediation of organic and metal contaminants with dissimilatory metal reduction. *J Indus Microbiol 14*:85–93.

Lovley DR (1997) Microbial Fe(III) reduction in subsurface environments. *FEMS Microbiol Rev 20*:305–315.

Lovley DR (1997) Potential for anaerobic bioremediation of BTEX in petroleum-contaminated aquifers. *J Indus Microbiol 18*:75–81.

Lovley DR and Anderson RT (2000) The influence of dissimilatory metal reduction on the fate of organic and metal contaminants in the subsurface. *Hydrogeol J 8*:77–88.

Lovley DR, Baedecker MJ, Lonergan DJ, Cozzarelli IM, Phillips EJP, and Siegel DI (1989) Oxidation of aromatic contaminants coupled to microbial iron reduction. *Nature 339*:297–299.

Lovley DR and Blunt-Harris EL (1999) Role of humics-bound iron as an electron transfer agent in dissimilatory Fe(III) reduction. *Appl Environ Microbiol 65*:4252–4254.

Lovley DR and Chapelle FH (1995) Deep subsurface microbial processes. *Rev Geophys 33*:365–381.

Lovley DR, Chapelle FH, and Phillips EJP (1990) Fe(III)-reducing bacteria in deeply buried sediments of the Atlantic Coastal Plain. *Geology 18*:954–957.

Lovley DR, Chapelle FH, and Woodward JC (1994a) Use of dissolved H_2 concentrations to determine the distribution of microbially catalyzed redox reactions in anoxic ground water. *Environ Sci Technol 28*:1205–1210.

Lovley DR, Coates JD, Blunt-Harris EL, Phillips EJP, and Woodward JC (1996a) Humic substances as electron acceptors for microbial respiration. *Nature 382*:445–448.

Lovley DR, Coates JD, Saffarini DA, and Lonergan DJ (1997) Dissimilatory iron reduction. In: Winkelman G and Carrano CJ (eds): *Iron and Related Transition Metals in Microbial Metabolism*. Switzerland: Harwood Academic Publishers, pp 187–215.

Lovley DR, Fraga JL, Blunt-Harris EL, Hayes LA, Phillips EJP, and Coates JD (1998) Humic substances as a mediator for microbially catalyzed metal reduction. *Acta Hydrochim Hydrobiol 26*:152–157.

Lovley DR, Fraga JL, Coates JD, and Blunt-Harris EL (1999a) Humics as an electron donor for anaerobic respiration. *Environ Microbiol 1*:89–98.

Lovley DR, Giovannoni SJ, White DC, Champine JE, Phillips EJP, Gorby YA, and Goodwin S (1993a) *Geobacter metallireducens* gen. nov. sp. nov., a microorganism capable of coupling the complete oxidation of organic compounds to the reduction of iron and other metals. *Arch Microbiol 159*:336–344.

Lovley DR, Kashefi K, Vargas M, Tor JM, and Blunt-Harris EL (2000) Reduction of humic substances and Fe(III) by hyperthermophilic microorganisms. *Chem Geol 169*:289–298.

Lovley DR and Lonergan DJ (1990) Anaerobic oxidation of toluene, phenol, and p-cresol by the dissimilatory iron-reducing organism, GS-15. *Appl Environ Microbiol 56*:1858–1864.

Lovley DR and Phillips EJP (1988a) Manganese inhibition of microbial iron reduction in anaerobic sediments. *Geomicrobiol J 6*:145–155.

Lovley DR and Phillips EJP (1988b) Novel mode of microbial energy metabolism: Organic carbon oxidation coupled to dissimilatory reduction of iron or manganese. *Appl Environ Microbiol 54*:1472–1480.

Lovley DR and Phillips EJP (1989) Requirement for a microbial consortium to completely oxidize glucose in Fe(III)-reducing sediments. *Appl Environ Microbiol 55*:3234–3236.

Lovley DR and Phillips EJP (1992a) Bioremediation of uranium contamination with enzymatic uranium reduction. *Environ Sci Technol 26*:2228–2234.

Lovley DR and Phillips EJP (1992b) Reduction of uranium by *Desulfovibrio desulfuricans*. *Appl Environ Microbiol 58*:850–856.

Lovley DR and Phillips EJP (1994) Reduction of chromate by *Desulfovibrio vulgaris* (Hildenborough) and its c_3 cytochrome. *Appl Environ Microbiol 60*:726–728.

Lovley DR, Phillips EJP, Gorby YA, and Landa ER (1991a) Microbial reduction of uranium. *Nature 350*:413–416.

Lovley DR, Phillips EJP, and Lonergan DJ (1989b) Hydrogen and formate oxidation coupled to dissimilatory reduction of iron or manganese by *Alteromonas putrefaciens*. *Appl Environ Microbiol 55*:700–706.

Lovley DR, Phillips EJP, and Lonergan DJ (1991b) Enzymatic versus nonenzymatic mechanisms for Fe(III) reduction in aquatic sediments. *Environ Sci Technol 25*:1062–1067.

Lovley DR, Roden EE, Phillips EJP, and Woodward JC (1993b) Enzymatic iron and uranium reduction by sulfate-reducing bacteria. *Marine Geol 113*:41–53.

Lovley DR, Stolz JF, Nord GL, and Phillips EJP (1987) Anaerobic production of magnetite by a dissimilatory iron-reducing microorganism. *Nature 330*:252–254.

Lovley DR, Widman PK, Woodward JC, and Phillips EJP (1993a) Reduction of uranium by cytochrome c_3 of *Desulfovibrio vulgaris*. *Appl Environ Microbiol 59*:3572–3576.

Lovley DR and Woodward JC (1996) Mechanisms for chelator stimulation of microbial Fe(III)-oxide reduction. *Chem Geol 132*:19–24.

Lovley DR, Woodward JC, and Chapelle FH (1994b) Stimulated anoxic biodegradation of aromatic hydrocarbons using Fe(III) ligands. *Nature 370*:128–131.

Lovley DR, Woodward JC, and Chapelle FH (1996b) Rapid anaerobic benzene oxidation with a variety of chelated Fe(III) forms. *Appl Environ Microbiol 62*:288–291.

Ludwig W, Bauer SH, Bauer M, Held I, Kirchhof G, Schulze R, Huber I, Spring S, Hartmann A, and Schleifer KH (1997) Detection and *in situ* identification of representatives of a widely distributed new bacterial phylum. *FEMS Microbiol Lett 153*:181–190.

Magnuson TS, Hodges-Myerson AL, and Lovley DR (2000) Characterization of a membrane-bound NADH-dependent Fe(3+) reductase from the dissimilatory Fe(3+)-reducing bacterium *Geobacter sulfurreducens*. *FEMS Microbiol Lett 185*:205–211.

Mortimer RJG and Coleman ML (1997) Microbial influence on the isotopic composition of diagenetic siderite. *Geochem Cosmochim Acta 61*:1705–1711.

Myers CR and Myers JM (1992) Localization of cytochromes to the outer membrane of anaerobically grown *Shewanella putrefaciens* MR-1. *J Bacteriol 174*:3429–3438.

Myers CR and Myers JM (1993) Ferric reductase is associated with the membranes of anaerobically grown *Shewanella putrefaciens* MR-1. *FEMS Microbiol Lett 108*:15–22.

Mycrs CR and Myers JM (1997) Cloning and sequencing of *cymA*, a gene encoding a tetraheme cytochrome c required for reduction of iron(III), fumarate, and nitrate by *Shewanella putrefaciens* strain MR-1. *J Bacteriol 179*:1143–1152.

Nealson KH and Saffarini D (1994) Iron and manganese in anaerobic respiration: Environmental significance, physiology, and regulation. *Ann Rev Microbiol 48*:311–343.

Nevin KP and Lovley DR (1999) Potential for nonenzymatic reduction of Fe(III) during microbial oxidation of organic matter coupled to Fe(III) reduction (submitted for publication).

Newman DK, Kennedy EK, Coates JD, Ahmann D, Ellis DJ, Lovley DR, and Morel FMM (1998) Dissimilatory As(V) and S(VI) reduction in *Desulfotomaculum aurpigmentum*, sp. nov. *Arch Microbiol 36*:380–388.

Oremland RS (1994) Biogeochemical transformations of selenium in anoxic environments. In: Frankenberger WTJ and Benson SN (eds): *Selenium in the Environment*. New York: Marcel Dekker, pp 389–419.

Pedersen K (1993) The deep subterranean biosphere. *Earth-Science Rev 34*:243–260.

Pedersen K, Arlinger J, Ekendahl S, and Hallbeck L (1996) 16S rRNA gene diversity of attached and unattached bacteria in boreholes along the access tunnel to the Aspo hard rock laboratory, Sweden. *FEMS Microbiol Ecol 19*:249–262.

Rooney-Varga JN, Anderson RT, Fraga JL, Ringelberg D, and Lovley DR (1999) Microbial communities associated with anaerobic benzene degradation in a petroleum-contaminated aquifer. *Appl Environ Microbiol 65*:3056–3063.

Rugge K, Hofstetter T, Haderlein SB, Bjerg P, Knudsen S, Zraunig C, Mosbaek H, and Christensen TH (1998) Characterization of predominant reductants in an anaerobic leachate-contaminated aquifer by nitroaromatic probe compounds. *Environ Sci Technol 32*:23–31.

Scott DT, McKnight DM, Blunt-Harris EL, Kolesar SE, and Lovley DR (1998) Quinone moieties act as electron acceptors in the reduction of humic substances by humics-reducing microorganisms. *Environ Sci Technol 32*:2984–2989.

Seeliger S, Cord-Ruwisch R, and Schink B (1998) A periplasmic and extracellular *c*-type cytochrome of *Geobacter sulfurreducens* acts as a ferric iron reductase and as an electron carrier to other acceptors or to partner bacteria. *J Bacteriol 180*:3686–3691.

Slobodkin A, Reysenbach A-L, Strutz N, Dreier M, and Wiegel J (1997) *Thermoterrabacterium ferrireducens* gen. nov., sp. nov., a thermophilic anaerobic dissimilatory Fe(III)-reducing bacterium from a continental hot spring. *Intern J System Bacteriol 47*:541–547.

Slobodkin A and Wiegel J (1997) Fe(III) as an electron acceptor for H_2 oxidation in thermophilic anaerobic enrichment cultures from geothermal areas. *Extremophiles 2*:106–109.

Synoeyenbos-West OL, Nevin KP, and Lovley DR (1999) Enrichment of *Geobacter* species in response to stimulation of Fe(III) reduction in sandy aquifer sediments. *Microbial Ecol 39*:153–167.

Utkin I, Woese C, and Wiegel J (1994) Isolation and characterization of *Desulfitobacterium dehalogenans* gen. nov., sp. nov., an anaerobic bacterium which reductively dechlorinates chlorphenolic compounds. *Internat J Syst Bacteriol 44*:612–619.

Vargas M, Kashefi K, Blunt-Harris EL, and Lovley DR (1998) Microbiological evidence for Fe(III) reduction on early Earth. *Nature 395*:65–67.

Walker JCG (1984) Suboxic diagenesis in banded iron formations. *Nature 309*:340–342.

Walker JCG (1987) Was the Archaean biosphere upside down? *Nature 329*:710–712.

8

MICROBIAL SULFUR CYCLING IN TERRESTRIAL SUBSURFACE ENVIRONMENTS

ETHAN L. GROSSMAN AND STEVEN DESROCHER

Department of Geology and Geophysics, Texas A&M University, College Station, Texas

Because of its reactivity and abundance, sulfur is one of the most important elements participating in oxidation and reduction reactions in the subsurface biosphere (Fig. 8.1). Sulfur is redox-sensitive because it occurs in a variety of oxidation states: -2, 0, 2, 4, and 6. Oxidized sulfur as sulfate ($SO_4^=$) might be the most important inorganic electron acceptor in the terrestrial subsurface. Other electron acceptors such as O_2, NO_3^-, Mn(IV), and Fe(III) typically occur in low concentrations in groundwaters or in sediments, and are often unavailable meters to tens of meters below the surface. Reduced sulfur (sulfide, FeS_2) is one of the most important electron donors in the subsurface. Sulfide minerals provide reduced iron and sulfur for iron- and sulfur-oxidizing chemoautotrophs in oxic environments. Aerobic oxidation of sulfide (-2) to sulfate ($+6$) occurs through intermediates such as thiosulfate ($S_2O_3^=$) and elemental sulfur (Nordstrom, 1982). The reaction results in the transfer of eight electrons and release of considerable energy. A major sink for oxygen in the subsurface, sulfide oxidation transfers electron-accepting capacity from molecule oxygen to sulfur compounds, which are subsequently carried to the

Subsurface Microbiology and Biogeochemistry, Edited by James K. Fredrickson and Madilyn Fletcher. ISBN 0-471-31577-X. Copyright 2001 by Wiley-Liss, Inc.

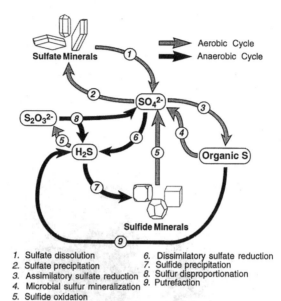

1. Sulfate dissolution
2. Sulfate precipitation
3. Assimilatory sulfate reduction
4. Microbial sulfur mineralization
5. Sulfide oxidation
6. Dissimilatory sulfate reduction
7. Sulfide precipitation
8. Sulfur disproportionation
9. Putrefaction

FIGURE 8.1 Generalized view of the sulfur cycle in the terrestrial subsurface.

deeper subsurface through groundwater flow. Clearly, sulfur plays a key role in the microbial ecology of the terrestrial subsurface.

Reduced and oxidized forms of sulfur are ubiquitous in the subsurface. Estimates of the proportion of oxidized and reduced sedimentary sulfur reservoirs vary from $0.1:1$ to nearly $2:1$ (Garrels and Lerman, 1984). Evaporitic salt beds furnish extensive sulfate deposits. Sandstones and shales, with an average sulfur content of 240 and 2400 ppm (Turekian and Wedepohl, 1961), respectively, provide disseminated sulfate and sulfide minerals. Sulfate is the second most abundant anion in seawater, with a concentration of 2710 ppm, and a key constituent in surface water and groundwater. Not surprisingly, sulfur-oxidizing and sulfate-reducing bacteria are widespread in the natural environment and are sometimes dominant community members in the terrestrial subsurface. Such bacteria mediate sulfur fluxes between reduced and oxidized reservoirs, which, when perturbed, can have global impacts on ocean chemistry and atmospheric O_2 and CO_2 levels (Garrels and Lerman, 1984; Berner, 1987).

Sulfur cycling has a prominent impact on the quality of natural waters. Sulfide oxidation in natural and perturbed systems produces acidic waters rich in iron, metals, and sulfate (Hem, 1989). Conversely, sulfate reduction can decrease the concentration of toxic metals in natural waters through precipitation of insoluble metal sulfides (Brierley and Brierley, 1996). But, in the absence of dissolved metals such as iron, sulfate reduction can result in noxious levels of H_2S gas.

Scientists investigate the sulfur cycle using geochemical and microbiological methods. Geochemists analyze the products of microbial sulfur cycling, the chemical and isotopic (^{34}S/^{32}S) compositions of water and sediment, and apply thermodynamic models to predict reaction progress and equilibrium states. When applied to the heterogeneous and complex environs of the subsurface, however, such thermodynamic models often prove inadequate. Microbiologists provide vital information regarding the rates of ongoing (rather than past) processes by quantitating the abundance, activity, spatial distribution, and nutritional needs of the ambient microbial population. In this chapter, we will review the microbial processes important in sulfur cycling, then summarize knowledge regarding the distribution, controls, rates, and impacts of these processes in the terrestrial subsurface.

1 IMPORTANT REACTIONS IN BIOGEOCHEMICAL SULFUR CYCLING

Within the E_h and pH of most natural waters, SO_4^{2-}, H_2S, HS^-, and elemental sulfur are the thermodynamically most stable sulfur species (Fig. 8.2A). These species form significant sulfur reservoirs in nature. Metastable sulfur species serve as intermediates in transitions between redox end-members, sulfide and sulfate. The thermodynamic stability fields of the metastable sulfur species, ignoring elemental sulfur and relatively inert sulfate, are shown in Fig. 8.2B. At pH 8, the sequence with increasing E_h is bisulfide (HS^-), thiosulfate ($S_2O_3^{2-}$), and sulfite (SO_3^{2-}) (Williamson and Rimstadt, 1992). The key reactions in the biogeochemical sulfur cycle are sulfide oxidation, sulfate reduction, and sulfur disproportionation (Fig. 8.1). These reactions are discussed below and in reviews by Goldhaber and Kaplan (1974), Nordstrom (1982), and Luther and Church (1992).

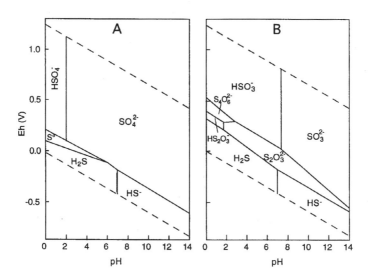

FIGURE 8.2 E_h-pH diagram for the thermodynamic stability of aqueous sulfur species in the S-O-H system at 25°C. (A) Thermodynamically stable species. (B) System neglecting elemental and sulfate sulfur (from Williamson and Rimstidt, 1992).

1.1 Sulfide Oxidation

The overall sulfide oxidation reaction is commonly represented by

$$FeS_2 + \frac{15}{4}O_2 + \frac{7}{2}H_2O \rightarrow Fe(OH)_3 + 2SO_4^{2-} + 4H^+ \qquad (1)$$

where FeS_2 is pyrite. The reaction product is typically not a pure ferric hydroxide phase, but a mixture of phases including goethite (α-FeOOH), ferrihydrite ($Fe_5OH_8 \cdot 4H_2O$), jarosite ($KFe_3(SO_4)_2(OH)_6$), and schwertmannite ($Fe_8O_8SO_4(OH)_6$) (Nordstrom and Southam, 1997). At low pH (< 4.5), Fe^{3+} oxidizes pyrite more rapidly than O_2 through the overall reaction

$$FeS_2 + 14Fe^{3+} + 8H_2O \rightarrow 15Fe^{2+} + 2SO_4^{2-} + 16H^+ \qquad (2)$$

(e.g., Nordstrom, 1982; Evangelou and Zhang, 1995). Ferric iron is regenerated by oxygen reduction:

$$Fe^{2+} + \frac{1}{4}O_2 + H^+ \rightarrow Fe^{3+} + \frac{1}{2}H_2O \qquad (3)$$

At low pH, reaction 3 occurs more slowly than reaction 2 and is thus the rate-limiting step. The Fe^{2+}-oxidation rate (reaction 3) can be accelerated by a factor of 10^6 through the actions of the iron-oxidizing bacteria *Thiobacillus ferrooxidans* (Singer and Stumm, 1970).

Low Fe^{3+} solubilities have led some researchers to suggest that O_2 is the primary electron acceptor for pyrite oxidation at circumneutral pH. However, experimental and theoretical studies argue that Fe^{3+} is the more important electron acceptor, even under those conditions (Luther, 1987; Moses and Herman, 1991). Nitrate may also serve as an electron acceptor for sulfide oxidation (Ehrlich, 1996). This mechanism may be important in groundwaters contaminated with fertilizers or sewage (Böttcher et al., 1992), but is relatively unimportant in pristine groundwaters where nitrate concentrations are typically low (< 10 mg L^{-1}; Davis and DeWiest, 1966).

Metastable intermediate sulfoxyanions are produced during sulfide oxidation, a consequence of the reaction involving the transfer of 15 electrons, with typically one or two electrons transferred per step (Nordstrom and Southam, 1997). Experimental studies in O_2-saturated water show that tetrathionate and sulfate are the dominant sulfoxyanions produced at pH 9, whereas thiosulfate and sulfite dominate at pH 6 (Goldhaber, 1983). When Fe^{3+} replaces oxygen as the electron acceptor, sulfoxyanions are not detected (Moses et al., 1987). Not surprisingly, metastable sulfoxyanions in the terrestrial subsurface occur in extremely low concentration.

As mentioned previously, the acidophile *T. ferrooxidans* catalyzes Fe^{2+} oxidation, increasing the concentration of Fe^{3+} available for pyrite oxidation. Predictably, *T. ferrooxidans* is common in acid mine waters (Nordstrom and Southam, 1997). *Leptospirillum ferrooxidans* can also be important in natural acidic environments. It can dominate at very low pH (0.3–0.7) and relatively high temperature (30–50°C)

(Schrenk et al., 1998). *Thiobacillus thiooxidans*, another acidophile, oxidizes sulfide-oxidation intermediates such as elemental sulfur, thiosulfate, and tetrathionate, but not Fe^{2+}. Surface chemistry studies show that pyrite surfaces are sulfur-rich due to the preferential dissolution of iron (Buckley and Woods, 1987). This surface enrichment of partially oxidized sulfur species may provide substrate for *T. thiooxidans* and explain the common association of that taxon with *T. ferroxidans* (Sasaki et al., 1995).

At the circumneutral pH of most aquifer systems, nonacidophiles could play a larger role in sulfide oxidation than their acidophilic counterparts. *Metallogenium*, a controversial mesophilic iron-oxidizing bacteria (Walsh and Mitchell, 1972), has been proposed to facilitate the transition to acidic environments (Nordstrom and Southam, 1997), but confusion caused by *Metallogenium*-like inorganic precipitates confounds the issue (Ghiorse, 1984; Ehrlich, 1996). *Thiobacillus thioparus* is a potentially important iron-oxidizer that shows optimum growth at neutral pH. In studies of mine tailings, the waste material from mining processes, *T. thioparus* was found to dominate at pH values above 5 (Blowes et al., 1995, 1998). On the other hand, *T. ferrooxidans* and *T. thiooxidans* dominate at pH values of 4 or less. These results led Blowes et al. (1998) to conclude that neutrophilic thiobacilli play a major role in sulfide oxidation at circumneutral pH. Other bacteria that show evidence for iron oxidation in neutral-pH environments include *Gallionella*, *Sphaerotilus*, and *Leptothrix* (Ehrlich, 1996; Tuhela et al., 1997). Another consideration is microenvironments. Southam and Beveridge (1992) recovered low numbers of acidophilic thiobacilli from circumneutral pH macroenvironments, which may reflect growth in acidic "nanoenvironments."

Acidophilic iron- and sulfur-oxidizing bacteria may be mesophilic or thermophilic. *Thiobacillus* sp. and *Leptospirillum* sp. are mesophiles with optimum temperatures of 30–35°C (Norris, 1990). Under normal geothermal gradients (2–3°C/100 m), this temperature range equates to terrestrial subsurface depths of 300–750 m. Moderate thermophiles such as *Leptospirillum thermoferrooxidans*, *Sulfobacillus thermosulfidooxidans*, and strain TH1/BC1 typically have optimum growth temperatures of 45–50°C. Extreme thermophiles, with optimum growth temperatures ranging from 65–80°C, include *Sulfolobus* sp. and *Acidianus* sp. (Norris, 1990; Nordstrom and Southam, 1997). These optimum growth temperatures translate to depths of 1500–3000 m.

1.2 Sulfate Reduction

Bacterial sulfate reduction (BSR) is the dominant origin of the reduced sulfur compounds in the terrestrial subsurface biosphere. The general reaction is

$$2CH_2O + SO_4^{2-} \rightarrow H_2S + 2HCO_3^- \tag{4}$$

where CH_2O is a representative organic compound. BSR may be assimilatory or dissimilatory. In assimilatory reduction, microorganisms take up and reduce sulfate in the process of amino acid and protein synthesis. Although assimilatory sulfate reduction is quantitatively more significant on a global scale than dissimilatory reduction (Postgate, 1984), sequestration of sulfur by this mechanism is of lesser significance to the geochemical sulfur cycle and will not be discussed further.

In dissimilatory sulfate reduction, sulfate is reduced during the microbial oxidation of organic matter without incorporating the majority of sulfur into biomass. This anaerobic respiration process releases reduced sulfur into the environment. Dissimilatory sulfate reduction is mediated by several bacterial species belonging to genera that bear little physiological similarity apart from their capacity to reduce sulfate (Table 8.1; Postgate, 1984). Sulfate-reducing bacteria (SRB) are strict anaerobes, being inhibited by oxygen. However, tolerance of O_2 has been widely reported (Hardy and Hamilton, 1981; Cypionka et al., 1985).

SRB are ubiquitous in most anaerobic environments, but their activity may be limited where environmental conditions are not optimal. They require highly reducing redox potentials, with E_h values less than -150 to -200 mV (Postgate, 1984). SRB exhibit a wide range of temperature optima, as well as a high degree of temperature adaptability. Similarly, salinity does not impose a significant limitation on their activity in most systems, as sulfate reducers are present in a wide range of osmotic conditions. Marine SRB have a salinity requirement of 20 g L^{-1} NaCl and 1.5 g L^{-1} $MgCl_2$ (Widdel, 1989), whereas freshwater and terrestrial strains exhibit no salinity requirements. Salt tolerances as high as 270 g L^{-1} NaCl were reported by Cord-Ruwisch et al. (1987). Reported pH tolerance varies within the range of approximately 4 (Benda, 1957) to 9.2 (Birnbaum and Wireman, 1984), with a circumneutral optimum (Widdel, 1989). This range is inclusive of most subsurface environments. An additional limitation on activity is dissolved sulfide concentration. O'Flaherty et al. (1998) report the growth inhibition of SRB at sulfide concentrations of 328–2059 mg L^{-1} in anaerobic sludges, with dependence on both pH and the substrate utilized. In subsurface systems containing soluble metals (e.g., iron, lead, or copper), dissolved sulfide is consumed by precipitation of insoluble metal sulfides.

TABLE 8.1 Substrates Utilized by SRB and Representative Reactions

Substrate	Reaction	Representative Organisms	References
H_2	$4H_2 + HSO_4^- \rightarrow$ $4H_2O + HS^-$ (5)	*Desulfobulbus propionicus* *Desulfovibrio baarsii*	
Formate	$4HCOO^- + SO_4^{2-} \rightarrow$ $4HCO_3^- + S^{2-}$ (6)	*Desulfobulbus propionicus* *Desulfovibrio baarsii*	Postgate (1984)
Lactate	$2CH_3CHOHCOO^- + SO_4^{2-} \rightarrow$ $2CH_3COO^- + 2HCO_3^- +$ $HS^- + H^+$ (7)	*Desulfovibrio* *Desulfotomaculum*	Voordouw (1995)
Acetate	$CH_3COO^- + SO_4^{2-} \rightarrow$ $2HCO_3^- + HS^-$ (8)	*Desulfobacter* *Desulfotomaculum*	Gebhardt et al. (1983) Schauder et al. (1986)

A variety of low-molecular-weight organic compounds may act as electron donors for BSR (Table 8.1), whereas H_2 is the sole inorganic electron donor. Organic species oxidized by SRB include formate (Postgate, 1984; Widdel, 1989), lactate (Voordouw, 1995), acetate and higher-carbon-number aliphatic acids (Gebhardt et al., 1983; Schauder et al., 1986). The utilization of hydrocarbons as electron donors for BSR has been contentious since its proposal by Rosenfeld (1947). Methane oxidation, described by

$$CH_4 + SO_4^{2-} \rightarrow HCO_3^- + HS^- + H_2O \tag{9}$$

is thermodynamically favorable, being exergonic at typical sedimentary temperatures. Furthermore, methane consumption in sulfate-reducing zones overlying methanogenic zones in marine sediments suggests methane oxidation via sulfate reduction (Reeburgh, 1980; Burns, 1998). However, isolation of a SRB capable of hydrocarbon utilization remains elusive. Recent studies suggest that methane oxidation is mediated by a sulfate reducer-methanogen consortium in which methanogens oxidize CH_4 (Zehnder and Brock, 1980) while under low H_2 partial pressure imposed by H_2-oxidation during BSR (Hoehler et al., 1994; Hansen et al., 1998).

1.3 Sulfur Disproportionation

Sulfur disproportionation constitutes a third metabolic pathway in addition to sulfate reduction and sulfide oxidation. In this process, thiosulfate or sulfite is converted to sulfate and sulfide according to the reactions:

$$S_2O_3^{2-} + H_2O \rightarrow SO_4^{2-} + HS^- + H^+ \tag{10}$$

and

$$4SO_3^{2-} + H^+ \rightarrow 3SO_4^{2-} + HS^- \tag{11}$$

for thiosulfate and sulfite disproportionation, respectively (Bak and Pfennig, 1987). Reactions 10 and 11 yield 21.9 and 58.9 kJ of energy per mole of thiosulfate or sulfite disproportionated, respectively, constituting the sole means of energy generation under anoxic conditions (Bak and Pfennig, 1987). The bacterium mediating these reactions was named *Desulfovibrio sulfodismutans* by the authors. In addition to this novel organism, Bak and Pfennig (1987) reported thiosulfate and sulfite disproportionation by *Desulfobacter curvatus* and sulfite disproportionation by *Desulfovibrio vulgaris*. More recently, the disproportionation of elemental sulfur by *Desulfocapsa thiozymogenes* (Bak, 1993) and *Desulfobulbus propionicus* (Lovley and Phillips, 1994) was reported. Thamdrup et al. (1993) describe the coupling of elemental sulfur disproportionation to iron or manganese reduction. Environmental requirements for bacterial disproportionation have not been fully constrained. Bak and Pfennig (1987) observed the growth of *D. sulfodismutans* in freshwater and brackish media with salinities less than $14\,g\,L^{-1}$ NaCl and $2\,g\,L^{-1}$ $MgCl_2 \cdot 6H_2O$. Growth temperatures range from 15–45°C, with optimal growth temperature at 35°C. As with sulfate reduction, pH requirements are circumneutral, with a range of 6.8–8.2 and an optimum between 7.2 and 7.5 (Bak and Pfennig, 1987).

2   REDOX CHEMISTRY OF SUBSURFACE SYSTEMS

Sulfur's role in the redox chemistry of the subsurface, as with other Earth systems, depends on the availability of other electron acceptors. Different electron acceptors produce different energy yields per mole organic substrate, and the dominant microorganism will be the one whose physiology generates the most energy. Oxygen reduction provides the greatest energy yield per mole of CH_2O, followed by nitrate reduction, Mn(IV)-reduction, Fe(III)-reduction, sulfate reduction, and finally CO_2-reduction (Berner, 1980; Stumm and Morgan, 1981). Alternatively, the redox succession can be viewed as a competition for electron donors (e.g., acetate, H_2), with the energetically favorable metabolisms able to function at much lower concentrations of these compounds (Lovley and Chapelle, 1995). These models explain the overall redox succession in aquifers (Fig. 8.3; Champ et al., 1979), although local distributions of redox reactions may be much more complex and heterogeneous (see later discussion).

The balance between oxygen influx and consumption controls dissolved oxygen concentrations (DO) in the subsurface. At and above the water table, atmospheric

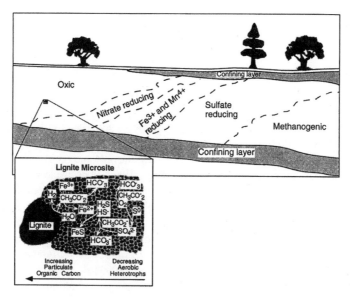

FIGURE 8.3   Redox sequence in an idealized aquifer system. Inset shows a hypothetical anaerobic microzone around a lignite clast (from Murphy et al., 1992).

oxygen provides a nearly limitless supply of electron acceptors for processes like respiration and sulfide oxidation. But below the water table, oxygen influx is restricted to infiltration with meteoric waters. The oxygen-carrying capacity of these waters is severely limited by the low solubility of oxygen. At temperatures from $30–0°C$, oxygen solubilities in freshwater range from $240–460 \, \mu mol \, L^{-1}(\mu M)$, with lower solubilities at higher temperatures. Dissolved oxygen is consumed in aquifers primarily through microbial respiration and secondarily through Fe^{2+}-oxidation. Where the rate of oxygen utilization by aerobic microorganisms exceeds the rate at which it can be replenished by diffusion and meteoric recharge, low dissolved oxygen concentrations will develop, allowing the proliferation of anaerobic microorganisms.

Nitrate and Mn(IV) are generally not important electron acceptors in the subsurface. Fe(III)-reduction can be important in some aquifer systems (Lovley et al., 1990), but readily reducible ferric iron is typically consumed in the upgradient portion of aquifers (Fig. 8.3; Champ et al., 1979; Edmunds et al., 1984). Sulfate reduction predominates over Fe-reduction in the downgradient portions of aquifers, where poorly crystalline ferric oxyhydroxides are absent (Chapelle and Lovley, 1992). Under increasingly reducing conditions where groundwater sulfate concentrations decrease to less than $1 \, mM$ (Beeman and Suflita, 1990; Zhang et al., 1998), methanogenesis emerges as the dominant terminal electron-accepting process (TEAP).

Differences in free-energy yields between these respiration processes results in zones dominated by a single process, a phenomenon known as *redox zonation* (Fig. 8.3). The distribution and size of redox zones in the subsurface are controlled by both geological and geochemical factors. Among the geological controls, the porosity and permeability of the porous medium are critical. Highly permeable aquifer systems will permit rapid meteoric water recharge on timescales that are short relative to the time required for organic substrate degradation. The net result of this rapid recharge is deepening of the aerobic zone. For example, the highly permeable sand and gravel Pauxtent Aquifer of Maryland maintains high dissolved oxygen contents up to 6 miles away from the recharge zone (Chapelle and Kean, 1985). Conversely, low-permeability aquifer systems are characterized by rapid oxygen depletion relative to recharge rates and will exhibit compression of the aerobic zone and development of anaerobic conditions at shallow depths (Martino et al., 1998). The Floridan Aquifer of South Carolina is one such system, becoming anaerobic within a few meters of the water table (Chapelle et al., 1988).

Aquifer lithology may determine the occurrence and size of redox zones by controlling the distribution of electron donors such as organic matter and pyrite. For example, fluvial deposits may have heterogeneously distributed lignite beds or carbonaceous clays, which provide microsites of enhanced microbial heterotrophy. This results in redox microzonation, with E_h progressively decreasing toward organic rich particles (Fig. 8.3; Murphy et al., 1992). Redox zonation may also develop at contacts between aquifer sands and fine-grained, organic-rich sediments. Impermeability limits the availability of electron acceptors in these fine-grained sediments, allowing excess production of organic acids through fermentation reactions (McMahon and Chapelle, 1991; Routh et al., 2001). Diffusive flux of these organic acids provides electron donors for heterotrophs in aquifers (McMahon and Chapelle, 1991).

The composition of the organic substrate undergoing degradation may also influence redox zonation. In laboratory incubations of aquifer sediments, von Gunten and Zobrist (1993) observed compression of redox zones with increasingly labile substrates. Incubations containing glucose as an electron donor exhibited significantly smaller scales of redox zonation relative to cellulose, a more recalcitrant organic substrate. Thus, the heterogeneous distribution and varied quality of organic matter in aquifer systems will produce a complex pattern of subsurface redox zonation.

3 SULFUR CYCLING IN THE TERRESTRIAL SUBSURFACE

3.1 Geochemical Approaches

Geochemical evidence of sulfur cycling in aquifers is typically based on the concentrations and isotopic compositions of dissolved sulfate, dissolved sulfides, and sulfide minerals. These data are supplemented by pH, dissolved oxygen, dissolved inorganic carbon (DIC), and DIC $^{13}C/^{12}C$ ratios, and are especially useful when collected along a flowpath. Unusually high sulfate contents and low pH (< 5) are evidence of sulfide oxidation. Carbonate minerals will buffer the acid-generating capability of sulfide oxidation and will result in DIC production (Houghton et al., 1984). Downgradient decreases in dissolved oxygen and increases in sedimentary sulfide provide supporting evidence of sulfide oxidation. Geochemical evidence for BSR is more subtle. One line of evidence is a downgradient decrease in dissolved sulfate content (Thorstenson et al., 1979). This alone, however, is inconclusive evidence because dilution with freshwater can produce the same trends. Anoxic (or low E_h), sulfidic water also argues for BSR, but H_2S is commonly undetectable in Fe^{2+}-rich waters because of FeS precipitation.

Stable sulfur isotopes can provide additional evidence of sulfur cycling in natural waters (see Nielsen, 1979; Pearson and Rightmire, 1980; Kaplan, 1983; Krouse and Grinenko, 1991 for reviews). Sulfur occurs as four stable isotopes: ^{32}S (95%), ^{33}S (0.8%), ^{34}S (4.2%), and ^{36}S (0.02%). Small mass differences in these isotopes result in isotopic fractionation, the preferential incorporation of one isotope over another in a particular compound or phase. For convenience, researchers measure and ratio the two most abundant sulfur isotopes, ^{32}S and ^{34}S. The $^{34}S/^{32}S$ ratios are expressed as per mil (‰) deviations from a standard using delta (δ) notation, defined by the equation

$$\delta^{34}S(‰) = \frac{R_{sample} - R_{std}}{R_{std}} \times 1000$$

where R_{sample} and R_{std} are the $^{34}S/^{32}S$ ratios of the sample and standard, respectively. Sulfur isotope ratios are reported relative to troilite (FeS) from the Cañon Diablo meteorite (CDT). Isotopic fractionation is described by the fractionation factors α, where $\alpha = R_A/R_B$, and A and B are different chemical species. For convenience, fractionation is also described by the enrichment factor ϵ, where ϵ (‰) = $(\alpha - 1) \times 1000$.

Under equilibrium conditions, oxidized compounds will be enriched in ^{34}S relative to reduced compounds. In addition, SRB preferentially reduce $^{32}SO_4^{2-}$ relative to $^{34}SO_4^{2-}$, resulting in a sulfide-sulfate ^{34}S enrichment factor ($\epsilon_{sulfide-SO4}$) of -19 to -45‰ (Chambers and Trudinger, 1979; Nielsen, 1979; Kaplan, 1983). High absolute $\epsilon_{sulfide-SO4}$ values may reflect amplification of fractionation through reoxidation of sulfide to S° and subsequent disproportionation to sulfide and sulfate (Canfield and Thamdrup, 1994). Experimental results for ^{34}S-fractionation during sulfide oxidation have been variable and contradictory, but generally attest to relatively small fractionation (Chambers and Trudinger, 1978; Thode, 1991).

Sulfate and sulfide mineral precipitation yields little (< 2‰) or no fractionation (Holser and Kaplan, 1966; Nielsen, 1979). Thus, mineral $\delta^{34}S$ values reflect the $\delta^{34}S$ of the dissolved sulfate or sulfide at the time of precipitation. Conversely, sulfate $\delta^{34}S$ values can be used to identify sulfate sources, distinguishing, for example, between dissolution of sulfate minerals and oxidation of sulfide minerals. As a consequence of isotopic fractionation during sulfate reduction, groundwater sulfate $\delta^{34}S$ increases as sulfate reduction progresses. As this process progresses, the $\delta^{34}S$ of dissolved sulfate can become unusually high, exceeding the $\delta^{34}S$ of typical evaporite (10–30‰) and sulfide (mostly -40 to 10‰) minerals.

Oxygen isotope ratios ($^{18}O/^{16}O$ reported at $\delta^{18}O$) in sulfates can also assist in understanding sulfur cycling in aquifer systems. Analogous to sulfur isotopes, SRB preferentially reduce ^{16}O relative to ^{18}O, resulting in a progressive increase in sulfate $\delta^{18}O$ values during BSR (Fritz et al., 1989). These sulfate $\delta^{18}O$ values can be preserved for hundreds to thousands of years because of the slow exchange rate between water and sulfate oxygen (Holt and Kumar, 1991). Oxygen isotopes have also been used to identify the electron acceptor involved in sulfide oxidation (Taylor et al., 1984; van Everdingen and Krouse, 1985). This is possible because in oxidation by molecular oxygen (Eq. 1), most of the oxygen in sulfate is supplied by O_2, whereas in oxidation by Fe^{3+}, oxygen is supplied by water. Laboratory and field studies show that both processes are significant, but oxidation by ferric iron usually dominates (Taylor and Wheeler, 1994).

3.2 Sulfide Oxidation in the Terrestrial Subsurface

Sulfide oxidation is a widespread environmental problem, both in the natural weathering processes that form acid sulfate soils, and in the weathering of mine tailings (e.g., Kittrick et al., 1982; Alpers and Blowes, 1994). Numerous studies have examined these surficial processes; few have examined sulfide oxidation in the terrestrial subsurface, despite the fact that some deep (100–1000 m) aquifers have waters with relatively high oxygen contents (60 μM) (Winograd and Robertson, 1982).

3.2.1 Geochemical Studies

Houghton et al.'s (1984) study of western North Dakota coal mines provides an example of the geochemical aspects of sulfide oxidation. The sulfur chemistry of undisturbed overburden sediments shows the removal of sulfide sulfur, with sedimentary sulfide comprising $< 5\%$ of the sulfur present in the oxygenated zone (< 5 m) compared with $> 50\%$ in the deeper, oxygen-free zone. In contrast, the distribution of sulfate sulfur is completely opposite. It dominates in the shallow interval and is overshadowed by sulfide sulfur in the deeper interval. Groundwater chemistry also shows distinctive trends. Sulfate concentrations and $\delta^{34}S$ values are < 1000 mg L^{-1} (< 10.4 mM) and about -5% at an undisturbed site, changing progressively to 3700 mg L^{-1} (38.5 mM) and -17% at the disturbed mine site. $\delta^{18}O$ values indicate that the sulfate in the undisturbed regions originates predominantly from sulfide oxidation by Fe^{3+} (eq. 2), whereas sulfate from the disturbed site is produced mostly from sulfide oxidation by O_2 (Eq. 1).

The impact of sulfide oxidation is also observed in natural subsurface sediments in the uplands coastal plain in Texas (Ulrich et al., 1998). In this case, the lack of carbonate minerals in these sands, silts, and mudstones allows pH values to show the impact of sulfide oxidation. As with the North Dakota study, the maximum impact of sulfide oxidation occurs at a depth of about 5 m, with sediment pH values less than 4 and sulfate contents greater than 50 μmol gdw^{-1} (Figs. 8.4A and 8.4B). Furthermore, sulfate from this 5-m interval has about the same $\delta^{34}S$ values as cooccurring sulfide (Fig. 8.4D). Lastly, pyritic sulfur contents are low and ferric iron contents

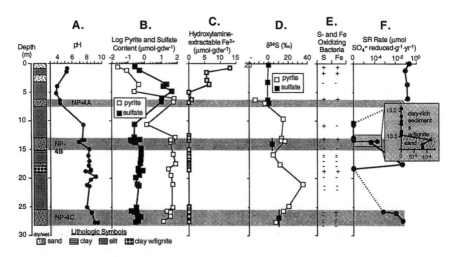

FIGURE 8.4   Sedimentology, microbiology, and sediment geochemistry of the NP-4 borehole in Eocene sediments from the uplands coastal plain in Texas. Stippled bands show screened intervals of wells (from Ulrich et al., 1998)

(hydroxylamine-extractable) are high above 5 m, reflecting the oxidation of pyritic iron (Figs. 8.4B and 8.4C).

Geochemical studies of sulfide oxidation in the deep subsurface are sparse, probably reflecting the small impact of this process away from the O_2-rich atmosphere of the unsaturated zone. From Eq. (1), water saturated with O_2 can produce a maximum of $0.25\,mM$ of sulfate. In contrast, waters impacted by shallow sulfide weathering can have sulfate concentrations exceeding 100 times that value (Houghton et al., 1984; Ulrich et al., 1998). Nevertheless, Fe- and S-oxidizers may still be present in some aquifers systems, enhancing microbial diversity.

3.2.2 *Microbiological Studies*

Researchers have recently begun to appreciate the importance of subsurface microorganisms in sulfide oxidation, leading to studies of the distribution and activity of iron- and sulfur-oxidizing bacteria in these environments. Hirsch and Rades-Rohkohl (1988) detected thousands of "iron-precipitating" and sulfur-oxidizing bacteria per gram dry weight (gdw^{-1}) at 10-m depth in sandy glacial sediments in Germany. Fredrickson et al. (1989) showed sulfur oxidizers at abundances of $100–1000/g$ in sediments (0–259 m) from the coastal plain of South Carolina. In the two boreholes yielding the greatest numbers of sulfur oxidizers, abundance correlated negatively with percent clay, suggesting that aquifer sands are the preferred environment for sulfur oxidation. This is not surprising, considering the importance of surface-derived oxygen in sulfide oxidation. E_h values in these sediments are mostly above 300 mV, and reduced sulfur species in pore waters are low in concentration or absent. SRB occur in low numbers, presumably living in anaerobic microenvironments (Jones et al., 1989). They may provide dissolved sulfide for sulfur oxidizers (Fredrickson et al., 1989).

The microbiology of the shallow aquifer system in the Texas coastal plain (Ulrich et al., 1998) mirrors the geochemistry discussed above. Sulfur- and iron-oxidizing bacteria are common in shallow, aerobic sediments (< 7 m) and scarce in deeper sediments (Fig. 8.4E). Similar results were obtained with DNA amplification of primers specific for *T. ferrooxidans* and *T. thiooxidans* (Martino, 1998). Based on denaturing gradient gel electrophoresis (DGGE) profiles of sulfur-oxidizing enrichments, the dominant microorganisms are closely related to *T. ferrooxidans and* T. intermedius. Amendment of sediment slurries indicates that these microorganisms can grow using elemental sulfur and thiosulfate, but not pyrite (Ulrich et al., 1998).

Clearly, Fe- and S-oxidizing microorganisms play a significant role in sulfide oxidation. This process "pumps" sulfate into the terrestrial subsurface, for use by SRB.

3.3 Sulfate Reduction in Aquifer Systems

It has long been recognized that microbial sulfate reduction occurs in the deep terrestrial subsurface. Studies of oil field brines in the 1920s revealed the occurrence of SRB in sulfide-bearing formation waters (ZoBell, 1958; Chapelle, 1993). Dissolved sulfide is also a common constituent in groundwater, attesting to the widespread influence of SRB in aquifer systems. In most occurrences, sulfides negatively affect groundwater quality, even at nontoxic abundances. Hydrogen sulfide is particularly problematic, producing detectable odors at concentrations of $0.03 \, mg \, L^{-1}$. Owing to water quality concerns, the distribution and origin of groundwater sulfides have been the focus of subsurface investigations (e.g., Thorstenson et al., 1979; Dockins et al., 1980).

3.3.1 Geochemical Evidence

The geochemical impact of sulfate reduction in a local flow system is exemplified in a study of a forested recharge area in Ontario, Canada (Robertson and Schiff, 1994). Robertson and Schiff found that dissolved sulfate contents decrease abruptly from values as high as $0.3 \, mM$ above 10–15 m depth, to values $< 0.01 \, mM$ below (Fig. 8.5A). Furthermore, the sulfate decrease corresponds to anoxic conditions and a faint odor of H_2S. Sulfur isotopes also provide strong evidence for sulfate reduction. Sulfate $\delta^{34}S$ values increase downward from $8 \pm 1‰$ in the recharge area to as high as 46‰ in the sulfate-depleted zone (Figs. 8.5B and 8.6A). If this $\delta^{34}S$ increase represents closed-system BSR with a constant $\epsilon_{sulfide-SO4}$, plotting $\delta^{34}S_{SO4}$ versus the fraction of sulfate remaining (from the original pool) should yield a straight line. As seen in Fig. 8.6B, this is the case for the Ontario site. The slope of the line suggests an $\epsilon_{sulfide-SO4}$ of −15.5‰. A similar approach by Strebel et al. (1990) yielded an $\epsilon_{sulfide-SO4}$ of −9.7‰ and an $\epsilon_{HCO3-SO4}$ for ^{18}O-fractionation of about −6‰.

Comprehensive studies of regional aquifers provide additional characterization of BSR in aquifer systems. Thorstenson et al. (1979) used geochemical trends to argue for sulfate reduction in the Fox Hills—Hell Creek Aquifer (North and South

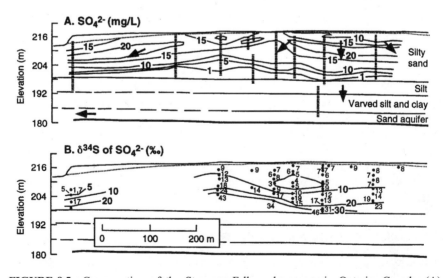

FIGURE 8.5 Cross-section of the Sturgeon Falls recharge are in Ontario, Canada. (A) Sulfate concentrations in $mg \, L^{-1}$. (B) Sulfur isotopic compositions of sulfate in ‰ versus CDT (modified from Robertson and Schiff, 1994).

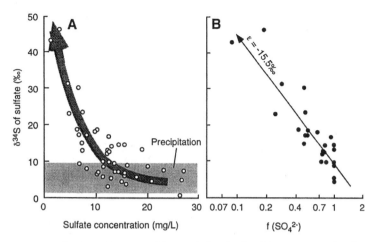

FIGURE 8.6 Sulfur isotopic composition and dissolved sulfate concentration for ground-waters from the Sturgeon Falls recharge area in Ontario, Canada. Arrows point in the direction of the trend with continued BSR. (A) Sulfate δ^{34}S value versus sulfate concentration. Stippled area denotes δ^{34}S range for meteoric precipitation. (B) Sulfate δ^{34}S value versus fraction of sulfate remaining (f) for the sulfate reducing zone. Regression line is δ^{34}S $= 9.6 - 15.5 \ln f$ (modified from Robertson and Schiff, 1994).

Dakota). Dissolved sulfate contents decrease from > 2 *mM* to < 0.1 *mM* within 50 km from the recharge area. Groundwaters from the Chalk Aquifer, Lincolnshire Limestone, and Sherwood Sandstone in Great Britain all show characteristic declines in E_h along the flowpath, with traces of dissolved sulfide in the distil portions of the aquifer (Edmunds et al., 1984). Oddly, the Chalk Aquifer and Sherwood Sandstone *increase* in dissolved sulfate downgradient, suggesting addition of sulfate from the dissolution of sulfate minerals. The sulfate increase, the lack of a decrease in DIC ^{13}C/^{12}C ratios, and low dissolved sulfide contents led Edmunds et al. (1984) to suggest that sulfate reduction is a quantitatively unimportant process in these aquifers. Nevertheless, SRB have been recovered from Chalk Aquifer groundwaters and sediments (Smith et al., 1976; Kimblin and Johnson, 1992).

3.3.2 *SRB Distribution and Abundance*

Researchers in the former USSR have been investigating SRB in the terrestrial subsurface for several decades (e.g., Ivanov, 1961), but only in the last 25 years has subsurface microbiology been the focus of interest in the United States and western Europe (Chapelle, 1993). Sulfate-reducing populations in aquifer systems are highly variable, controlled by the complex interplay of redox potential, temperature, salinity, substrate availability, electron acceptor concentrations, and medium porosity. The potential of microorganisms in bioremediation led to detailed microbiological studies of the siliciclastic sediments (sands, silts, clays) of the Atlantic coastal plain by the U.S. Geological Survey and the U.S. Department of Energy. Chapelle et al. (1987) showed that BSR was common in these sediments, finding activity in 7 of 19 cores recovered from 14- to 182-m depth in Maryland. In a study of deeper (≤ 300 m) Atlantic coastal plain sediments, Jones et al. (1989) reported SRB abundances ranging from < 1 cell g^{-1} sediment to $> 10^5$ cells g^{-1} sediment. They observed an inverse correlation between SRB population and clay content, illustrating the influence of lithology on microbial environment and, consequently, abundance. BSR activities were recorded in 3 of 20 subsurface samples from these Atlantic coastal plain sediments (Shanker et al., 1991). The highest BSR rates occurred at depth (290 m, 416 m) in sediments with high pore-water sulfate contents.

Martino et al. (1998) investigated SRB distribution in a shallow, low-permeability aquifer system in the Texas upland coastal plain. Supporting the results of Jones et al., SRB are abundant (10^5–10^6 gdw^{-1}) within aquifer sands (27- to 31-m depth), and sparse or absent in fine-grained aquitard sediments. Johnson and Wood (1992) also noted that the greatest concentration of heterotrophic bacteria in the Basal Sands of the London Basin was in the sandy sediments, but these sediments also contained the greatest concentration of metabolizable organic carbon.

The occurrence of SRB in indurated sedimentary rocks, environments less porous and hospitable than coastal plain aquifers, has also been noted. Ivanov (1961) found that SRB are widely distributed in porous and fractured limestone, but are less abundant or absent in less porous rock. Krumholz et al. (1997) used a sensitive autoradiography technique to quantify BSR in Cretaceous shales and sandstones in New Mexico. As with coastal plain sediments, SRB activity is mainly limited to sandstones. The highest activity appears to be associated with sandstone–shale boundaries, supporting the hypothesis that fermenting bacteria in organic-rich shales provide electron donors for SRB in sandstones (McMahon and Chapelle, 1991). Krumholz et al. also discovered that addition of ground shale to sandstone-ground-water slurries increased SRB activity in proportion to the amount added, further evidence of the ecological advantages of sand–shale interfaces.

SRB are also found in igneous aquifers. Groundwaters from fractured granitic rocks in Sweden (129–680 m) yield SRB abundances of up to 10^5 cells mL^{-1} (Pedersen and Ekendahl, 1990). These SRB likely account for the > 1 mg L^{-1} concentrations of dissolved sulfide. Groundwaters from basaltic aquifers have also produced SRB (Stevens et al., 1993).

3.3.3 *Rates of BSR*

BSR rates in anaerobic aquifer systems have been quantified by mass balance and flux calculations (e.g., Chapelle and McMahon, 1991; Jakobsen and Postma, 1994), and *in situ* (Jakobsen and Postma, 1994) and microcosm (Phelps et al., 1994; Ulrich et al., 1998) radiotracer experiments. Microcosm experiments often yield rates that are several orders of magnitude greater than the geochemical calculations (Chapelle and Lovley, 1990; Phelps et al., 1994), presumably because of stimulation caused by sampling and slurrying. These results are best used only for examining relative rates between samples. Geochemical rate calculations summarized in Jakobsen and Postma (1994) are based on groundwater chemistry gradients and age data from studies by Thorstenson et al. (1979), Robertson et al. (1989), and Chapelle and McMahon (1991), among others.

BSR rates calculated for aquifer systems in this manner range between 0.7×10^{-5} and 2.3×10^{-2} mmol SO_4^{2-} L^{-1} yr^{-1}.

BSR rates determined from *in situ* radiotracer experiments are closer to those based on geochemical models, presumably because disturbance is minimized. $^{35}SO_4^{2-}$-amended aquifer sediments from Rømø, Denmark yield BSR rates that are at least an order of magnitude greater than the geochemical determinations, with the maximum for each site falling between 0.5×10^{-1} and 4.5 mmol SO_4^{2-} L^{-1} yr^{-1} (Fig. 8.7; Jakobsen and Postma, 1994). It is not yet known how well these rates represent natural processes. The discrepancy between *in situ* experiments and geochemical models, in part, reflects differences of scale. Geochemical models average rates over a wide variety of subsurface environments, whereas *in situ* experiments focus on zones of enhanced BSR activity (Jakobsen and Postma, 1994). Uncertainties notwithstanding, studies clearly show that BSR rates in aquifers are substantially less than those in marine and lacustrine systems. This reflects lower reactivity of organic carbon in most aquifers.

3.3.4 Environmental and Ecologic Influences on BSR

As described in previous sections, the environmental factors and substrate requirements strongly control the distribution and rate of bacterial sulfate reduction in aquifer systems. Jakobsen and Postma (1994) and Schulte et al. (1997) contend that the principal control on BSR activity is the availability of organic carbon as an electron donor, with sulfate concentration of secondary importance. Even when sulfate contents in groundwaters are low, BSR may be supported by diffusion of sulfate from aquitards (Chapelle and McMahon, 1991). On the other hand, both BSR rates in Atlantic coastal plain sediments (Shanker et al., 1991) and SRB numbers in Edwards aquifer groundwaters (Zhang, 1994) correlated positively with sulfate contents. Likewise,

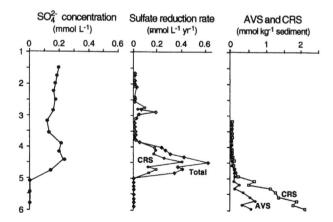

FIGURE 8.7 Sulfate concentration, sulfate reduction rates, acid volatile sulfide (AVS), and chromium-reducible sulfide (CRS) in aquifer sediment from the island of Rømø, Denmark. Note that BSR rates are at a maximum just above the SO_4^{2-}-concentration decline and CRS accumulation (from Jakobsen and Postma, 1994).

BSR activity is high in sulfate-rich sediments associated with sulfide oxidation (see later discussion).

In several observations of BSR in aquifer systems, an association of high-sulfide groundwaters with lignite deposits has been noted (Thorstenson et al., 1979; Dockins et al., 1980). It is likely that lignite constitutes a source of organic electron donors for sulfate reduction, being enzymatically degraded by anaerobic micro-organisms to produce complex, water-soluble humic substances. The role of SRB in this lignite degradation is unclear. Stoner et al. (1993) report that SRB cultures decrease the carbon content of water-soluble, lignite-derived material (suggesting some degree of lignite degradation), but apparently do not significantly depolymer-ize the lignite matrix. SRB can degrade aliphatic and aromatic functional groups (Bak and Widdel, 1986; Aeckersberg et al., 1991), making it likely that they can degrade such structures in the lignite matrix. This contention is supported by laboratory studies demonstrating degradation of coal-tar-derived polyaromatic structures by SRB as the sole carbon source for sulfate reduction (Bedessem et al., 1997).

As with marine sediments, methane has been proposed as a possible electron donor for BSR in aquifer systems (Kelly et al., 1985). Following a natural gas well blowout, Kelly and co-workers observed lower SO_4^{2-} concentrations and higher S^{2-} concentrations in groundwaters receiving CH_4, relative to pristine wells. Comple-mentary laboratory experiments in which groundwaters were amended with CH_4 showed a sulfide concentration increase of eight orders of magnitude, although concentrations remained low. A similar increase in sulfide content was not recorded in controls receiving N_2 rather than CH_4. As with studies of BSR in marine systems, these results are equivocal with regard to anaerobic methane oxidation by SRB. It is conceivable that other microbes, such as methanogens, acted in concert with sulfate reducers to oxidize methane, as postulated for anoxic marine sediments by Hoehler et al. (1994). Alternatively, it is possible that an electron donor other than CH_4 was utilized in the perturbed sediments. Molecular hydrogen, a common natural gas constituent usable by SRB (Voordouw, 1995), could have served as an electron donor. Alternatively, hydrocarbon oxidation by aerobic microorganisms (Widdel, 1989) in the perturbed aquifers may have produced suitable electron donors (e.g., carboxylic acids) that entered the BSR zone through diffusion or mass flow.

An interconnection of sulfate reduction and sulfide oxidation, hypothesized in Fredrickson et al. (1989), is reported in both disturbed and pristine subsurface systems. Fortin et al. (1995) and Fortin and Beveridge (1997) describe the relation-ships between oxidation and reduction of sulfur in mine tailings. Although such deposits artificially introduce reduced sulfur species into an oxic environment, they present a system of microbial sulfur reactions analogous to those in undisturbed subsurface ecosystems. Anoxic conditions arise deep in the tailings through the consumption of oxygen by microbial metabolism and sulfide oxidation reactions (Fortin et al., 1995). Sulfide oxidation by *Thiobacillus* sp. produces SO_4^{2-}. It infiltrates into the deeper anoxic zone of the tailings piles, providing electron acceptors for BSR. In fine-grained deposits such as those described by Fortin and Beveridge (1997), aggregation of tailing particles limits water infiltration and promotes the development of anaerobic microsites within the otherwise oxic zone. Limited SRB populations, apparently of the genera *Desulfovibrio* and *Desulfotomaculum*, are viable in the centers of such microsites, promoting the reduction of sulfate within a predominantly sulfide-oxidizing environment.

As with the inorganic sulfur species, organic electron donors may be cycled between sulfide oxidizers and sulfate reducers in tailings, and by extension in undisturbed systems. Fortin et al. (1995) postulate that lysis of dead and decaying cells of autotrophic sulfide-oxidizing bacteria liberates organic carbon that is used by SRB as an electron donor for metabolism.

The coupling of sulfur cycling processes in undisturbed subsurface systems has received little attention. Ulrich et al. (1998) examined the spatial relationships and interactions between sulfide oxidation and sulfate reduction in Texas uplands coastal plain sediments. Like previous studies of mine tailing deposits, these authors identified viable SRB populations and high BSR activity in oxic, sulfate-rich environments dominated by sulfide oxidation. Sulfide oxidation serves to promote anaerobic conditions and provide sulfate for SRB in the otherwise oxic zone. Figure 8.8 summarizes sulfur cycling in this system. Sulfate-reduction activity exhibits a high degree of heterogeneity throughout the intervals studied. It is largely confined to sands, especially in the vicinity of low-permeability clay and lignite layers (Figs. 8.4F and 8.9). It is likely that these organic-rich layers provide electron donors for SRB (Ulrich et al., 1998; Routh et al., 2001) and low permeability to support the development of anaerobic microsites within a predominantly oxic system (Fig. 8.8).

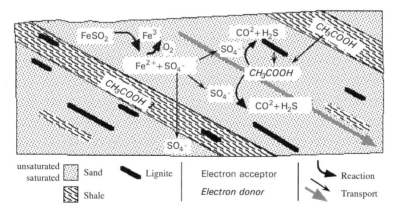

Pyrite Oxidation and Fermentation Reactions
Enhance Sulfate Reduction in the Substance

FIGURE 8.8 General model for microbial sulfur cycling in the terrestrial subsurface. Pyrite oxidation occurs near the water table with ferric iron as the electron acceptor. Iron oxidizers mediate the reoxidation of Fe^{2+}, supplying ferric iron for the abiotic oxidation of pyrite. Sulfate generated by pyrite oxidation is used by SRB in shallow anaerobic microsites and deeper anoxic waters. These SRB oxidize organic acids generated by fermenters in lignitic sediments and in organic-rich clays.

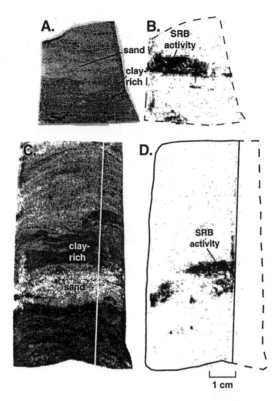

FIGURE 8.9 Core segments (left) and complementary autoradiographic images (right) of sulfate reduction activity in Eocene sediments from the Texas uplands coastal plain (NP-5, 25.8 m; A, B) and (NP-5, 18.6 m; C, D) (from Ulrich et al., 1998).

This system is similar to that described by McMahon and Chapelle (1991) for the Atlantic coastal plain aquifers.

3.3.5 Microbial Impacts on Sediment Chemistry Previous discussion has shown that microorganisms actively cycle sulfur in the terrestrial subsurface and greatly influence *groundwater* chemistry. Their influence on *sediment* chemistry, on the other hand, varies by the process. Microbially enhanced sulfide oxidation has profound effects on sediment. As noted in Fig. 8.4B, Fe- and S-oxidizers can be effective in removing pyrite and adding amorphous iron oxyhydroxides and gypsum. In addition, the low pH values generated by sulfide oxidation dissolve carbonate minerals and alter silicate minerals such as chlorite and smectite (Dixon et al., 1982).

The influence of SRB on sediment chemistry is much more subtle and difficult to demonstrate. Chapelle and McMahon (1991) cite pyrite and calcite cementation at clay–sand contacts as evidence of BSR. This cementation is consistent with their model of enhanced BSR activity at such contacts, where fermenters in shales provide organic acids for SRB in sands (McMahon and Chapelle, 1991). Pyrite $\delta^{34}S$ values

(-36 to $-21‰$) are consistent with the reduction of modern aquifer sulfate, but do not rule out ancient marine sulfate as a source. Furthermore, the $^{13}C/^{12}C$ ratios of calcite cements suggest that only $\sim 11\%$ of the carbon comes from oxidation of organic matter (McMahon et al., 1992). Kimblin and Johnson (1992) attribute framboidal pyrite in fissures in the Chalk Aquifer to recent BSR activity. They noted textural differences between pyrite in fissures versus that in the matrix. In addition, they found SRB in fissured chalk, but not in unfissured chalk. A single-sulfur isotopic analysis of framboidal pyrite is consistent with precipitation in modern waters, but not conclusive. Also, the recent age of the fissures has been vigorously contested (Elliot and Younger, 1994). In a related study, the sulfur isotopic compositions of pyritic sulfur from the Texas Yegua Aquifer were too high ($> 40‰$) to be attributed to modern microbial processes, despite the fact that SRB in the aquifer sediments numbered 10^5 gdw^{-1} (Martino et al., 1998). Results to date are inconclusive, but suggest that pyrite precipitation associated with BSR in aquifer systems (late diagenesis) is minor compared with pyrite precipitation occurring during early diagenesis in the marine or marginal-marine environment. Clearly, the identification and characterization of recent BSR-related mineralization require development of better methods or investigation of simpler, better-constrained aquifer systems.

3.4 Sulfate Reduction in Deep Subsurface Systems

Sulfate reduction in deep formation waters is attributed to both biotic and abiotic processes. Abiotic or thermochemical sulfate reduction (Orr, 1977) is of concern only in systems exceeding temperature thresholds on the order of $140°C$ (Worden et al., 1995) to $210°C$ (Machel et al., 1995). Below this temperature range, sulfate reduction is principally bacterially mediated. The majority of thermophilic SRB have been isolated from marine geothermal habitats. However, recent work has recognized the role of thermophilic strains in deep subsurface systems.

Thermophilic SRB comprise both eubacteria and archaea. Eubacterial thermophiles include the genera *Desulfotomaculum* and *Desulfovibrio*, with the species *Desulfotomaculum nigrificans*, *Desulfotomaculum* sp. strains T90A and T93B, and *Desulfovibrio thermophilus* having been isolated from deep formation waters in sedimentary basins (Rozanova and Nazina, 1980; Rosnes et al., 1991). Archaean SRB include the species *Archaeoglobus fulgidus*, occurring in volcanic geothermal systems (Rosnes et al., 1991). *Desulfotomaculum* exhibits sporulation in old or thermally stressed cells, permitting its survival under subsurface conditions where growth is not possible.

Temperature, pH, and salinity, as well as the concentrations of electron donors and acceptors, control occurrence of thermophilic SRB in the subsurface. Optimal growth temperatures vary between eubacterial and archaean varieties. Eubacterial SRB characteristic of deep subsurface and geothermal systems exhibit peak growth within the range of $55–70°C$ (Rosnes et al., 1991; Sanchez et al., 1993), whereas active reduction by *Desulfovibrio thermophilus* was reported at $85°C$ (Rozanova and Nazina, 1980). The thermophilic sulfate-reducing archaea *Archaeoglobus fulgidus* is tolerant of still higher temperatures, with optimum growth at 83°C and a maximum growth temperature of 92°C (Rosnes et al., 1991). Acidity optima for thermophilic sulfate reducers are circumneutral, with *Desulfotomaculum* sp. growing within a pH range of 6.5–7.5 (Rosnes et al., 1991). A wide range of formation water salinities is tolerated by thermophilic SRB, with *Desulfotomaculum* growth in the laboratory over the entire range of 0–1200 mg L^{-1} NaCl (Rosnes et al., 1991). *In situ* growth and metabolism of SRB have been reported over a total dissolved solids range of 11,000–18,000 mg L^{-1} (Rozanova and Nazina, 1980).

Sulfate-reducing thermophiles exhibit a considerable degree of versatility with regard to electron donor utilization (Nazina et al., 1995). In laboratory enrichment experiments, Rosnes et al. (1991) obtained growth of thermophilic *Desulfotomaculum* species on media containing butanol, propanol, acetate, lactate, butyrate, and a valerate-caproate-caprylate mixture. These strains were capable of growth in the absence of sulfate, relying on the fermentation of pyruvate.

The environmental and substrate requirements for thermophilic SRB are frequently met in mixing zones between waters of disparate chemistry. Frequently, the injection of freshwater or seawater into petroleum reservoirs during hydrocarbon recovery stimulates bacterial sulfate reduction. Such is the case in the Mykhpay and Taliskoe oil fields of Siberia, where seawater injection cools the reservoir and supplies sulfate and organic acid electron donors, permitting the growth of mesophilic SRB (Nazina et al., 1995). BSR by thermophilic eubacteria is enhanced in the Russian Uzen oil field, where injection of heated (60–85°C) sulfate-rich (3 g L^{-1} SO$_4^{2-}$) seawater provides the necessary temperature, substrate, and electron acceptor requirements for active growth. The lower salinities developed in such mixing zones may also serve to promote sulfate reduction in the subsurface, as both mesophilic and thermophilic SRB exhibit improved growth at relatively low salinities (Rosnes et al., 1991). Although descriptions of *in situ* stimulation of BSR are lacking, it is conceivable that mixing low-salinity meteoric groundwaters with sedimentary basin fluids may enhance this process. Circumstantial evidence for such an enhancement has been reported in the Alberta Basin, in which areas of H$_2$S-rich formation waters appear to correlate with a mixing zone of meteoric groundwater and saline formation waters along the eastern basin margin (Cody and Hutcheon, 1994).

SRB and sulfate reduction in deep formation waters have been quantified by Nazina et al. (1995). In the mixing zone between injection and formation waters, thermophilic SRB populations of 10^5 cells mL^{-1} were reported. Similar numbers of sulfate reducers were reported in the Talinskoe and Mykhpay oil fields, with a range of 10^4–10^5 cells mL^{-1} in the waters surrounding injection wells. Sulfate-reduction rates vary by three orders of magnitude in these petroleum reservoirs. In the Mykhpay field, a maximum mesophilic sulfide production rate of 1.2 µg S^{2-} L^{-1} day^{-1} was observed, contrasting with a maximum rate of 1.8 × 10^{-3} µg S^{2-} L^{-1} day^{-1} in the Talinskoe oil field. In the vicinity of producing oil wells in the Mykhpay oil field, sulfate-reduction rates are one to three orders of magnitude lower than those in waters in the injection water zone, with reported rates of 0.001–0.48 µg S^{2-} L^{-1} day^{-1} (Nazina et al., 1995).

4 FUTURE DIRECTIONS

In the past two decades, understanding of the microbial sulfur cycle in the terrestrial subsurface has greatly advanced. Geochemical models incorporate a better appreciation of the role aquitards play in controlling microbial activity in aquifers, and of the interrelationship between microorganisms in different lithologies. Microbiological studies reveal the fine-scale heterogeneity of microbial activity and lithologic dependence on a variety of scales. Present and future studies will make wider use of molecular methods to characterize microbial populations on increasingly finer scales. Factors controlling the spatial distribution of BSR activity will be better defined through the broader application of *in situ* radiotracer techniques. Efforts will be made to characterize the fermenting population in aquitards and to explain the absence of SRB in such environments. Research will explore the role of sulfur disproportionation in microbial sulfur cycling in the subsurface. Finally, continued collaboration between microbiologists and geologists will quantify the role of BSR in sulfide mineralization within aquifer systems. These efforts will lead to better prediction of microbial distributions and activities in the terrestrial subsurface, and to more accurate models for the fate and transport of organic contaminants and toxic metals in our aquifers.

REFERENCES

Aeckersberg F, Bak F, and Widdel F (1991) Anaerobic oxidation of saturated hydrocarbons to CO_2 by a new type of sulfate-reducing bacterium. *Arch Microbiol 156*:5–14.

Alpers CN and Blowes DW (eds) (1994) *Environmental Geochemistry of Sulfide Oxidation.* Washington, DC: American Chemical Society.

Bak F (1993) Fermentation of inorganic sulfur compounds by sulfate-reducing bacteria. In Guerrero R and Pedrós-Alió C (eds): *Trends in Microbial Ecology.* Barcelona, Spain: Spanish Society for Microbiology, pp 75–78.

Bak F and Pfennig N (1987) Chemolithotrophic growth of *Desulfovibrio sulfodismutans* sp. nov. by disproportionation of inorganic sulfur compounds. *Arch Microbiol 147*:184–189.

Bak F and Widdel F (1986) Anaerobic degradation of phenol and phenol derivatives by *Desulfobacterium phenolicum* sp. nov. *Arch Microbiol 146*:177–180.

Bedessem ME, Swoboda-Colberg NG, and Colberg PJS (1997) Naphthalene mineralization coupled to sulfate reduction in aquifer-derived sediments. *FEMS Microbiol Lett 152*:213–218.

Beeman RE and Suflita JM (1990) Environmental factors influencing methanogenesis in a shallow anoxic aquifer: A field and laboratory study. *J Indus Microbiol 5*:45–58.

Benda I (1957) Mikrobiologische untersuchungen über das auftreten von schwefelwasserstoff in den anaeroben zonen des hockmoores. *Arch Mikrobiol 27*:337–374.

Berner RA (1980) *Early Diagenesis: A Theoretical Approach.* Princeton, NJ: Princeton University Press.

Berner RA (1987) Models for carbon and sulfur cycles and atmospheric oxygen: Application to Paleozoic geologic history. *Amer J Sci 287*:177–196.

Birnbaum SJ and Wireman JW (1984) Bacterial sulfate reduction and pH: Implications for early diagenesis. *Chem Geol 43*:143–149.

Blowes DW, Lortie L, Gould WD, Jambor JL, and Hanton-Fong CJ (1998) Geochemical, mineralogical, and microbiological characterization of a sulphide-bearing carbonate-rich gold-mine tailing impoundment, Joutel, Québec. *Appl Geochem 13*:687–705.

Blowes DW, Al T, Lortie L, Gould WD, and Jambor JL (1995) Microbiological, chemical, and mineralogical characterization of the Kidd Creek Mine Tailings Impoundment, Timmins Area, Ontario. *Geomicrobiol J 13*:13–31.

Böttcher J, Strebel O, and Kölle W (1992) Redox conditions and microbial sulfur reactions in the Fuhrberger Feld sandy aquifer. In Matthess G, Frimmel F, Hirsch P, Schulz HD, and Usdowski H-E (eds): *Progress in Hydrogeochemistry*. Berlin: Springer-Verlag, pp 219–226.

Brierley CL and Brierley JA (1996) Microbiology for the metal mining industry. In Hurst CJ, et al. (eds): *Manual of Environmental Microbiology*, 1st ed. Washington, DC: American Society for Microbiology Press, pp 830–841.

Buckley AN and Woods RW (1987) The surface oxidation of pyrite. *Appl Surf Sci 27*:437–452.

Burns SJ (1998) Carbon isotopic evidence for coupled sulfate reduction-methane oxidation in Amazon Fan sediments. *Geochim Cosmochim Acta 62*:797–804.

Canfield CE and Thamdrup B (1994) The production of ^{34}S-depleted sulfide during bacterial disproportionation of elemental sulfur. *Science 266*:1973–1975.

Chambers LA and Trudinger PA (1979) Microbiological fractionation of stable sulphur isotopes. A review and critique. *Geomicrobiol J 1*:249–293.

Champ DR, Gulens J, and Jackson RE (1979) Oxidation-reduction sequences in ground water flow systems. *Can J Earth Sci 16*:12–23.

Chapelle FH (1993) *Ground-water Microbiology and Geochemistry*. New York: John Wiley & Sons.

Chapelle FH and Kean TM (1985) Hydrogeology, digital solute-transport simulation, and geochemistry of the Lower Cretaceous aquifer system near Baltimore, Maryland. Maryland Geological Survey Report of Investigations no. 43.

Chapelle FH and Lovley DR (1990) Rates of microbial metabolism in deep coastal plain aquifers. *Appl Environ Microbiol 56*:1865–1874.

Chapelle FH and Lovley DR (1992) Competitive exclusion of sulfate reduction by Fe(III)-reducing bacteria: A mechanism for producing discrete zones of high-iron ground water. *Ground Water 30*:29–36.

Chapelle FH and McMahon PB (1991) Geochemistry of dissolved inorganic carbon in a Coastal Plain aquifer. 1. Sulfate from confining beds as an oxidant in microbial CO_2 production. *J Hydrol 127*:85–108.

Chapelle FH, Morris JT, McMahon PB, and Zelibor JL (1988) Bacterial metabolism and the δ^{13}C composition of ground water, Floridan aquifer system, South Carolina. *Geology 16*:117–121.

Chapelle FH, Zelibor JL, Grimes DJ, and Knobel LL (1987) Bacteria in deep coastal plain sediments of Maryland: A possible source of CO_2 to groundwater. *Water Resources Res 23*:1625–1632.

Cody JD and Hutcheon IE (1994) Regional water and gas geochemistry of the Mannville Group and associated horizons, southern Alberta. *Bull Can Petrol Geol 42*:449–464.

Cord-Ruwisch R, Kleinitz W, and Widdel F (1987) Sulfate-reducing bacteria and their activities in oil production. *J Petrol Technol*:97–106.

Cypionka H, Widdel F, and Pfennig N (1985) Survival of sulfate-reducing bacteria after oxygen stress, and growth in sulfate-free oxygen-sulfide gradients. *FEMS Microbiol Ecol 31*:39–45.

Davis SN and DeWiest RJM (1966) *Hydrogeology*. New York: John Wiley & Sons.

Dixon JB, Hossner LR, Senkayi AL, and Egashira K (1982) Mineralogical properties of lignite overburden as they relate to mine spoil reclamation. In Kittrick JA, Fanning DS, and Hossner LR (eds): *Acid Sulfate Weathering*. Madison, WI: Soil Science Society of America Special Publication 10, pp 169–191.

Dockins WS, Olson GJ, McFeters GA, and Turbak SC (1980) Dissimilatory bacterial sulfate reduction in Montana groundwater. *Geomicrobiol J 2*:83–98.

Edmunds WM, Miles DL, and Cook JM (1984) A comparative study of sequential redox processes in three British aquifers. In Eriksson E (ed): *Hydrochemical Balances of Freshwater Systems*. Wallingford: IAHS-AISH Publication No. 150, pp 55–70.

Ehrlich HL (1996) *Geomicrobiology*, 3rd ed. New York: Marcel Dekker.

Elliot T and Younger PL (1994) Recent localised sulphate reduction and pyrite formation in a fissured chalk aquifer—comments. *Chem Geol 14*131–136.

Evangelou VP and Zhang YL (1995) A review: Pyrite oxidation mechanisms and acid mine drainage prevention. *Crit Rev Environ Sci Technol 25*:141–199.

Fortin D and Beveridge TJ (1997) Microbial sulfate reduction within sulfidic mine tailings: Formation of diagenetic Fe sulfides. *Geomicrobiol J 14*:1–21.

Fortin D, Davis B, Southam G, and Beveridge TJ (1995) Biogeochemical phenomena induced by bacteria within sulfidic mine tailings. *J Indus Microbiol 14*:178–185.

Fredrickson JK, Garland TR, Hicks RJ, Thomas JM, Li SW, and McFadden KM (1989) Lithotrophic and heterotrophic bacteria in deep subsurface sediments and their relation to sediment properties. *Geomicrobiol J 7*:53–66.

Fritz P, Basharmal GM, Drimmie RJ, Ibsen J, and Qureshi RM (1989) Oxygen isotope exchange between sulphate and water during bacterial reduction of sulphate. *Chem Geol* (Isotope Geoscience Section) *79*:99–105.

Garrels RM, and Lerman A (1984) Coupling of the sedimentary sulfur and carbon cycles—an improved model. *Amer J Sci 284*:989–1007.

Gebhardt NA, Linder D, and Thauer RK (1983) Anaerobic acetate oxidation to CO_2 by *Desulfobacter postgatei*. 2. Evidence from ^{14}C-labelling studies for the operation of the citric acid cycle. *Arch Microbiol 136*:230–233.

Ghiorse WC (1984) Biology of iron- and manganese-depositing bacteria. *Ann Rev Microbiol 38*:515–550.

Goldhaber MB (1983) Experimental study of metastable sulfur oxyanion formation during pyrite oxidation at pH 6–9 and 30°C. *Amer J Sci 283*:193–217.

Goldhaber MB and Kaplan IR (1974) The sulfur cycle. In Goldberg ED (ed): *The Sea*, Vol 4. New York: John Wiley & Sons, pp 569–655.

Hansen LB, Finster K, Fossing H, and Iversen N (1998) Anaerobic methane oxidation in sulfate depleted sediments: Effects of sulfate and molybdate additions. *Aquatic Microbial Ecol 14*:195–204.

Hardy JA and Hamilton WA (1981) The oxygen tolerance of sulfate-reducing bacteria isolated from North Sea water. *Curr Microbiol 6*:259–262.

Herm JD (1989) Study and interpretation of the chemical characteristics of natural water, 3rd ed. Washington, DC: U.S. Government Printing Office, U.S. Geological Survey Water-Supply Paper 2254.

Hirsch P and Rades-Rohkohl E (1988) Some special problems in the determination of viable counts of groundwater microorganisms. *Microbial Ecol 16*:99–113.

Hoehler TM, Alperin MJ, Albert DB, and Martens CS (1994) Field and laboratory studies of methane oxidation in an anoxic marine sediment: Evidence for a methanogen-sulfate reducer consortium. *Global Biogeochem Cycles 8*:451–463.

Holser WT and Kaplan IR (1966) Isotope geochemistry of sedimentary sulfates. *Chem Geol 1*:939–135.

Holt BD and Kumar R (1991) Oxygen isotope fractionation for understanding the sulphur cycle. In Krouse HR and Grinenko VA (eds): *Stable Isotopes: Natural and Anthropogenic Sulphur in the Environment*. New York: John Wiley & Sons, pp 27–64.

Houghton RL, Koob RD, and Groenewold GH (1984) Sulfur cycle in western North Dakota coal mines. In Hitchon B and Wallick EI (eds): *Proceedings First Canadian/American Conference on Hydrogeology: Practical Applications of Ground Water Geochemistry*. Worthington, OH: National Water Well Association, pp 306–313.

Ivanov MV (1961) Microbiological studies of Carpathian sulfur deposits. *Microbiology* [English trans.] *30*:428–430.

Jakobsen R and Postma D (1994) *In situ* rates of sulfate reduction in an aquifer (Rømø, Denmark) and implications for the reactivity of organic matter. *Geology 22*:1103–1106.

Johnson AC and Wood M (1992) Microbial potential of sandy aquifer material in the London Basin. *Geomicrobiol J 10*:1–13.

Jones RE, Beeman RE, and Suflita JM (1989) Anaerobic metabolic processes in the deep terrestrial subsurface. *Geomicrobiol J 7*:117–130.

Kaplan IR (1983) Stable isotopes of sulfur, nitrogen and deuterium in recent marine environments. In: Arthur MA, Anderson TF, Kaplan IR, Veizer J, and Land LS (eds): *Stable Isotopes in Sedimentary Geology*, SEPM Short Course No 10. Tulsa, OK: Society of Economic Paleontologists and Mineralogists, pp 2-1 to 2-108.

Kelly WR, Matisoff G, and Fisher JB (1985) The effects of a gas well blow out on groundwater chemistry. *Environ Geol Water Sci 7*:205–213.

Kimblin RT and Johnson AC (1992) Recent localised sulphate reduction and pyrite formation in a fissured Chalk aquifer. *Chem Geol 100*:119–127.

Kittrick JA, Fanning DS, and Hossner LR (1982) *Acid Sulfate Weathering*. Madison, WI: Soil Science Society of America Special Publication 10.

Krouse HR and Grinenko VA (1991) *Stable Isotopes: Natural and Anthropogenic Sulphur in the Environment*. New York: John Wiley & Sons.

Krumholz LR, McKinley JP, Ulrich GA, and Suflita JM (1997) Confined subsurface microbial communities in Cretaceous rock. *Nature 386*:64–66.

Lovley DR and Chapelle FH (1995) Deep subsurface microbial processes. *Rev Geophys 33*:365–381.

Lovley DR, Chapelle FH, and Phillips EJP (1990) Fe(III)-reducing bacteria in deeply buried sediments of the Atlantic Coastal Plain. *Geology 18*:954–957.

Lovley DR and Phillips EJP (1994) Novel processes for anaerobic sulfate production from elemental sulfur by sulfate-reducing bacteria. *Appl Environ Microbiol 60*:2394–2399.

Luther GW III (1987) Pyrite oxidation and reduction: Molecular orbital theory considerations. *Geochim Cosmochim Acta 51*:3193–3199.

Luther GW III and Church TM (1992) An overview of the environmental chemistry of sulphur in wetland systems. In Howarth RW, Stewart JWB, and Ivanov MV (eds): *Sulphur Cycling on the Continents, Scope 48*. Chichester: John Wiley & Sons, pp 125–144.

Machel HG, Krouse HR, and Sassen R (1995) Products and distinguishing criteria of bacterial and thermochemical sulfate reduction. *Appl Geochem 10*:373–389.

Martino DP (1998) Geomicrobiology and molecular biology of S- and Fe-oxidizing bacteria in subsurface sediments associated with the Yegua Formation. PhD dissertation, Texas A&M University, College Station.

Martino DP, Grossman EL, Ulrich GA, Burger KC, Schlichenmeyer JL, Suflita JM, and Ammerman JW (1998) Microbial abundance and activity in a low-conductivity aquifer system in East-Central Texas. *Microbial Ecol 35*:224–234.

McMahon PB and Chapelle FH (1991) Microbial production of organic acids in aquitard sediments and its role in aquifer geochemistry. *Nature 349*:233–235.

McMahon PB, Chapelle FH, Falls WF, and Bradley PM (1992) Role of microbial processes in linking sandstone diagenesis with organic-rich clays. *J Sediment Petrol 62*:1–10.

Moses CO and Herman JS (1991) Pyrite oxidation at circumneutral pH. *Geochim Cosmochim Acta 55*:471–482.

Moses CO, Nordstrom DK, Herman JS, and Mills AL (1987) Aqueous pyrite oxidation by dissolved oxygen and by ferric iron. *Geochim Cosmochim Acta 51*:1561–1571.

Murphy EM, Schramke JA, Fredrickson JK, Bledsoe HW, Francis AJ, Sklarew DS, and Linehan JC (1992) The influence of microbial activity and sedimentary organic carbon on the isotope geochemistry of the Middendorf aquifer. *Water Resources Res 28*:723–740.

Nazina TN, Ivanova AE, Borzenkov IA, Belyaev SS, and Ivanov MV (1995) Occurrence and geochemical activity of microorganisms in high-temperature, water-flooded oil fields of Kazakhstan and Western Siberia. *Geomicrobiol J 13*:181–192.

Nielsen H (1979) Sulfur isotopes. In Jäger E and Hunziker J (eds): *Lectures in Isotope Geology*. Berlin: Springer-Verlag, pp 283–312.

Nordstrom DK (1982) Aqueous pyrite oxidation and the consequent formation of secondary iron minerals. In Kittrick JA, Fanning DS, Hossner LR (eds): *Acid Sulfate Weathering*. Madison, WI: Soil Science Society of America Special Publication 10, pp 37–56.

Nordstrom DK and Southam G (1997) Geomicrobiology of sulfide mineral oxidation. In Banfield JF and Nealson KH (eds): *Geomicrobiology: Interactions between Microbes and Minerals*. Washington, DC: Mineralogical Society of America, Vol 35, pp 361–390.

Norris PR (1990) Acidophilic bacteria and their activity in mineral sulfide oxidation. In Ehrlich HL and Brierley CL (eds): *Microbial Mineral Recovery*. New York: McGraw-Hill, pp 3–27.

O'Flaherty V, Mahony T, O'Kennedy R, and Colleran E (1998) Effect of pH on growth kinetics and sulphide toxicity thresholds of a range of methanogenic, syntrophic and sulphate-reducing bacteria. *Proc Biochem 33*:555–569.

Orr WL (1977) Geologic and geochemical controls on the distribution of hydrogen sulfide in natural gas. In Campos R and Goni J (eds): *Advances in Organic Geochemistry*. pp 572–597.

Pearson FJ Jr and Rightmire CT (1980) Sulphur and oxygen isotopes in aqueous sulphur compounds. In Fritz P and Fontes JC (eds): *Handbook of Environmental Isotope Geochemistry, Vol 1: The Terrestrial Environment A*. Amsterdam: Elsevier, pp 227–258.

Pedersen K and Ekendahl S (1990) Distribution and activity of bacteria in deep granitic groundwaters of southeastern Sweden. *Microbial Ecol 20*:37–52.

Phelps TJ, Murphy EM, Pfiffner SM, and White DC (1994) Comparison between geochemical and biological estimates of subsurface microbial activities. *Microbial Ecol 28*:335–349.

Postgate JR (1984) *The Sulphate Reducing Bacteria*, 2nd ed. London: Cambridge University Press.

Reeburgh WS (1980) Anaerobic methane oxidation: Rate depth distributions in Skan Bay sediments. *Earth Planet Sci Lett 47*:345–352.

Robertson WD and Schiff SL (1994) Fractionation of sulphur isotopes during biogenic sulphate reduction below a sandy forested recharge area in south-central Canada. *J Hydrol 158*:123–134.

Robertson WD, Cherry JA, and Schiff SL (1989) Atmospheric sulfur deposition 1950–1985 inferred from sulfate in groundwater. *Water Resources Res 25*:1111–1123.

Rosenfeld WD (1947) Anaerobic methane oxidation of hydrocarbons by sulfate-reducing bacteria. *J Bacteriol 54*:664–665.

Rosnes JT, Torsvik T, and Lein T (1991) Spore-forming thermophilic sulfate-reducing bacteria isolated from North Sea oil field waters. *Appl Environ Microbiol 57*:2302–2307.

Routh J, Grossman EL, Ulrich GA, and Suflita J (2001) Volatile organic acids in the Yegua formation, east-central Texas. *Appl Geochem* v. 16, in press.

Rozanova EP and Nazina TN (1980) Occurrence of thermophilic sulfate-reducing bacteria in oil-bearing strata of Apsheron and Western Siberia. *Microbiology 48*:1113–1117.

Sanchez G, Marin A, and Vierma L (1993) Isolation of thermophilic bacteria from a Venezuelan oil field. In Premuzic E and Woodhead A (eds): *Microbial Enhancement of Oil Recovery—Recent Advances. Developments in Petroleum Science*, Vol 34. Amsterdam: Elsevier, pp 383–389.

Sasaki K, Tsunekawa M, Ohtsuka T, and Konno H (1995) Confirmation of a sulfur-rich layer on pyrite after oxidative dissolution by Fe(III) ions around pH 2. *Geochim Cosmochim Acta 59*:3155–3158.

Schauder R, Eikmanns B, Thauer RK, Widdel F, and Fuchs G (1986) Acetate oxidation to CO_2 in anaerobic bacteria via a novel pathway not involving reactions of the citric acid cycle. *Arch Microbiol 145*:162–172.

Schrenk MO, Edwards KJ, Goodman RM, Hamers RJ, and Banfield JF (1998) Distributions of *Thiobacillus ferrooxidans* and *Leptospirillum ferrooxidans*: Implications for generation of acid mine drainage. *Science 279*:1519–1522.

Schulte U, Strauß H, Bergmann A, and Obermann P (1997) Isotopenverhältnisse der schwefel- und kohlenstoffspezies aus sedimenten und tiefen Grundwässern der Niederrheinischen Bucht. *Grundwasser 3*(97):103–109.

Shanker R, Kaiser J-P, and Bollag J-M (1991) Microbial transformation of heterocyclic molecules in deep subsurface sediments. *Microbial Ecol 22*:305–316.

Singer PC and Stumm W (1970) Acidic mine drainage: The rate-determining step. *Science 167*:1121–1123.

Smith DB, Downing RA, Monkhouse RA, Otlet RL and Pearson FJ (1976) The age of groundwater in the Chalk of the London Basin. *Water Resources Res 12*:392–404.

Southam G and Beveridge TJ (1992) Enumeration of Thiobacilli within pH-neutral and acidic mine tailings and their role in the development of secondary mineral soil. *Appl Environ Microbiol 58*:1904–1912.

Stevens TO, McKinley JP, and Fredrickson JK (1993) Bacteria associated with deep, alkaline anaerobic groundwaters in southeast Washington. *Microbial Ecol 25*:35–50.

Stoner DL, Miller KS, Polman JK, and Wright RB (1993) Modification of organosulfur compounds and water-soluble coal-derived material by anaerobic microorganisms. *Fuel 72*:1651–1656.

Strebel O, Böttcher J, and Fritz P (1990) Use of isotope fractionation of sulfate-sulfur and sulfate-oxygen to assess bacterial desulfurication in a sandy aquifer. *J Hydrol 121*:155–172.

Stumm W and Morgan JJ (1981) *Aquatic Chemistry*, 2nd ed. New York: John Wiley & Sons.

Taylor BE and Wheeler MC (1994) Sulfur- and oxygen-isotope geochemistry of acid mine drainage in the western United States. In Alpers CN and Blowes DW (eds): *Environmental Geochemistry of Sulfide Oxidation*. Washington, DC: American Chemical Society, pp 481–514.

Taylor BE, Wheeler MC, and Nordstrom DK (1984) Isotope composition of sulphate in acid mine drainage as measure of bacterial oxidation. *Nature 308*:538–541.

Thamdrup B, Finster K, Hansen JW, and Bak F (1993) Bacterial disproportionation of elemental sulfur coupled to chemical reduction of iron or manganese. *Appl Environ Microbiol 59*:101–108.

Thode HG (1991) Sulphur isotopes in nature and the environment: An overview. In Krouse HR and Grinenko VA (eds): *Stable Isotopes: Natural and Anthropogenic Sulphur in the Environment*. New York: John Wiley & Sons, pp 1–26.

Thorstenson DC, Fisher DW, and Croft MG (1979) The geochemistry of the Fox Hills—Basal Hell Creek aquifer in southwestern North Dakota and northwestern South Dakota. *Water Resources Res 15*:1479–1498.

Tuhela L, Carlson L, and Tuovinen OH (1997) Biogeochemical transformations of Fe and Mn in oxic groundwater and well water environments. *J Environ Sci Health A32*:407–426.

Turekian KK and Wedepohl KH (1961) Distribution of the elements in some major units of the Earth's crust. *Geol Soc Amer Bull 72*:175–192.

van Everdingen RO and Krouse HR (1985) Isotope composition of sulphates generated by bacterial and abiological oxidation. *Nature 315*:395–396.

von Gunten U and Zobrist J (1993) Biochemical changes in groundwater-infiltration systems: Column studies. *Geochim Cosmochim Acta 57*:3895–3906.

Voordouw G (1995) The genus *Desulfovibrio*: The centennial. *Appl Environ Microbiol 61*:2813–2819.

Ulrich GA, Martino D, Clemence K, Grossman EL, Ammerman JW, and S Sufflita JM (1998) S-cycling in the terrestrial subsurface: Commensal interactions, spatial scales, and microbial heterogeneity. *Microbial Ecol 36*:141–151.

Walsh F and Mitchell R (1972) An acid-tolerant iron-oxidizing *Metallogenium*. *J Gen Microbiol 72*:369–376.

Widdel F (1989) Microbiology and ecology of sulfate and sulfur-reducing bacteria. In Zehnder AJB (ed): *Biology of Anaerobic Microorganisms*. New York: John Wiley & Sons, pp 469–585.

Williamson MA and Rimstidt JD (1992) Correlation between structure and thermodynamic properties of aqueous sulfur species. *Geochim. Cosmochim Acta 56*:3867–3880.

Winograd IJ and Robertson FN (1982) Deep oxygenated groundwater-anomaly or common occurrence. *Science 216*:1227–1230.

Worden RH, Smalley PC, and Oxtoby NH (1995) Gas souring by thermochemical sulfate reduction at 140°C. *Amer Assoc Petrol Geol Bull 79*:854–863.

Zehnder AJB and Brock TD (1979) Methane formation and methane oxidation by methanogenic bacteria. *J Bacteriol 37*:420–432.

Zhang C (1994) Microbial geochemistry of groundwater in deep aquifers, central and east-central Texas. PhD dissertation, Texas A&M University, College Station.

Zhang C, Grossman EL, and Ammerman JW (1998) Factors influencing methane distribution in Texas ground water. *Ground Water 36*:58–66.

ZoBell CE (1958) Ecology of sulfate reducing bacteria. *Producers Monthly 22*:12–29.

9

INTRINSIC BIOREMEDIATION OF ORGANIC SUBSURFACE CONTAMINANTS

EUGENE L. MADSEN

Department of Microbiology, Cornell University, Ithaca, New York

1 INTRODUCTION

1.1 Overview

This chapter presents the latest developments pertinent to microbially mediated destruction of organic contaminants in subsurface environments. Because most of the relevant scientific studies have been restricted to shallow aquifers, the deep subsurface habitat may not be directly addressed in the literature discussed here. Nonetheless, fundamental principles of intrinsic bioremediation apply equally in the deep and shallow subsurface, as well as to virtually all the Earth's habitats.

The chapter begins by (1) defining intrinsic bioremediation; (2) discussing problems of subsurface contamination by organic compounds; (3) describing the subsurface habitat; and (4) placing naturally occurring biodegradation reactions in evolutionary and disciplinary perspectives. Next, intrinsic bioremediation's transition from science to technology is described. The final discussions address recent literature that provides critical evidence for verifying intrinsic bioremediation in the field and intrinsic bioremediation's quantitative future.

Subsurface Microbiology and Biogeochemistry, Edited by James K. Fredrickson and Madilyn Fletcher. ISBN 0-471-31577-X. Copyright 2001 by Wiley-Liss, Inc.

1.2 Definition: Natural Attenuation

Natural attenuation happens in the real world to contaminant compounds in soils, sediments, and aquifers. The U.S. Environmental Protection Agency (U.S. EPA) defines natural attenuation as "a variety of physical, chemical, or biological processes that, under favorable conditions, act without human intervention to reduce the mass, toxicity, mobility, volume or concentration of contaminants in soil or groundwater. These processes include biodegradation, dispersion, dilution, sorption, volatilization, and/or chemical or biological stabilization, transformation, or destruction of contaminants. Monitored natural attenuation is appropriate as a remedial approach only when it can be demonstrated capable of achieving a site's remedial objectives within a time frame that is reasonable compared to that offered by other methods" (U.S. EPA, 1998). These multifaceted attenuation processes occur regardless of human wishes or actions. They are natural and inescapable. However, "natural attenuation" is not well understood because this term is used in many different contexts and conveys different types of information depending on the background and orientation of the audience and speakers. To scientists and engineers, natural attenuation is usually the above, U.S. EPA-defined *series of complex processes* that reduce the concentrations of contaminant compounds. But, to owners of contaminated sites and environmental regulatory officers trying to select the optimum way to comply with EPA cleanup regulations, natural attenuation can be a *remedial action technology* that may or may not be effective in reaching targeted decontamination goals. For individuals concerned with public health, natural attenuation can be a *risk management strategy* that contains contaminants by preventing their migration. And finally, for a skeptical community living adjacent to toxic waste sites, natural attenuation can be seen as a "do-nothing" excuse for avoiding costly engineered measures to clean up a site (for a full discussion of natural attenuation's many meanings, see NRC, 2000).

1.3 Definition: Intrinsic Bioremediation

In 1993, the National Research Council (NRC, 1993) coined the term *intrinsic bioremediation* to mean: "A type of bioremediation that manages the innate capabilities of naturally-occurring microbes to degrade contaminants without taking any engineering steps to enhance the process." Accordingly, in this chapter, intrinsic bioremediation is defined as the microbiological component of natural attenuation. Intrinsic bioremediation (and its mechanistic metabolic basis, biodegradation) of subsurface contaminants is usually the most important of the natural attenuation processes because microbial enzymatic reactions routinely break intramolecular carbon–carbon (and other) bonds. Contaminant compounds are, thus, permanently altered, detoxified, and often converted completely to carbon dioxide. Microbial enzymatic destruction contrasts with other attenuation processes that merely dilute or influence transport behavior of contaminants. It must be recognized, however, that not all biodegradation reactions are beneficial. For instance, during anaerobic dechlorination of perchloroethylene (PCE), this compound (a suspected carcinogen) passes through the often stable intermediate, vinyl chloride (VC), a proven carcinogen (Bouwer, 1994; McCarty, 1997; U.S. EPA, 1998).

2   THE PROBLEM:˷SUBSURFACE CONTAMINANTS

Approximately 8 million synthetic and naturally occurring organic compounds have been widely disseminated in the nineteenth and twentieth centuries (Swoboda-Colberg, 1995; Lenhard et al., 1995; Wackett, 1996). Many of these compounds play a vital role in a variety of industrial processes, including production of fuels, solvents, pesticides, and food additives. Furthermore, a broad array of wastes have been generated by commercial, industrial, and military activities. Table 9.1 lists many classes of organic chemicals that have contaminated the subsurface, along with examples of contaminants belonging to each class, some of the common industrial sources or applications, mechanisms of microbe–contaminant interaction, types of contaminant alteration, susceptibility to microbial transformation, and status of intrinsic bioremediation technology. The table is organized by chemical classes because predictable types of abiotic and biotic reactions are expected within each class under particular subsurface conditions. It is important to note that Table 9.1 is not an exhaustive survey of environmental contaminants—only some of the most common are listed. Furthermore, regulatory agencies such as the U.S. EPA may occasionally overlook compounds that warrant closer scrutiny.

Compounds that are most susceptible to microbial metabolism occur naturally, have a simple molecular structure, are water-soluble, exhibit no sorptive tendencies, are nontoxic, and serve as a growth substrate for aerobic or anaerobic microorganisms. By contrast, compounds that are resistant to microbial metabolism exhibit properties such as a complex molecular structure, low water solubility, strong sorptive interactions, toxicity, or do not support the growth of microorganisms (Spain, 1997). Table 9.1 presents a broad perspective on how chemical and microbiological properties jointly affect prospects for bioremediation. There are exceptions to the classification scheme. However, the trends are clear. Bioremediation treatment technology has been established only for certain classes of petroleum hydrocarbons, whereas the technologies for treating all other classes in Table 9.1 are still emerging [see NRC (2000) for further details]. One class of compounds on the verge of progressing to "established" status in Table 9.1 is chlorinated aliphatics [such as PCE and TCE (trichloroethylene)]. As will be elaborated later in this chapter, hindsight, in combination with understanding the mechanisms of microbe–contaminant interactions and the contaminants' hydrogeological settings, provides a basis for assessing the utility of intrinsic bioremediation technology.

3   THE SUBSURFACE AND INTRINSIC BIOREMEDIATION

3.1   The Subsurface Habitat

As discussed in a variety of previous reviews and books (Amy and Haldeman, 1997; Chapelle, 1993; Fredrickson and Fletcher, 1999; Ghiorse, 1997; Ghiorse and

TABLE 9.1   Categories of Subsurface Contaminants, Their Frequency, Sources, Mechanisms of Microbe–contaminant Interaction, Types of Contaminant Alteration, Susceptibility to Microbiological Transformation, and Status of Intrinsic Bioremediation Technology

Chemical Classes[a]	Example Compounds[a]	Examples of Industrial Sources or Applications	Mechanisms of Microbe–contaminant Interaction[b]	Types of Contaminant Alteration[c]	Susceptibility to Microbiological Transformation[d]	Status of Intrinsic Bioremediation[e]
Organic Hydrocarbons						
Low MW	Benzene, toluene, xylenes, (BTEX), alkanes, gasoline	Crude oil, refined fuels, solvents	CEA	M	A1, A2, AN2	Es
High MW	Polycyclic aromatic hydrocarbons (PAHs), alkanes	Creosote, coal tar, crude oil, dyestuffs	CEA, CoA	M, P, IP	A1, A2, A4, AN4	Em
Oxygenated Hydrocarbons						
Low MW	Alcohols, ketones, esters, ethers, phenols, methyl-tert-butyl ether (MTBE)	Fuel oxygenates, solvents, paints, pesticides, adhesives, pharmaceuticals, fermentation products, detergents	CoA, CoN, CEA	P, M	A2–A5, AN3	Em
Halogenated Aliphatics						
Highly chlorinated	Tetrachloroethylene (PCE), trichloroethylene (TCE), 1,1,1-trichloroethane, carbon tetrachloride	Dry cleaning, degreasing, solvents	EA, CoA, CoN	P, M	A3–5, AN2	Em

Less chlorinated	Dichloroethane, dichloroethene, vinyl chloride	Solvents, pesticides, landfills, biodegradation byproducts	EA, CEA, CEN	P, M	A2, A3, A5, AN2	Em
Halogenated Aromatics						
Highly chlorinated	Pentachlorophenol (PCP), polychlorinated biphenyls (PCBs), polychlorinated dioxins (TCDD)	Wood treatment, insulators, heat exchangers, byproducts of chemical synthesis and combustion	EA, CoN	P, M, IP	A4, AN2	Em
Less chlorinated	Dichlorobenzene, PCBs	Solvents, pesticides	EA, CEA	P, M	A2, A1, AN4	Em
Nitroaromatics	Trinitrotoluene (TNT), Dinitrotoluene (DNT)	Explosives, plastics	CoN, CoA, CEA	P, M, IP	A4, AN4, A2	Em

[a] Compound abbreviations: MW=molecular weight; TCDD=2,3,7,8-tetrachlorodibenzo-p-dioxin; RDX=1,3,5-trinitrohexahydro-s-triazine.

[b] CEA=carbon and energy source under aerobic conditions; CEN=carbon and energy source under anaerobic conditions; EA=electron acceptor under anaerobic conditions; CoN=cometabolized under anaerobic conditions; CoA=cometabolized under aerobic conditions.

[c] M=mineralized to CO_2; P=partial alteration of carbon skeleton; IP = immobilized by polymerization or soil binding reactions.

[d] The alphanumeric entries for each compound provide a rating of susceptibility to microbial transformation (1–6 see below) under aerobic (A), anaerobic (AN) conditions. 1=readily mineralizable as growth substrate; 2=biodegradable under narrow range of conditions; 3=metabolized partially when second substrate is present (cometabolized); 4=resistant; 5=insufficient information.

[e] Es=established; Em=emerging.

Wilson, 1988; Madsen and Ghiorse, 1993; Madsen, 1995; Fredrickson and Onstott, 1996; Pedersen, 1993, 1997; White et al., 1998), the subsurface habitat is remarkable in its physical and chemical complexity, its inextricable ties to hydrogeology, its inaccessibility, and its impact on groundwater (hence, drinking-water) supplies. But, perhaps the most significant aspect of the subsurface is that microorganisms are, with few exceptions (e.g., Longly, 1981), the exclusive inhabitants. Microorganisms colonize and adapt to all biosphere habitats, including the terrestrial subsurface, as governed by the available metabolic resources (Madsen, 1998b). Aside from being located beneath the zone of soil formation (pedogenesis), the terrestrial subsurface resists most generalizations. The habitat's structure (geology, stratigraphy, climate, hydrologic regime, hydrologic fluxes, and geochemical characteristics) governs the nutritional resources present in every subsurface location around the globe. The microorganisms in each of these unique subhabitats respond to a variety of nutritional and physiological parameters that include temperature, pH, moisture, availability of energy sources (e.g., organic carbon, CH_4, H_2, NH_3, etc.), final electron acceptors (O_2, NO_3^-, Mn^{4+}, Fe^{3+}, SO_4^{2-}, CO_2), and other nutrients. Some subsurface habitats may be proximate to surface recharge, feature high fluxes of dissolved energy sources (e.g., carbon), be frequently recolonized by soil micro-organisms, and contain metabolically active, growing microbial communities. However, given the depth of the subsurface habitat (e.g., thousands of meters), the prevalence of low-permeable strata, and the immense ages of some geological formations (and groundwaters residing therein), many subsurface habitats are extremely oligotrophic and, perhaps, hostile to microbial life. Under such circumstances, selective pressures on native microbial communities are probably for survival and dormancy, not resource exploitation.

3.2 An Evolutionary Perspective on Intrinsic Bioremediation

In an evolutionary sense, microorganisms (primarily heterotrophic microorganisms) are the recycling agents responsible for biosphere maintenance. In order to survive and proliferate, microorganisms exploit thermodynamically favorable chemical reactions and, in so doing, derive carbon and energy from deceased biomass. As a result of microbial decay processes, essential nutrients (e.g., nitrogen and phosphorous) present in the biomass of one generation of microbiota are available to the next.

Figure 9.1 emphasizes the importance of natural microbial processes in maintaining the biosphere. At the center of Fig. 9.1 is a timeline that, simply but usefully, divides all of history into two periods. Before the twentieth century, naturally occurring biodegradation processes (intrinsic bioremediation) were adequate in recycling the organic materials on the surface of the Earth. A myriad of microbial processes digested different types of biomass derived, directly or indirectly, from photosynthesis, so that organic substances seldom accumulated to cause environmental pollution.* Therefore, few, if any, remedies were needed. In contrast, in the

---

* Of course, under certain circumstances, the microbial precursors to petroleum have accumulated. In some portions of the Earth's crust, tars and oils have formed and later were released into the biosphere. In the timeframe of the scenario described here, however, these accumulations and releases of organic compounds were transient or of minor toxicological impact.

Naturally occurring biodegradation processes in the biosphere

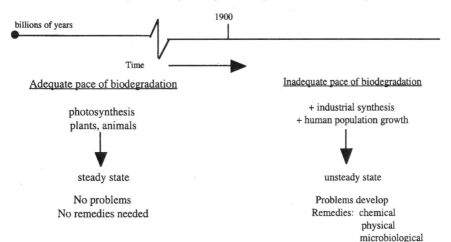

FIGURE 9.1 Historical perspective on the effectiveness of intrinsic bioremediation processes in maintaining the biosphere (from Madsen, 1998a).

twentieth and twenty-first centuries, two major developments have occurred: (1) Human beings developed the ability to synthesize and disseminate large quantities of industrial chemicals, and (2) the human population increased dramatically. The environmental stresses implicit in natural resource exploitation by large human populations, combined with the production and utilization of industrial products, have become apparent today. Our awareness of environmental pollution problems has increased because of concomitant improvements in analytical chemistry, epidemiology, and toxicological expertise. Thus, developments in twentieth-century civilization have caused production and dispersal rates of industrial and other toxins to outpace natural biodegradation processes. An unsteady state has resulted, and threats to human health and ecosystem function have become common in the latter part of the twentieth century (British Medical Association, 1991; Paigen and Goldman, 1987; Johnson, 1997; Sever, 1997).

Globally, the often slow pace of natural biodegradation processes has failed to prevent problems of environmental contamination. But this failure is a human perception, judged within the timeframe of human activities, transactions, and lives. In some localities, contaminants reside in geographic, hydrogeologic, and political/economic settings that allow the pace of these naturally occurring biodegradation reactions to be adequate for human needs. These settings allow humans to be patient. In these instances, intrinsic bioremediation can be an effective, and often preferred, site decontamination process.

3.3 Disciplines Related to Intrinsic Bioremediation and Its Reliability

Figure 9.2 depicts relationships between key disciplines and applications in biodegradation research as they developed from the distant past through the twentieth century. Empirical aspects of biodegradation have been practiced for thousands of years. At the center of the concentric circles of Fig. 9.2 is "empiricism, the distant past." The center represents microbial process management practices, developed by peoples worldwide, that have ranged from food fermentations to retting of flax to recover its fiber to preparation of silage in agriculture to sewage management (Postgate, 1992). With the advent of sophisticated chemical and biological inquiry in the nineteenth and twentieth centuries, the microbiological foundations of many sanitation and biocatalysis practices were established as knowledge of physiology, biochemistry, molecular biology, and genetic engineering (the four rings in Fig. 9.2) gradually accrued. In addition to time, three arrows (each representing applied disciplines) are superimposed on the concentric rings of Fig. 9.2. These applied disciplines are civil and environmental engineering (essential for the development of sanitation and sewage treatment technologies), biocatalysis (essential for modifying many materials in industrially valuable ways), and intrinsic bioremediation. Only the latter falls within the scope of this chapter.

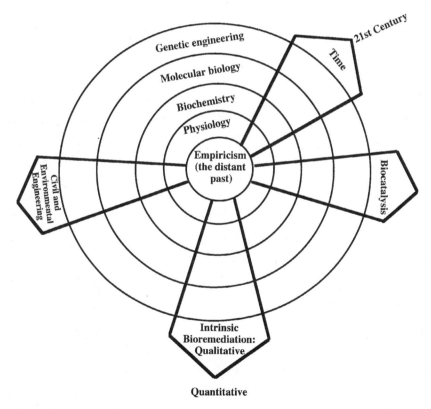

FIGURE 9.2 Historical relationships between fundamental disciplines of biodegradation (concentric circles) and applied disciplines (civil and environmental engineering, biocatalysist, and intrinsic bioremediation).

Intrinsic bioremediation technology is based on discoveries in microbial metabolism of environmental pollutants, made largely in laboratory experiments since the early part of this century (Alexander, 1981, 1999; Gibson, 1984; Dagley, 1977; Madsen, 1991; Young and Cerniglia, 1995). Laboratory-based studies, often using individual microbial cultures, constitute the essential scientific foundation for physiological, biochemical, and molecular biological knowledge of biodegradation reactions. But laboratory biodegradation experiments, even when employing recently gathered environmental samples, can be ecologically irrelevant, revealing what may be, but not necessarily what actually is, occurring in field sites. The details of problems that can arise from laboratory-imposed incubation of environmental samples, and extrapolating from laboratory data to the field, are discussed briefly below; they have been extensively discussed previously (Brockman et al., 1998; Madsen, 1991, 1996, 1998b; Phelps et al., 1994).

As pointed out in the above discussion of the evolutionary role of microorganisms in the biosphere, microbial processes have served and continue to serve both natural habitats and engineering systems (e.g., sewage treatment plants). In both, it must be recognized that temporary nutrient deficiencies or imbalances in interactions between populations within microbial communities may cause metabolic inefficiencies, or lapses in the reliability of microbial processes. For most microbially mediated processes (e.g., decaying crop residue in agricultural fields, recycling of leaf litter in forest soils, fermenting biomass in agricultural silage, nitrogen and phosphorus removal by municipal sewage treatment plants), inefficiencies caused by system imbalance are of little consequence. However, imbalances and/or failure of microbial systems carrying out biodegradation processes may have dire consequences: When toxic environmental pollutants migrate freely, they may jeopardize human health or ecosystem function. For this reason, intrinsic bioremediation is making the transition from a qualitative to a quantitative foundation (Fig. 9.2; see discussion below).

4 EVIDENCE FOR VERIFYING INTRINSIC BIOREMEDIATION: PAST AND FUTURE DEVELOPMENTS IN INTRINSIC BIOREMEDIATION AND NATURAL ATTENUATION

In 1993, the National Research Council (NRC) published the book, "*In situ* Bioremediation: When Does it Work?". It provided three criteria for documenting microbial metabolism of organic contaminants *in situ* in the subsurface: (1) documented loss of contaminants from the site, (2) laboratory assays showing that microorganisms from site samples have the *potential* to transform the contaminants under the expected site conditions, and (3) one or more pieces of information showing that the biodegradation potential is *actually realized* in the field.

It is important to realize that the third criterion, field verification of biodegradation, is a site- and contaminant-specific measure crucial for developing confidence in intrinsic bioremediation. Without this, the technology is based partially on faith, rather than fact. The need for assurances that intrinsic bioremediation can reliably safeguard public and ecological health is driving the technology toward quantitative measures, sustainability, and contingency plans (in case of system failure).

4.1 Over-acceptance of Natural Attenuation and a Proliferation of Government and Industrial Protocols

Despite the fact that Raymond and co-workers devised and implemented engineered, hydrologically controlled, *in situ* bioremediation strategies for subsurface hydro-carbons in the 1970s (Raymond et al., 1976, 1977), nearly two decades passed before intellectual and technological developments verified that microorganisms in subsurface habitats were, in fact, carrying out the expected biodegradation processes (Heitzer and Sayler, 1993; McDonald and Rittmann, 1993; Madsen, 1991; Madsen et al., 1991; Salanitro, 1993; Shannon and Unterman, 1993). The 1990s was a decade in which both the technological and regulatory communities were receptive to passive microbiological solutions to groundwater cleanup. As testimony to the attention directed to *in situ* remediation technologies (both passive and engineered, featuring biotic and abiotic processes), witness the five highly influential and ever-growing International *In situ* and On Site-Bioremediation Symposia sponsored by the Battelle Memorial Institute, U.S. DOE, U.S. EPA, and others (e.g., Hinchee et al., 1995).

During the 1990s, intrinsic bioremediation gained widespread acceptance as a remedial action for cleaning up contaminated sites. Evidence of its popularity is provided by U.S. EPA records showing that, in 1995, natural attenuation was the leading remedial action (47%) for cleaning up groundwater contaminated by leaking underground fuel storage tanks. Furthermore, in the same year natural attenuation was the second leading remedy (28%) for soil contaminated by fuel stored underground (U.S. EPA, 1997a).

Although it was somewhat gratifying for environmental microbiologists to witness society's recognition of microbial processes, concern arose that permits granted for intrinsic bioremediation were sometimes inappropriate (NRC, 2000). Microbial processes are remarkable and have served the biosphere well, but they are neither infallible nor respectful of anthropocentric timeframes.

Several factors contributed to the perhaps overzealous acceptance of intrinsic bioremediation. A growing number of empirical examples, especially of BTEX compounds from underground storage tanks in the shallow aerobic subsurface, indicated that plumes of fuel hydrocarbons routinely stopped advancing and later receded (Rice et al., 1995). Another factor was the realization that pump-and-treat technology, while effective at contaminant containment, was ineffective at accel-erating cleanup (Mackay and Cherry, 1989). In addition, intrinsic bioremediation is perceived to incur lower costs—although when properly implemented, the monitor-ing and analytical fees for intrinsic bioremediation are substantial. And finally, an inadvertent, yet somewhat insidious relaxation of intrinsic bioremediation quality control standards was caused by the proliferation of many technical protocols that extended but deviated from the original guidelines advanced by the NRC (NRC, 2000). A representative list of 12 such protocols appears in Table 9.2. These

TABLE 9.2 Description of Key Protocols for Natural Attenuation of Subsurface Contaminants

Organization	Title	Authors	Comments	Reference
U.S. Air Force (USAF)	Technical protocol for implementing intrinsic remediation with long-term monitoring for natural attenuation of fuel contamination dissolved in groundwater	Weidemeier, Wilson, Kampbell, Miller, Hansen	An excellent, comprehensive "how to" manual that thoroughly draws on latest scientific principles and practices. Two large volumes (574 pp.) consisting of a primary document plus six appendices with extensive case studies.	USAF (1995)
U.S. Air Force	Technical protocol for evaluating natural attenuation of chlorinated solvents in groundwater	Weidemeier, Swanson, Montoux, Kinzie, Wilson, Kampbell, Hansen, Haas, Chapelle	Clear, thorough presentation of physiological principles and field practices underlying reductive dechlorination of PCE and TCE. (93 pp.)	USAF (1997)
U.S. Navy (USDON)	Technical guidelines for evaluating monitored natural attenuation at naval and Marine Corps facilities	Weidemeier and Chapelle	Covers both petroleum hydrocarbons and chlorinated solvents. Draws heavily on the two U.S. Air Force studies above. (42 pp. plus two case study appendices)	USDON (1998)
U.S. EPA	Use of monitored natural attenuation at Superfund, RCRA corrective action, and underground storage tank sites		An interim draft that clarifies EPA policy for sites under the Office of Solid Waste Emergency Response. (24 pp.)	U.S. EPA (1997b)
U.S. EPA	Technical protocol for evaluating natural attenuation of chlorinated solvents in groundwater	Weidemeier, Swanson, Montoux, Gordon, Wilson, Wilson, Kampbell, Haas, Miller, Hansen, Chapelle	A comprehensive "how to" manual for data collection and analysis to evaluate remediation of water contaminated with a mixture of fuels and chlorinated solvents. (77 pp., plus three appendices totalling 154 pp.)	U.S. EPA (1998)

(continued)

TABLE 9.2 (*continued*)

Organization	Title	Authors	Comments	Reference
American Petroleum Institute (API)	Methods for measuring indicators of intrinsic bioremediation: guidance manual		The focus is on quality control for field measurements. (80 pp.)	API (1997)
Chevron Research	Protocol for monitoring intrinsic bioremediation of hydrocarbons in groundwater	Buscheck, O'Reilly	Succinct restatement of U.S. Air Force protocol for fuel contaminants, emphasizing geochemical parameters. (20 pp.)	Chevron (1995)
Chevron Research	Protocol for monitoring natural attenuation of chlorinated solvents in groundwater	Buscheck, O'Reilly	A succinct restatement of U.S. Air Force protocol for chlorinated solvents, text emphasizes geochemical parameters. (26 pp.)	Chevron (1997)
American Society for Testing Materials (ASTM)	Standard guide for remediation of groundwater by natural attenuation of petroleum release sites		Developed to meet the need for a standardized approach in government and industry for evaluating natural attenuation of fuels. (85 pp.)	ASTM (1977)
Dept. of Energy (DOE)	Site screening and technical guidance for monitored natural attenuation of inorganic and organic materials at DOE sites	Brady, Spalding, Krupka, Waters, Zhang, Barns, Brady	Software package from Sandia National Laboratory that produces "natural attenuation factors" of questionable validity.	U.S. DOE (1998)
Remediation Technology Development Forum (RTDF)	Natural attenuation of chlorinated solvents in groundwater: principles and practices		A question-and-answer format to explain microbial processes and the development of site conceptual models.	RTDF (1997)
Minnesota Pollution Control Agency	Draft guidelines: natural attenuation of chlorinated solvents in groundwater		Captures essential technical points from the U.S. Air Force protocol on attenuation. (24 pp.)	MPCA (1997)

protocols, largely aimed at standardized implementation of "natural attenuation technology," have been published by the U.S. government (Air Force, Navy, EPA, DOE), state government (Minnesota), trade (ASTM, API, RTDF; the latter is now synonymous with the Interstate Technical and Regulatory Work Group) and corporate (Chevron) organizations. One key issue to note in Table 9.2 is that the U.S. Air Force (actually the Air Force Center for Environmental Excellence) spearheaded the two most comprehensive and influential protocols: one for petroleum hydrocarbons (entry #1, Table 9.2) and one for chlorinated solvents (entry #2, Table 9.2). Key authors for these Air Force documents participated in drafting many subsequent documents for other organizations. Furthermore, virtually all the other protocols directly reference the U.S. Air Force documents. Many of the protocols in Table 9.2 build on the prior three NRC criteria for intrinsic bioremediation and adhere to staunch scientific standards. However, other protocols (some authored by industrial or trade organizations, perhaps biased toward eliminating environmental liabilities) may fail to maintain rigorous standards of scientific quality that are essential for protecting human health and ecosystem integrity.

4.2   Linking Field Observations to Biodegradation Processes

Because the active agents in bioremediation are microorganisms, it is essential that proposed cleanup processes make sense to the microorganisms, themselves, in the physiological and thermodynamic context where they reside. Contexts for bioremediation processes range from shallow aerobic sands to clay-rich lacustrine deposits to deep anaerobic poorly sorted paleosols to fractured granite or basalt. Regardless of the particular context, each must be scrutinized as a *habitat for microbial metabolism* where individual cells can develop into populations and complex ecological communities whose fundamental physiological needs for ATP generation, carbon assimilation, terminal electron acceptors, other inorganic and organic nutrients, and dynamic intercellular interactions (competition, synergism, interspecies hydrogen transfer, commensalism, predation, parasitism, etc.) demand constantly improving sets of hypotheses aimed at refining our understanding of both the microorganisms and the processes they effect. Once the types of organic contaminants and local geochemical constraints have been established, and fundamental thermodynamic, nutritional, and ecological bases for the particular metabolic functions are initially conjectured, then a series of hypotheses will naturally unfold that provide a means for documenting the bioremediation process of interest on a site- and case-specific basis.

   Table 9.3 contains four examples of contaminants in subsurface field sites whose physiological contexts dictate how microorganisms can metabolize the offending organic compounds. Knowledge from laboratory-based (using environmental samples, mixed cultures, and/or pure cultures) assays provides the biochemical basis for mechanisms operating in the field. Answers to key questions, such as "Are the contaminants suitable carbon and electron sources?", "Which physiological electron acceptors (O_2, NO_3^-, Fe^{3+}, Mn^{4+}, SO_4^{2-}, CO_2) are required coreactants?", "Are the contaminants, themselves, final electron acceptors?", and "What com-

TABLE 9.3 Examples of Contaminated Sites, Hypothesized Key Bioremediation Processes, and the Corresponding Field and Laboratory Measurements That Allow Site- and Case-specific Verification of Microbiological Destruction of Contaminants

Example Sites	Hypothesized Key Bioremediation Processes	Supportive Field and Laboratory Measurements
Aerobic subsoil contaminated with petroleum products	Heterotrophic microorganisms grow using petroleum components as carbon and energy sources (Atlas and Cerniglia, 1995; Rifai et al., 1995b; Wilson and Jawson, 1995). Metabolism in this context relies on O_2 both in the attack of aliphatic and aromatic compounds, and as a final electron acceptor in respiratory chains.	• Coincident depletion of petroleum components and O_2 in the field. • Corresponding production of CO_2. • High numbers of petroleum-degrading aerobic heterotrophs inside but not outside the contaminated areas. • If petroleum has a distinctive $^{13}C/^{12}C$ ratio, this should be reflected in the CO_2. • Adding nitrogen or phosphorus fertilizer to contaminated zones may relieve nutrient limitation and hence enhance contaminant loss compared to unfertilized controls. • Genes involved in catabolism of petroleum components should be expressed in high abundance inside but not outside the contaminated zone.
Anaerobic aquifer contaminated with perchloroethylene	Dehalorespiring bacteria use chlorinated aliphatic compounds as final electron acceptors (Maymó-Gatell et al., 1997; McCarty, 1997). Dechlorination reactions are governed by: complex microbial and chemical interactions that generate physiological electron sources (especially H_2 gas); the presence of alternative electron acceptors (NO_3^-, Mn^{4+}, Fe^{3+}, SO_4^{2-}, CO_2) in site sediments and waters; and ecological and physiological competition between the microorganisms carrying out the metabolism that links electron donors and acceptors (Bouwer, 1994; Chapelle et al., 1997; Fennel et al., 1997; Yager et al., 1997; Yang and McCarty, 1998).	• Dechlorinated daughter products, TCE, DCE, VC within contaminant plume. • Products of complete detoxification, such as ethene, should be inside and not outside the plume. • Adaptation of site microorganisms to dehalorespiration can be documented by finding dechlorination activity in site samples from inside but not outside the contaminant plume. • Immunological or PCR-based data demonstrating the presence of dehalorespiring enzymes, genes, and characteristic bacteria inside but not outside the plume.

(continued)

TABLE 9.3 (*continued*)

Example Sites	Hypothesized Key Bioremediation Processes	Supportive Field and Laboratory Measurements
Aerobic aquifer contaminated with TCE	TCE destruction is achieved cometabolically by aerobic microorganisms supplied with a primary carbon source. Oxygenase enzymes (involved in metabolizing primary substrates such as methane, propane, toluene, and phenol) fortuitously convert TCE to unstable compounds that spontaneously hydrolyze to nontoxic and/or readily biodegradable components (Hopkins et al., 1993; Pfiffner et al., 1997; Semprini and McCarty, 1992).	• Microcosms prepared with site samples consume TCE only when supplied with both O_2 and the primary substrate. • In recirculating field site waters, the TCE loss is enhanced only when both O_2 and primary substrate are supplied. • Assays for the oxygenase enzymes, genes, and appropriate metabolites (e.g., *trans*-dichloro-ethylene oxide) reveal high abundances inside but not outside the contaminated zone.
Anaerobic aquifer contaminated with jet fuel	Aromatic fuel components, especially toluene, serve as growth substrates for anaerobic microorganisms that utilize sulfate as a final electron acceptor.	• Microcosms containing site sediments incubated under sulfate-reducing conditions produce $^{14}CO_2$ from ^{14}C-labeled toluene and benzene (Chapelle et al., 1996). • Sulfate is depleted along the groundwater flowpath in the field site. • Dissolved inorganic carbon (e.g., CO_2) increases along the flowpath in the field sites (Chapelle et al., 1996). • Contaminant plume has ceased advancing despite a constantly dissolving reservoir of jet fuel (Chapelle et al., 1996). • A solute transport model accounts for dispersion, flow velocity, and adsorption, and produces biodegradation rate estimates that are consistent with microcosm estimates (Chapelle et al., 1996).

peting reactions may slow or prevent the desired biodegradation?", provide a framework that launches a broad array of possible assays that can argue for or against the successful establishment of a given biodegradation process (Madsen, 1997, 1998a). The sites listed in Table 9.3 range from aerobic soil to aerobic and anaerobic aquifers. The assays range from field-based oxygen probes, to measurements of populations of contaminant-degrading bacteria, to laboratory biodegradation assays, to molecular biological assays of DNA and RNA. The next section of this chapter elaborates on the reasoning and detailed analyses required to generate and test hypotheses that allow bioremediation to be verified. Related aspects of bioremediation and biodegradation have been the subject of numerous books (Alexander, 1999; Crawford and Crawford, 1997; Norris et al., 1994; NRC, 1993, 2000; Skipper and Turco, 1995; Young and Cerniglia, 1995).

4.3 Twelve Strategies for Verifying Intrinsic Bioremediation of Organic Contaminants in Subsurface Field Sites

Three sources of uncertainty must be confronted and overcome when demonstrating that microorganisms are the active agents of pollutant loss in bioremediation projects:

1. We must acknowledge that extrapolation from laboratory-based metabolic activity assays to the field is usually unwise because of the propensity of microorganisms in field samples to respond to laboratory-imposed physiological conditions that are unlikely to perfectly match those in the field (Madsen, 1996, 1998b; Phelps et al., 1994).

2. The spatial heterogeneity of field sites may impede or completely prevent trends in the behavior of environmental contaminants from being discerned (Wilson and Jawson, 1995; Chapelle et al., 1996).

3. The action of a multitude of both abiotic and biotic processes may simultaneously contribute to pollutant attenuation (Chapelle et al., 1996; Madsen, 1991).

To contend with these challenges and simultaneously comply with the 1993 NRC guidelines, a variety of strategies have been developed for verifying the success of pollutant-destroying microbial processes in subsurface field sites (Madsen, 2000). These (gathered from the recent literature and codified in Table 9.4) are simple, logical linkages between biodegradation mechanisms and field observations introduced in Section 4.2 and Table 9.3. The strategies that appear in Table 9.4 are firmly based in the physiological principles that distinguish biotic from abiotic contaminant attenuation processes. Four of the strategies involve tracers (internal conservative, added conservative, added radioactive, and added stable isotopic) that either account for or circumvent problems arising from abiotic changes in field concentrations of contaminants and related metabolites. Five of the strategies in Table 9.4 rely on detailed prior knowledge of specific microbiological processes (stable isotope fractionation, detection of intermediary metabolites, metabolic adaptation, *in situ* respiration, and gradients of coreactants and/or products) that are manifest as

TABLE 9.4   Strategies for Obtaining Evidence for Field Expression and Biodegradation Potential (after Madsen, 2000)

Type of Strategy	Principles and Examples	References
Internal conservative tracers	Assess loss of certain compounds relative to the persistence of less biodegradable, but similarly transported, compounds. Examples include using ratios of straight- to branched-chain alkanes (C^{17}/pristane, C^{18}/phytane), and ratios of other compounds to vanadium or hopane in crude oil; ratios of lower to higher chlorinated congeners in PCB mixtures; trimethyl benzene can serve as a conservative tracer in BTEX plumes; ratios of nonchlorinated to chlorinated aromatics in mixed solvents; selective metabolism of one stereoisomer of particular pesticides (e.g., α-chlorocyclohexane).	Bragg et al. (1994), Harkness et al. (1993), Jackson et al. (1996), Kampbell et al. (1996), Ludwig et al. (1992), Prince et al. (1994), Pritchard and Costa (1991), Sasaki et al. (1998), Tett et al. (1994), Wiedemeier et al. (1995), Williams et al. (1997), Wilson and Jawson (1995), Zipper et al. (1996)
Added conservative tracers	In some field sites, contaminant mixtures may lack internal tracers but be amenable to the addition of materials that provide a baseline measure of various transport processes. Examples include He to assess O_2 loss or CO_2 production in groundwater, propane to assess toluene loss from a stream, and bromide to assess groundwater flow.	Kim et al. (1995), NRC (1993), Robertson (1994)
Added radioactive tracers	In rare instances, regulatory authorities have allowed intentional field release of radioactive (e.g., ^{14}C-labeled) pollutants in field sites. Subsequent recovery of $^{14}CO_2$, ^{14}C metabolites, and ^{14}C parent compounds provide definitive proof of metabolic and other field processes.	Führ et al. (1998), Lee et al. (1985)
Added stable isotopic tracers	Pollutant compounds that are nonradioactive, but isotopically labeled with deuterium or ^{13}C, have been released in field sites. Subsequent stable isotopic analyses of field samples for labeled CO_2, metabolites, and/or the parent compound provide proof of metabolic and other field processes.	Thierrin et al. (1993, 1995)

(continued)

TABLE 9.4 *(continued)*

Type of Strategy	Principles and Examples	References
Stable isotopic fractionation patterns	CO_2 has different $^{13}C/^{12}C$ ratios depending on the $^{13}C/^{12}C$ signature of the substrates respired and the ^{13}C-enriching process of methanogenesis. When site-specific signatures of both inorganic and organic carbon reservoirs have been characterized, the relative contribution of pollutant biodegradation to the pool of CO_2 can be discerned. The radioactive (^{14}C) component of CO_2 is also revealing because petroleum contaminants contain no ^{14}C.	Aggarwal and Hinchee (1991), Conrad et al. (1997), Grossman (1997), Jackson et al. (1996), Landmeyer et al. (1996)
Detection of intermediary metabolites	When sufficient biochemical knowledge of pollutant biodegradation has accrued, particular metabolites can be targeted using a combination of careful sampling and analytical chemistry. Detection of stable (dead end) metabolites and transient metabolites (indicative of "real-time" biodegradation) have been reported. The metabolites include *trans*-dichloroethylene oxide, dihydrodiols of aromatic compounds, DDE, and hydroxylated pesticides.	Beller and Spormann (1997), Beller et al. (1995), Flanagan and May (1993), Lerch et al. (1995), Semprini et al. (1990, 1995), Wilson and Madsen (1996), Wilson et al. (1990)
Microbial metabolic adaptation	Naturally occurring microbial communities that grow in response to pollutant exposure have predictable characteristics relative to adjacent unexposed communities. Adaptation is reflected in laboratory or field measure of pollutant metabolism (qualitative or quantitative); numbers of specific pollutant degraders; and enhanced concentrations of protozoan predators of bacteria inside but not outside contaminant plumes.	Madsen et al. (1991), Yager et al. (1997)

Molecular biological indicators	Based on molecular biological characterization of pure cultures capable of pollutant metabolism, a variety of assays consistent with established genetic sequences and their expression can be devised. These include PCR amplification of structural genes, mRNA extracted from field sites, reverse-transcriptase-PCR detection of mRNAs, nucleic acid sequencing, immunodetection or enzymes and metabolites, and 16S rRNA analysis of the composition of microbial communities.	Brockman (1995), Burlage (1997), Fleming et al. (1993), Ogram et al. (1995), Selvaratnam et al. (1995), Shields and Francesconi (1996), Wilson et al. (1999), Zhou et al. (1997)
Gradients of coreactants and/or products	Ongoing *in situ* metabolism of pollutants consumes physiological final electron acceptors and generates metabolic endproducts that reflect site-specific pollutant metabolism. Chemical gradients in field sites should be apparent using measures that include O_2, NO_3^-, Mn^{4+}, Fe^{3+}, $SO_4^=$, CO_2, NO_2^-, N_2O, Mn^{2+}, Fe^{2+}, CH_4, H_2, pH, and alkalinity.	Barker et al. (1996), Chapelle et al. (1996), Rifai et al. (1995a), Wilson et al. (1990)
In situ rates of respiration	A subset of the entry immediately above that has been effectively applied to engineered bioremediation of subsurface sites involves cessation of an oxygen (or air) sparging regime, followed by insertion of an oxygen probe that documents real-time oxygen consumption. This respiratory activity should be high inside but not outside the contaminated area. The conserved gas, He, can be included in the sparging step to account for diffusional O_2 loss.	Davis et al. (1995), Deyo et al. (1993), Hickey (1995), Li (1995)

(*continued*)

TABLE 9.4 (*continued*)

Type of Strategy	Principles and Examples	References
Mass balances of contaminants, coreactants, and products (total expressed assimilative capacity)	Under well-defined hydrogeologic regimes, fluxes of water, contaminants, and physiological electron donors or acceptors can be quantified in a cross-sectional analysis between site sampling stations. The stoichiometry of all appropriate aerobic, anaerobic, isotopic fractionation, and inorganic equilibria reactions can serve to predict and distinguish biotic from abiotic processes and to identify contributions from a variety of microbiological groups.	Cho et al. (1997), Höhener et al. (1998), Rifai et al. (1995b), Semprini et al. (1995), Williams et al. (1997), Wolfe et al. (1994)
Computer modeling that incorporates transport and reaction stoichiometries of electron donors and acceptors	This approach considers quantitative aspects of fluid flow, dilution, sorption, volatilization, mixing, microbial growth, and metabolic reaction stoichiometries to achieve an integrated and predictive tool for understanding all processes influencing the fate of pollutant compounds. This approach resembles the entry immediately above, but is implemented at a larger scale and uses more sophisticated computations.	Bekins et al. (1993), Chiang et al. (1989), Lang et al. (1997), Rifai et al. (1995b), Semprini and McCarty (1991, 1992), McNab and Dooher (1998)

observable geochemical changes in the field. The molecular biological strategy in Table 9.4 is an elegant, emerging approach that is constantly strengthened by genetic links forged between information from pure cultures and real-world mixed microbial populations. The molecular microbiology linkages are limited by the relatively small database of genetic sequences pertinent to pollutant metabolism, and uncertain metabolic diversity that arises when genes of unrelated lineage may have converged on the same metabolic function. The final two strategies in Table 9.4 (computer modeling and mass balances of contaminants, reactants, and products) attempt to quantitatively account for both transport and metabolic processes within entire field sites or along distinct transects therein.

4.4   Reliability, Robustness, Sustainability, Quantification

All the entries in Table 9.4 represent sound approaches for verifying intrinsic bioremediation—they provide evidence in support of microbial metabolism of pollutants in the field. Yet, all but the final two entries are largely qualitative. To put this in perspective, it is important to recall the precise wording of the U.S. EPA's definition of natural attenuation (Section 1.2). The wording stipulates "favorable conditions" and "achieving a site's remedial objectives." A qualitative demonstration that a particular metabolic process (be it metabolism of benzene, vinyl chloride, or molasses) is necessary, but not sufficient to effect a true "remedy." In order to achieve remedial objectives, the rates and extent of reaction(s) must be adequate. For instance, Yager et al. (1997) documented the occurrence of ethene inside, but not outside, a plume of TCE-contaminated groundwater—but the ethene was present only in trace quantities. Thus, although complete detoxification (dechlorination) of TCE was documented *in situ*, the percent of total TCE converted to nontoxic products was not addressed. This latter quantitative goal is elusive in many field investigations of intrinsic remediation, but it is crucial for establishing the technology as a reliable and effective one.

Detailed understanding of the physiological and biochemical mechanisms by which specific contaminants are metabolized in specific geochemical settings is also crucial for establishing the reliability of intrinsic bioremediation technology. Only if the mechanisms of microbe–contaminant interaction (Table 9.1) are known can insightful hypotheses be formulated in developing conceptual paradigms that link contaminant biodegradation to field observations (see Section 4.2). As noted in Section 2 and Table 9.1, petroleum hydrocarbons is the only class of contaminants for which intrinsic bioremediation is established. The reason is that petroleum hydrocarbons serve as carbon and energy sources for microbial growth. Furthermore, this microbial growth is usually sustainable because a wide range of terminal electron acceptors (e.g., O_2, NO_3^-, $SO_4^=$, Fe^{3+}, Mn^{4+}, CO_2) can be coupled to hydrocarbon oxidation and are present (or replenished by diffusion) in most subsurface environments. Thus, because reactants are unlikely to be limiting, petroleum hydrocarbon metabolism is a reliable process.

Now, let us compare petroleum hydrocarbon metabolism to reductive dechlorination of PCE. The most robust mechanism by which PCE is metabolized is via "dehalorespiration"—PCE is used as a terminal electron acceptor. This process requires both an external electron donor (often a cocontaminant such as petroleum hydrocarbons) and a properly balanced anaerobic microbial food chain that directs electron flow toward PCE [and its daughter products, TCE, DCE (dichloroethylene) and DC], and not toward other potential electron acceptors (Fennell et al., 1997; Yang and McCarty, 1998). Thus, the "favorable conditions" (mentioned above as a stipulation in the U.S. EPA's definition) for intrinsic bioremediation of PCE are found relatively rarely. If and when intrinsic bioremediation of PCE is successful, assessing its sustainability requires careful site characterization to assure that all reactants are both properly distributed and in sufficient supply.

As indicated in Fig. 9.2 and Section 3.3, recent and future advancements in intrinsic bioremediation aspire toward quantification, as well as robustness and sustainability. The final two entries in Table 9.4 provide a means toward that end. In the studies cited, the site characterization procedures were sufficiently thorough to allow formulation of equations that account for masses of contaminants, hydrologic flow, transport parameters, and biodegradation reaction stoichiometries, so that site geochemistry and pollutant metabolism were fully integrated. Although quantification of intrinsic bioremediation is still rather rare (restricted primarily to sites that have received extensive research and funding), quantification is likely to be required in the future; hence, it will become increasingly common.

The final quality control steps that need to be included in all intrinsic bioremediation plans are long-term monitoring and contingencies. These plans are used to monitor the contaminant plume over time and to verify that intrinsic bioremediation (and often other components of natural attenuation) is occurring at rates sufficient to attain site-specific remediation objectives within the timeframe predicted. In addition, the long-term monitoring plan should be designed to evaluate behavior of the plume over time, verify that exposure to contaminants does not occur, verify that natural attenuation byproducts pose no additional risks, determine actual (rather than predicted) attenuation rates for refining predictions of remediation timeframe, and to document when site-specific remediation objectives have been attained. The long-term monitoring plan should be developed based on site characterization data, analysis of potential exposure pathways, and the results of solute fate and transport modeling. It often includes two types of monitoring wells. Long-term monitoring wells are needed to determine if the behavior of the plume is changing. Performance evaluation wells are necessary to confirm that contaminant concentrations meet regulatory acceptance levels, and to trigger engineering procedures to manage potential expansion of the plume, should system failure occur.

5 SUMMARY: INTRINSIC BIOREMEDIATION'S LIMITATIONS AND PROMISE

The biosphere has been well served by microbiological processes (see Section 3.2). However, twentieth century commercial and political systems may not always be. Bioremediation technology may perform suboptimally (by human standards) for many reasons. Microorganisms are capable of partial transformations that actually increase the toxicity of organic compounds (Alexander, 1999). Also, microorganisms may not physiologically respond to particular contaminant resources in specific settings because of intricate mechanisms that control gene expression (e.g., inducers, promoters, catabolite repression) and because enzymes involved in substrate uptake and attack may have low affinity constants. Furthermore, the contaminants, themselves, may present major barriers to efficient cleanup when they are toxic or are present as mixtures of compounds whose physiological requirements for biodegradation are incompatible with one another (for instance, compounds susceptible only to anaerobic metabolism may be mixed with compounds whose attack requires oxygen). In addition, site complexity and heterogeneity may impair (1) characterization of the contaminant problem (which compounds are where?); (2) understanding of ambient physiological processes carried out by native microorganisms; and (3) effectiveness of bioremediation efforts. Moreover, uncertainties regarding contaminant bioavailability have major impacts both on the health risks posed by the contaminants and their susceptibility to microbial metabolism. Also, rates of microbial processes may simply not conform to human expectations. Finally, it must be recognized that the biochemical reactions catalyzed by microorganisms exist in response to evolutionary pressures that include resource exploitation, survival, dormancy, and competition. Given this evolutionary history, it is uncertain that past selective pressures, the legacies of which are recorded in each microbial genome, have shaped microorganisms for performance considered optimal by humans devising bioremediation systems.

Despite the above-mentioned limitations of intrinsic bioremediation, its future as a technology is bright in a wide variety of pollution-control applications. This positive outlook is based on the fact that (1) advances in the diverse disciplines that shape bioremediation (e.g., biochemistry, microbial physiology, molecular biology, genetic engineering, chemical engineering, analytical chemistry, hydrogeology) are accelerating; (2) empirical observations of successful field case studies of intrinsic bioremediation are constantly accruing; and (3) the abilities to quantify the field processes that sustain intrinsic bioremediation are advancing.

Acknowledgments During preparation of this manuscript, the author's laboratory research was supported by NSF, NIEHS, New York State Agricultural Experiment Station, the U.S. EPA, Textron, Inc., and the United States Geological Survey. Expert manuscript preparation by P. Lisk was greatly appreciated.

REFERENCES

Aggarwal PK and Hinchee RE (1991) Monitoring *in situ* biodegradation of hydrocarbons by using stable carbon isotopes. *Environ Sci Technol 25*:1178–1180.

Alexander M (1981) Biodegradation of chemicals of environmental concern. *Science 211*:132–138.

Alexander M (1999) *Biodegradation and Bioremediation.* 2nd Ed. New York: Academic Press.

Amy PS and Haldeman DL (1997) *Microbiology of the Terrestrial Deep Subsurface*. Boca Raton, FL: CRC Press.

API (1997) *Methods for Measuring Indicators of Intrinsic Bioremediation: Guidance Manual*. Washington, DC: American Petroleum Institute, Health and Environmental Sciences Dept., Publication No 46548, API Publishing Services.

ASTM (1997) *Standard Guide for Remediation of Ground Water by Natural Attenuation at Petroleum Release Sites*. E1943-98, ASTM, West Conshohocken, PA.

Atlas RM and Cerniglia CE (1995) Bioremediation of petroleum pollutants. *Bioscience* 45:332–338.

Barker GW, Raterman KT, Fisher JB, Corgan JM, Trent TL, Brown DR, Kemp NP, and Sublette KL (1996) A case study of the natural attenuation of gas condensate hydrocarbons in soil and groundwater *Appl Biochem Biotechnol* 57/58:791–802.

Bekins BA, Godsy EM, and Goerlitz DR (1993) Modeling steady-state methanogenic degradation of phenols in groundwater. *J Contam Hydrol* 14:279–294.

Beller HR and Spormann AM (1997) Anaerobic activation of toluene and o-xylene by addition to fumarate in denitrifying strain. *J Bacteriol* 179:670–676.

Beller HR, Ding WH, and Reinhard M (1995) Byproducts of anaerobic alkylbenzene metabolism useful as indicators of *in situ* bioremediation. *Environ Sci Technol* 29:2864–2870.

Bouwer EJ (1994) Bioremediation of chlorinated solvents using alternative electron acceptors. In Norris RD, Hinchee RE, Brown R, McCarty PL, Semprini L, Wilson JT, Kampbell G, Reinhard M, Bouwer DJ, Borden RC, Vogel TM, Thomas JM, and Ward CH (eds): *Handbook of Bioremediation*, Boca Raton, FL: Lewis Publishers, pp 149–175.

Bragg JR, Prince RC, Harner EJ, and Atlas RM (1994) Effectiveness of bioremediation for the Exxon Valdez oil spill. *Nature* 368:413–418.

British Medical Association (1991) *Hazardous Wastes and Human Health*. New York: Oxford University Press.

Brockman FJ (1995) Nucleic-acid-based methods for monitoring the performance of *in situ* bioremediation. *Molecular Ecology* 4:567–578.

Brockman FJ, Li SW, Fredrickson JK, Ringelberg DB, Kieft TL, Spadoni CM, White DC, and McKinley JP (1998) Post-sampling changes in microbial community composition and activity in a subsurface paleosol. *Microbial Ecol* 36:152–164.

Burlage RS (1997) Emerging technologies: Bioreceptors, biosensors, and microprobes. In Hurst DJ, Knudsen GR, McInerney MJ, Stetzenbach LD, and Walter MV (eds): *Manual of Environmental Microbiology*. Washington, DC: ASM Press, pp 115–123.

Chapelle FH (1993) *Ground-Water Microbiology and Geochemistry*. New York: John Wiley & Sons.

Chapelle FH, Bradley PM, Lovley DR, and Vroblesky DA (1996) Measuring rates of biodegradation in a contaminated aquifer using field and laboratory methods. *Ground Water* 34:691–698.

Chapelle FH, Vroblesky DA, Woodward JC, and Lovley DR (1997) Practical considerations for measuring hydrogen concentrations in groundwater. *Environ Sci Technol* 31:2873–2877.

Chevron (1995) *Protocol for Monitoring Intrinsic Bioremediation in Groundwater*, Buscheck T and O'Reilly K (eds). Chevron Research and Technology Company, Health, Environment, and Safety Group, Richmond, CA.

Chevron (1997) *Protocol for Monitoring Natural Attenuation of Chlorinated Solvents in Groundwater*, Buscheck T and O'Reilly K (eds). Chevron Research and Technology Company, Health, Environment, and Safety Group, Richmond, CA.

Chiang CY, Salanitro PJ, Chai EY, Colthart JD, and Klein CL (1989) Aerobic biodegradation of benzene, toluene, and xylene in a sandy aquifer—data analysis and computer modeling. *Ground Water 27*:823–834.

Cho JS, Wilson JT, DiGiulio DC, Vardy JA, and Choi W (1997) Implementation of natural attenuation at a JP4 jet fuel release after active remediation. *Biodegradation 8*:265–273.

Conrad ME, Daley PF, Fischer ML, Buchanan BB, Leighton T, and Kashgarian M (1997) Combined ^{14}C and δ^{13}C monitoring of *in situ* biodegradation of petroleum hydrocarbons. *Environ Sci Technol 31*:1463–1469.

Crawford RL and Crawford DL (eds) (1997) *Bioremediation: Principles and Applications*. New York: Cambridge University Press.

Dagley S (1977) Microbial degradation of organic compounds in the biosphere. In Scott FJ (ed): *Survey of Progress in Chemistry*. New York: Academic Press, pp 121–170.

Davis GB, Johnston CD, Patterson BM, Barber C, Bennett M, Sheehy A, and Dunbavan M (1995) Monitoring bioremediation of weathered diesel NAPL using oxygen depletion profiles. In Hinchee RE, Douglas GS, and Ong SK (eds): *Monitoring and Verification of Bioremediation*. Columbus, OH: Battelle Press, pp 115–133.

Deyo GB, Robbins GA, and Binkhorst GK (1993) Use of portable oxygen and carbon dioxide detectors to screen soil gas for subsurface gasoline contamination. *Ground Water 31*:598–604.

Fennell DE, Gossett JM, and Zinder SH (1997) Comparison of butyric acid, ethanol, lactic acid, and propionic acid, as hydrogen donors for the reductive dechlorination of tetrachlorethene. *Environ Sci Technol 31*:918–926.

Flanagan WP and May RJ (1993) Metabolite detection as evidence for naturally occurring aerobic PCB biodegradation in Hudson River sediment. *Environ Sci Technol 27*:2207–2212.

Fleming JT, Sanseverino J, and Sayler GS (1993) Quantitative relationship between naphthalene catabolic gene frequency and expression in predicting PAH degradation in soils. *Environ Sci Technol 27*:1068–1074.

Fredrickson JK and Fletcher M (1999) *Deep Subsurface Microbiology and Biogeochemistry*. New York: John Wiley & Sons.

Fredrickson JK and Onstott TC (1996) Microbes deep inside the Earth. *Scien Amer 275*:68–73.

Führ F, Burauel P, Dust M, Mittelstaedt W, Putz T, Reinken G, and Stork A (1998) In Fuhr F, Hance RJ, Plimmer JR, and Nelson JO (eds): The Lysimeter concept: Environmental behavior of pesticides. ACS Symposium series 669, American Chemical Society, Washington, DC, pp 1–29.

Gibson DT (1984) *Microbial Degradation of Organic Compounds*. New York: Marcel Dekker.

Ghiorse WC (1997) Subterranean life. *Science 275*:789.

Ghiorse WC and Wilson JT (1988) Microbial ecology of the terrestrial subsurface. *Adv Appl Microbiol 33*:107.

Grossman EL (1997) Stable carbon isotopes as indicators of microbial activity in aquifers. In Hurst DJ, Knudsen GR, McInerney MJ, Stetzenbach LD, and Walter MV (eds): *Manual of Environmental Microbiology*. Washington, DC: ASM Press, pp 565–576.

Harkness MR, McDermott JB, Abramowicz DA, Salvo JJ, Flanagan WP, Stephen ML, Mondello FJ, May FJ, Lobos JG, Carroll KM, Brennan MJ, Bracco AA, Fish KM, Warner GL, Wilson PR, Dietrich DK, Lin DT, Morgan CB, and Gately WL (1993) *In situ* stimulation of aerobic PCB biodegradation in Hudson River sediments. *Science* 259:503–507.

Heitzer A and Sayler GS (1993) Monitoring the efficacy of bioremediation. *TIB Tech 11*:334–343.

Hickey WJ (1995) *In situ* respirometry: Field method and implications for hydrocarbon biodegradation in subsurface soils. *J Environ Qual 24*:583–588.

Hinchee RE, Wilson TI, and Downey DC (eds) (1995) *Intrinsic Bioremediation*. Columbus, OH: Battelle Press.

Höhener P, Hunkeler D, Hess A, Bregnard T, and Zeyer J (1998) Methodology for the evaluation of engineered *in situ* bioremediation: Lessons from a case study. *J Microbiol Meth 32*:179–192.

Hopkins GD, Munakata J, Semprini L, and McCarty PL (1993) Trichloroethylene concentration effects on pilot field-scale *in-situ* groundwater bioremediation by phenol-oxidizing microorganisms. *Environ Sci Technol 27*:2542–2547.

Jackson AW, Pardue JH, and Araujo R (1996) Monitoring crude oil mineralization in salt marshes: Use of stable carbon isotope ratios. *Environ Sci Technol 30*:1139–1144.

Johnson BL (1997) Hazardous waste: Human health effects. *Toxicol Indus Health 13*:121–143.

Kampbell DH, Wiedemeier TH, and Hansen JE (1996) Intrinsic bioremediation of fuel contamination in a ground water at a field site. *J Hazard Mat 49*:197–204.

Kim H, Hemond HF, Krumholz LR, and Cohen BA (1995) *In situ* biodegradation of toluene in a contaminated stream. 1. Field studies. *Environ Sci Technol 29*:108–116.

Landemeyer JE, Vroblesky DA, and Chapelle FH (1996) Stable carbon isotope evidence of biodegradation zonation in a shallow jet-fuel contaminated aquifer. *Environ Sci Technol 30*:1120–1128.

Lang MM, Roberts PV, and Semprini L (1997) Model simulations in support of field scale design and operation of bioremediation based on cometabolic degradation. *Ground Water 35*:565–573.

Lee K, Wong CS, Cretney WJ, Whitney FA, Parsons TR, Lalli CM, and Wu J (1985) Microbial response to crude oil and corexist 9527: SEAFLUXES enclosure study. *Microbial Ecol 11*:337–351.

Lenhard RJ, Skeen RS, and Brouns TM (1995) Contaminants at US DOE sites and their susceptibility to bioremediation. In Skipper HD and Turco RF (eds): *Bioremedia-tion Sciences and Applications*. Madison, WI: SSSA, Special Publications No 43, pp 157–172.

Lerch RN, Donald WW, Li Y-X, and Alberts EE (1995) Hydroxylated atrazine degradation products in a small Missouri stream. *Environ Sci Technol 29*:2759–2768.

Li DX (1995) Continuous bioventing monitoring using a new sensor technology. In Hinchee RE, Douglas GS, and Ong SK (eds): *Monitoring and Verification of Bioremediation*. Columbus, OH: Battelle Press, pp 115–133.

Longly G (1981) The Edwards aquifer: Earth's most diverse groundwater ecosystem? *Internat J Speleol 11*:123–138.

Ludwig P, Huehnerfuss H, Koenig WA, and Gunkel W (1992) Gas chromatographic separation of the enantiomers of marine pollutants part 3. Enantioselective degradation of alpha hexachlorocyclohexane and gamma hexacholorocyclohexane by marine microorganisms. *Mar Chem 38*:13–23.

Mackay CM and Cherry JA (1989) Groundwater-contamination: Pump-and-treat remediation. *Environ Sci Technol 23*:630–636.

Madsen EL (1991) Determining *in situ* biodegradation: Facts and challenges. *Environ Sci Technol 25*:1662–1673.

Madsen EL (1995) Impacts of agricultural practices on subsurface microbial ecology. In Sparks D (ed): *Advances in Agronomy*, Vol 54. San Diego, CA: Academic Press, pp 1–67.

Madsen EL (1996) A critical analysis of methods for determining the composition and biogeochemical activities of soil microbial communities *in situ*. In Stotzky G and Bollag J-M (eds): *Soil Biochemistry*. New York: Marcel Dekker, pp 287–370.

Madsen EL (1997) Theoretical and applied aspects of bioremediation: The influence of microbiological processes on organic compounds in field sites. In Burlage R (ed): *Techniques in Microbial Ecology*. NYC: Oxford University Press, pp 354–407.

Madsen EL (1998a) Method for determining biodegradability. In Hurst DJ, Knudsen GR, McInerney MJ, Stetzenbach LD, and Walter M (eds): *Manual of Environmental Microbiology*. Washington, DC: ASM Press, pp 709–720.

Madsen EL (1998b) Epistemology of environmental microbiology. *Environ Sci Technol 32*:429–439.

Madsen EL and Ghiorse WC (1993) Ground water microbiology: Subsurface ecosystem processes. In Ford T (ed): *Aquatic Microbiology: An Ecological Approach*. Cambridge, MA: Blackwell Scientific Publications, pp 167–213.

Madsen EL, Sinclair JL, and Ghiorse WC (1991) *In situ* biodegradation: Microbiological patterns in a contaminated aquifer. *Science 252*:820–833.

Madsen EL (2000) Verifying bliomediation: How do I know if it is taking place? In JJ Valdes (ed) *Bioremediation* Kluwer, Dordrecht, pp 101–122.

Maymo-Gatell X, Chien YT, Gossett JM, and Zinder SH (1997) Isolation of a bacterium that reductively dechlorinates tetrachloroethene to ethene. *Science 276*:1568–1571.

McCarty PL (1997) Breathing with chlorinated solvents. *Science 276*:1521–1522.

McDonald JA and Rittmann BE (1993) Performance standards for *in situ* bioremediation. *Environ Sci Technol 27*:1974–1979.

McNab WW and Dooher BP (1998) Uncertainty analyses of fuel hydrocarbon biodegradation signatures in groundwater by probabilistic modeling. *Ground Water 36*:691–698.

MPCA (Minnesota Pollution Control Agency) (1999) *Natural Attenuation of Chlorinated Solvents in Ground Water Minnesota Pollution Control Agency*. St. Paul, MN: Site Response Section, working draft.

NRC (1993) *In situ Bioremediation: When Does It Work?* Washington, DC: National Academy Press.

NRC (2000) *Natural Attenuation for Groundwater Remediation*, Washington, DC: National Academy Press.

Norris RD, Hinchee RE, Brown R, McCarty PL, Semprini L, Wilson JT, Kampbell DG, Reinhard M, Bouwer DJ, Borden RC, Vogel TM, Thomas JM, and Ward CH (1994) *Handbook of Bioremediation*. Boca Raton, FL: Lewis Publishers.

Ogram A, Sun W, Brockman FJ, and Fredrickson JK (1995) Isolation and characterization of RNA from low-biomass deep-subsurface sediments. *Appl Environ Microbiol 61*:763–768.

Paigen B and Goldman LR (1987) Lessons from Love Canal New York: The role of the public and the use of birth weight growth and indigenous wildlife to evaluate health risk. In Andelman JB and Underhill DW (eds): *Health Effects from Hazardous Waste Sites*. Chelsea, MI: Lewis Publishers, pp 177–192.

Pedersen K (1993) The deep subterranean biosphere. *Earth Sci Rev 34*:243–260.

Pedersen K (1997) Microbial life in deep granitic rock. *FEMS Microbiol Rev 20*:399–414.

Pfiffner SM, Palumbo AV, Phelps TJ, and Hazen TC (1997) Effects of nutrient dosing on subsurface methanotrophic populations and trichloroethylene degradation. *J Indus Microbiol Biotechnol 18*:204–212.

Phelps TJ, Murphy EM, Pfiffner SM, and White DC (1994) Comparison between geochemical and biological estimates of subsurface microbial activities. *Microbial Ecol 28*:335–349.

Postgate J (1992) *Microbes and Man*. Cambridge, MA: Cambridge University Press.

Prince RC, Elmendorf DL, Lute JR, Hsu CS, Halth CE, Senius JD, Dechert GJ, Douglas GS, and Butler EL (1994) $17\alpha(H)$-$21\beta(H)$-hopane as a conserved internal marker for estimating the biodegradation of crude oil. *Environ Sci Technol 28*:142–145.

Pritchard PH and Costa CF (1991) EPA's Alaska oil spill bioremediation project. *Environ Sci Technol 25*:372–379.

Raymond RL, Hudson JO, and Jamison VW (1976) Oil degradation in soil. *Appl Environ Microbiol 31*:522–535.

Raymond RL, Jamison VW, and Hudson JO (1977) Beneficial stimulation of bacterial activity in groundwater containing petroleum hydrocarbons. *Amer Inst Chem Eng Symp Series 73*(1666):390–404.

Rice DW, Gorse RD, Michaelson JC, Dooher BP, MacQueen DH, Cullen SJ, Kastenberg WE, Everett LG, and Marino MA (1995) California leaking underground fuel tank (LUFT) historical case analyses.

Rifai HS, Borden RC, Wilson JT, and Ward CH (1995a) Intrinsic bioattenuation for subsurface restoration. In Hinchee RE, Wilson JT, and Downey DC (eds): *Intrinsic Bioremediation*. Columbus, OH: Battelle Press, pp 1–29.

Rifai HS, Newell CJ, Miller RN, Taffinder S, and Rounsaville M (1995b) Simulation of natural attenuation with multiple electron acceptors. In Hinchee RE, Wilson JT, and Downey DC (eds): *Intrinsic Bioremediation*. Columbus, OH: Battelle Press, pp 53–58.

Robertson WD (1994) Chemical fate and transport in a domestic septic system: Site description and attenuation of dichlorobenzene. *Environ Toxicol Chem 13*:183–19.

RTDF (1997) *Natural Attenuation of Chlorinated Solvents in Groundwater: Principles and Practices. Industrial Members of the Bioremediation of Chlorinated Solvents Consortium of the Remediation Technologies Development Forum (RTDF)*.

Salanitro JP (1993) The role of bioattenuation in the management of aromatic hydrocarbon plumes in aquifers. *Ground Water Moniter Remed 13*:150–161.

Sasaki T, Maki H, Ishihara M, and Harayama S (1998) Vanadium as an internal marker to evaluate microbial degradation of crude oil. *Environ Sci Technol 32*:3618–3621.

Selvaratnam S, Schoedel BA, McFarland BL, and Kulpa CF (1995) Application of reverse transcriptase PCR for monitoring expression of the catabolic *dmpN* gene in a phenol-degrading sequencing batch reactor. *Appl Environ Microbiol 61*:3981–3985.

Semprini L and McCarty PL (1991) Comparison between model simulations and field results for *in-situ* biorestoration of chlorinated aliphatics. Part 1. Biostimulation of methanotrophic bacteria. *Ground Water 29*:365–374.

Semprini L and McCarty PL (1992) Comparison between model simulations and field results for *in-situ* biorestoration of chlorinated aliphatics. Part 2. Cometabolic transformations. *Ground Water 30*:37–44.

Semprini L, Kitanidis PK, Kampbell DH, and Wilson JT (1995) Anaerobic transformation of chlorinated aliphatic hydrocarbons in a sand aquifer based on spatial chemical distribution. *Water Resources Res 31*:1051–1062.

Semprini L, Roberts PV, Hopkins GD, and McCarty PL (1990) A field evaluation of *in-situ* biodegradation of chlorinated ethenes: Part 2. Results of biostimulation and biotransformation experiments. *Ground Water 28*:715–727.

Sever LE (1997) Environmental contamination and health effects: What is the evidence. *Toxicol Indust Health 13*:145–161.

Shannon MJR and Unterman R (1993) Evaluating bioremediation: Distinguishing fact from fiction. *Ann Rev Microbiol 47*:715–738.

Shields MS and Francesconi SC (1996) Molecular techniques in bioremediation. In Crawford RL and Crawford DL (eds): *Bioremediation: Principles and Applications*. New York: Cambridge University Press, pp 341–390.

Skipper HD and Turco RF (eds) (1995) *Bioremediation Science and Applications*. Madison, WI: Soil Science Society of America.

Spain J (1997) Synthetic chemicals with potential for natural attenuation. *Bioremediation J 1*:1–9.

Swoboda-Colberg NG (1995) Chemical contamination of the environment: Sources, types, and fate of synthetic organic chemicals. In Young LY and Cerniglia CE (eds): *Microbial Transformation and Degradation of Toxic Organic Chemicals*. New York: Wiley-Liss, pp 27–74.

Tett VA, Willetts AJ, and Lappin-Scott HM (1994) Enantioselective degradation of the herbicide mecoprop [2-(2-methyl-4-chlorophenoxy)propionic acid] by mixed and pure bacterial cultures. *FEMS Microbiol Ecol 14*:191–199.

Thierrin J, Davis GB, and Barber C (1995) A ground-water tracer test with deuterated compounds for monitoring *in situ* biodegradation and retardation of aromatic hydrocarbons. *Ground Water 33*:469–475.

Thierrin J, Davis GB, Barber C, Patterson BM, Pribac F, Power TR, and Lambert M (1993) Natural degradation rates of compounds and naphthalene in a sulphate-reducing groundwater environment. *Hydrol Sci J 38*:309–322.

USAF (1995) *Technical Protocol for Implementing Intrinsic Remediation with Long-Term Monitoring for Natural Attenuation of Fuel Contamination in Groundwater*, Vols I and II, Wiedemeier R, Wilson JT, Kampbell DH, Miller RN, and Hanson JE (eds). San Antonio, TX: Air Force Center for Environmental Excellence, Technology Transfer Division, Brooks Air Force Base.

USAF (1997) *Technical Protocol for Evaluating Natural Attenuation of Chlorinated Solvents in Groundwater*, Wiedemeier R, Swanson MA, Montoux DE, Gordon EK, Wilson JT, Wilson BH, Kampbell DH, Hansen JE, Haas P, and Chapelle FJ (eds). San Antonio, TX: Air Force Center for Environmental Excellence, Technology Transfer Division, Brooks Air Force Base.

USDON (1998) *Technical Guidelines for Evaluating Monitored Natural Attenuation at Nova and Marine Corps Facilities*, Wiedemeier TH and Chapelle FH (eds).

U.S. DOE (1998) *Screening and Technical Guidance for Monitored Natural Attenuation at DOE Sites*, Brady PV, Spalding BP, Krupka KM, Waters RD, Zhang P, Borns DJ, and Brady WD (eds).

U.S. EPA (1997a) *Cleaning Up the Nation's Waste Sites: Markets and Technology Trends*, 1996 ed. Washington, DC: EPA, Office of Solid Waste and Emergency Response, EPA 542-R-96-005.

U.S. EPA (1997b) *Use of Monitored Natural Attenuation at Superfund, RCRA Corrective Action, and Underground Storage Tank Sites.* Washington, DC: EPA, Office of Solid Waste and Emergency Response. OSWER Monitored Natural Attenuation Policy, OSWER Directive 9200.4-17.

U.S. EPA (1998) *Technical Protocol for Evaluating Natural Attenuation of Chlorinated Solvents in Groundwater,* Wiedemeier TH, Swanson MA, Montoux DE, Gordon EK, Wilson JT, Wilson BH, Kampbell DH, Haas PE, Miller RN, Hansen JE, and Chapelle FH (eds). Washington, DC: EPA/600/R-98/128.

Wackett LP (1996) Co-metabolism: Is the emperor wearing any clothes? *Curr Opin Biotechnol* 7:321–325.

White DC, Phelps TJ, and Onstott TC (1998) What's up down there? *Curr Opin Biotechnol* 1:286–290.

Wiedemeier TH, Swanson MA, Wilson JT, Kampbell DH, Miller RN, and Hansen JE (1995) Patterns of intrinsic bioremediation at two U.S. Air Force bases. In Hinchee RE, Wilson JT, and Downey DC (eds): *Intrinsic Bioremediation.* Columbus, OH: Battelle Press, pp 31–51.

Williams RA, Shuttle KA, Kunkler JL, Madsen EL, and Hooper SW (1997) Intrinsic bioremediation in a solvent-contaminated alluvial groundwater. *J Indust Microbiol Biotechnol* 18:177–188.

Wilson MS, Bakermans C, and Madsen EL (1999) *In situ*, real-time catabolic gene expression: Extraction and characterization of naphthalene dioxygenase mRNA transcripts from groundwater. *Appl Environ Microbiol* 65:80–87.

Wilson JT and Jawson MD (1995) Science needs for implementation of bioremediation. In Skipper HD and Turco RF (eds): *Bioremediation Science and Applications.* Madison, WI: Soil Science Society of America, pp 293–303.

Wilson MS and Madsen EL (1996) Field extraction of a unique intermediary metabolite indicative of real time *in situ* pollutant biodegradation. *Environ Sci Technol* 30:2099–210.

Wilson BH, Wilson JT, Kampbell DH, Bledsoe BE, and Armstrong JM (1990) Biotransformation of monoaromatic and chlorinated hydrocarbons at an aviation gasoline spill site. *Geomicrobiol J* 8:225–240.

Wolfe DA, Hameedi MJ, Galt JA, Watabayashi G, Short J, O'Claire C, Rice S, Michel J, Payne JR, Braddock J, Hanna S, and Sale D (1994) The fate of the oil spilled from the Exxon Valdez. *Environ Sci Technol* 28:561A–568A.

Yager RM, Bilotta SE, Mann CL, and Madsen EL (1997) Metabolic adaptation and *in situ* attenuation of chlorinated ethenes by naturally occurring microorganisms in a fractured dolomite aquifer near Niagara Falls, NY. *Environ Sci Technol* 31:3138–3147.

Yang Y and McCarty PL (1998) Competition for hydrogen within a chlorinated solvent dehalogenating anaerobic mixed culture. *Environ Sci Technol* 32:3591–3597.

Young LY and Cerniglia CE (eds) (1995) *Microbial Transformation and Degradation of Toxic Organic Chemicals.* New York: Wiley-Liss.

Zhou J, Palumbo AV, and Tiedje JM (1997) Sensitive detection of a novel class of toluene-degrading denitrifiers, *Azoarcus tolulyticus,* with small-subunit rRNA primers and probes. *Appl Environ Microbiol* 63:2384–2390.

Zipper C, Nickel K, Angst W, and Kohler HP (1996) Complete microbial degradation of both enantiomers of the chiral herbicide mecoprop [(RS)-2-(4-chloro-2-methylhenoxy)propionic acid] in an enantioselective manner by *Sphingomonas herbicidovorans* sp. nov. *Appl Environ Microbiol* 62:4318–4322.

FUTURE TRENDS

≡ 10

NUCLEIC ACID ANALYSIS OF SUBSURFACE MICROBIAL COMMUNITIES: PITFALLS, POSSIBILITIES, AND BIOGEOCHEMICAL IMPLICATIONS

DARRELL P. CHANDLER and FRED J. BROCKMAN

Pacific Northwest National Lab, Richland, Washington

1 INTRODUCTION

The integration of geology, hydrology, geochemistry, and microbiology has significantly enhanced our understanding of deep ($> 30\,\mathrm{m}$) subsurface microorganisms and their role in subsurface geochemical processes, exemplified by the preceding chapters of this book. Most early investigations of subsurface ecosystems relied heavily on cultivation techniques, enrichment cultures, and mesocosm studies using varied electron donors and acceptors to isolate, characterize, and identify subsurface microorganisms and metabolic processes (Albrechtsen and Winding, 1992; Amy, 1997; Balkwill and Ghiorse, 1985; Balkwill et al., 1989; Balkwill, 1989; Bock et al., 1994; Beloin et al., 1988; Balkwill and Boone, 1997; Bone and Balkwill, 1988;

Subsurface Microgeobiology and Biogeochemistry, Edited by James K. Fredrickson and Madilyn Fletcher.
ISBN 0-471-31577-X. Copyright 2001 by Wiley-Liss, Inc.

Boone et al., 1995; Brockman et al., 1998; Colwell, 1989; Francis et al., 1989; Fredrickson et al., 1989, 1995a, 1995b; Fuller et al., 1995; Ghiorse and Balkwill, 1983; Ghiorse and Wilson, 1988; Haldeman and Amy, 1993; Haldeman et al., 1993, 1994; Hazen et al., 1991; Hicks and Fredrickson, 1989; Jones et al., 1989; Kieft et al., 1995, 1997; Madsen and Bollag, 1989; Phelps et al., 1989, 1994; Sinclair and Ghiorse, 1989; Stevens and Holbert, 1995). These studies have significantly enhanced our understanding of deep subsurface ecology, but suffer from some well-recognized limitations and biases. For example, many microorganisms cannot be successfully cultivated in the laboratory; laboratory-based measurements of microbial activity can be orders of magnitude greater or smaller than *in situ* activities; and geochemical and physical changes often occur in the sample as a result of drilling and coring, and before microbiological analyses are initiated.

These limitations led to the development of culture-independent techniques for the analysis of *in situ* microbial communities and involve extraction, purification, and analysis of nucleic acids and membrane lipids *directly* from the sample of interest. It is through the *synthesis* of culture-independent, community-level studies with geochemical, geo-hydrological, and other microbiological data that our understanding of subsurface microbial communities and biogeochemical processes will flourish. In this context, lipids are a particularly powerful and sensitive technique for determining *in situ* viability, community structure, and nutritional status (reviewed in White and Ringelberg, 1997). Signature lipid biomarkers are unique (or relatively unique) to specific taxonomic groupings (e.g., eukaryote, gram negative versus gram positive) and certain functional groups (e.g., sulfate reducers, methanotrophs). In some cases, genus-level signatures are possible (e.g., Sphingomonas). On the other hand, nucleic acids provide an inherent evolutionary history from which to infer taxonomic and ecological properties of interest, in addition to identifying a microorganism at the species or strain level and/or specific metabolisms, enzymes, or other cellular functions. In this chapter, then, we will examine nucleic acid analysis specifically within the context of deep subsurface and other low-biomass ecosystems.

The purpose of this chapter, however, is *not* to provide an exhaustive description of nucleic acid techniques in environmental microbiology or subsurface microbial ecology, as numerous review articles and book chapters have already been written on the subject (e.g., Sayler and Layton, 1990; Ogram and Sayler, 1988; Hugenholtz and Pace, 1996; Ward et al., 1992, 1995; Stahl, 1995, 1997; Steffan and Atlas, 1991; Olson and Tasi, 1992; Brockman, 1995). Instead, the objectives of this chapter are to (1) briefly introduce general strategies for culture-independent analysis of community nucleic acids, (2) elucidate some of the challenges, limitations, and pitfalls of culture-independent techniques and data interpretation, (3) introduce some possible technological solutions for overcoming these pitfalls, and (4) discuss the potential implications of technology developments for a more comprehensive understanding of biogeochemical processes in (*in situ* and *ex situ*) subsurface materials.

2 OVERVIEW

The way in which nucleic acid analyses are routinely applied in environmental microbiology is outlined in Fig. 10.1 and is generally relevant for the analysis of deep subsurface environments. When used to analyze community nucleic acids, it is generally accepted that specific DNAs provide evidence for the presence/absence of microorganisms and metabolic potential within the environment; specific rRNAs are an indicator of cell viability that can be correlated with general metabolic activity of the corresponding microorganisms in the environment; and specific mRNAs provide evidence for expression of metabolic processes or functions (Stahl, 1997; Power et al., 1998; Stapleton et al., 1998; Olson and Tasi, 1992). Within this framework, culture-independent nucleic acid analysis has significantly contributed to the understanding of *in situ* microbial diversity, activity, and response to environmental stimuli.

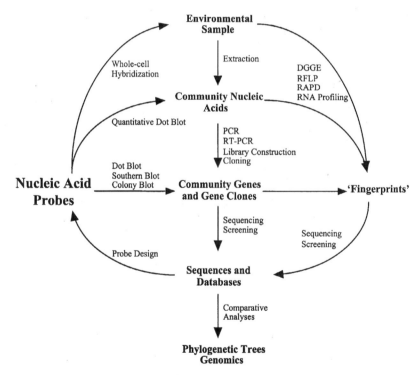

FIGURE 10.1 Common strategies for the analysis of uncultivated microorganisms and nucleic acids isolated from environmental samples. DGGE, denaturing gradient gel electrophoresis; RFLP, restriction fragment length polymorphism; RAPD, randomly amplified polymorphic DNA; PCR, polymerase chain reaction; RT-PCR, reverse transcriptase polymerase chain reaction. Numerous variations on these basic principles and techniques are also abundant in the literature.

For example, deep granitic groundwaters (Ekendahl et al., 1994; Pedersen et al., 1996a, 1996b) were characterized through 16S rDNA cloning and sequence analysis, and 16S rRNA hybridizations have been used to characterize the population structure of alkaline basalt aquifer microbial communities (Fry et al., 1997) Microbial communities in deep ocean sediments (Parkes et al., 1994), subsurface clays (Boivin-Jahns et al., 1996), and buried paleosols (Chandler et al., 1997a, 1997b, 1998) have also been characterized through direct nucleic acid extraction and analysis of 16S rDNA sequences. Taxon-specific 16S rRNA probes were used to probe microbial activity via dot-blot hybridizations against total RNA (Ogram et al., 1995) and *in situ* hybridizations of cells extracted from lacustrine cores (Fredrickson et al., 1995b). Functional gene probes have also been used to assess and characterize the progress of bioremediation (Stapleton et al., 1998). Advances in total RNA extraction from low-biomass sediments have been extended to detect mRNAs associated with toluene and trichloroethylene degradation (Ogram et al., 1995), and to identify novel Crenarchaeal clone-specific activity in a deep subsurface paleosol via RT-PCR analysis of rRNA (Chandler et al., 1998). However, the conclusions drawn from community nucleic acids are intimately tied to technical difficulties of the analysis and to our understanding of the environmental context in which the techniques are applied.

3 PITFALLS

Every analytical technique has its own strengths and weaknesses that must be understood and appreciated for the technique to be properly implemented and interpreted. Ecological studies, in general, further compound inherent limitations of analytical techniques due to our imperfect understanding of all environmental variables and their effect on microorganisms or the analytical method. The extent to which the challenges and limitations of environmental molecular biology are recognized or considered to be problematic varies considerably among scientists. Molecular biologists studying environmental samples, for example, often do not recognize or appreciate some of the technical or ecological issues associated with representative sampling, chemical and genetic complexity of *in situ* microbial communities and samples, and storage effects (to name a few). On the other hand, ecologists lacking extensive background in molecular biology often do not recognize or appreciate the analytical pitfalls of molecular biology, especially techniques involving multitemplate amplification by polymerase chain reaction (PCR). When limitations are recognized, they are frequently ignored for practical reasons, for example, the need to publish field-relevant studies instead of developing new molecular methodologies specifically within an environmental context. In this section, we discuss some of the pitfalls of nucleic acid analysis in subsurface ecology and biogeochemistry studies, with an emphasis on sample integrity and representativeness, biomass considerations, PCR at low target concentrations, and extrapolating from sequences to the environment.

3.1 Biomass Considerations

Relative to soils, eutrophic lakes, and bioreactors, the subsurface is a rather barren and inhospitable place. Subsurface environments generally exhibit very low fluxes of one or more nutrients (electron donor, electron acceptor, micronutrients), exceptionally slow rates of microbial metabolism and growth, and biomass concentrations of 10^4–10^6 cells g^{-1} (Kieft and Phelps, 1997; Kieft et al., 1993, 1995; Phelps et al., 1989, 1994; Balkwill, 1989; Balkwill et al., 1989; Fredrickson et al., 1989, 1995a,b; Sinclair and Ghiorse, 1989; Brockman et al., 1992, 1997). Typical bioloads in subsurface materials are therefore 10^3–10^5 times lower concentrations than a stationary-phase *E. coli* culture, and 10^4–10^6 times lower than many bioreactors. Stated another way, to achieve the same level of biomass typically available in 1 g of surface soil, 1 mL of *E. coli* culture or bioreactor requires up to *1 metric ton* (1–1,000 kg) of subsurface material! Natural exceptions to the low-biomass "dilemma" include subsurface systems with moderate to high levels of easily degradable contaminants, high-flux regions or large water-conducting fractures, locales adjacent to natural high-flux point sources of nutrients, and regions being stimulated by engineered bioremediation.

For the *in situ* analysis of deep subsurface microorganisms, then, the fundamental impediment to all bioanalytical techniques stems from the practical limitations of low or insufficient biomass for direct measurement and observation. To take advantage of culture-independent techniques in subsurface systems, there are two obvious solutions to the low-biomass problem: increase sample size and/or concentrate cells prior to nucleic acid extraction. The practical upper limit on sample size for a single nucleic acid extraction is ca. 100 g. More common is to extract a single 10-g sample or pool the extracts from 10×1-g samples. At best, this strategy for increasing sample size represents only a 10- to 100-fold improvement over indigenous biomass levels. And while it is common to filter and concentrate hundreds of liters of ocean, lake, and stream water, groundwater is not normally a suitable surrogate for subsurface sediments because the great majority of microorganisms remain attached to sediment grains; those that *are* present in groundwater may not represent (or poorly represent) the dominant, attached *in situ* community.

The low-biomass dilemma is ultimately a problem of detection sensitivity for the analytical *process*, as opposed to the analytical technique; PCR, for example, can be used to amplify and detect single copy genes. Table 10.1 illustrates the impact of biomass level on the ability to analyze subsurface environments with DNA or RNA detection methods illustrated in Fig. 10.1. To determine if nucleic acid techniques can detect a target microorganism (or assemblage of related target microorganisms) present at 1% of the total population, we have made a number of (reasonable) assumptions regarding indigenous biomass levels, sample processing and analytical procedures, and detection limits. Table 10.1 illustrates the finding that even under conditions of relatively high biomass (10^6–10^7 cells g^{-1}), the amount of nucleic acid recovered from subsurface sediments is frequently below the detection limit ($\geq 10^6$ copies, or 1 attomole) for direct probing strategies (Fig. 10.1). PCR, then, is an

TABLE 10.1 Theoretical detection limits of indigenous (nonseeded) microorganisms with DNA or rRNA targets utilizing hybridization or PCR methods, assuming that 1% of the total population is composed of the target organism. (−) indicates below detection, (+) indicates hybridization at the detection limit of the assay, (10^1) indicates potential signal at 10× to < 100× above the detection limit, (10^2) indicates detection at 100× to < 1000× above detection limit, etc. NA=not applicable. Assumptions used to estimate theoretical detection limits are: 10 g solid, 10 L groundwater concentrated by filtration to 10 mL with 20% efficiency, and 1 g sand or coarse silt for FISH; one copy of DNA and 1000 copies of rRNA per cell; 10 and 50% extraction/purification efficiency in solid and groundwater, respectively; 10% of final sample volume utilized for hybridization or PCR; no dilution of purified template required to relieve PCR inhibition; 10^6 molecule detection limit for hybridization; 10^2 molecule detection limit for PCR and 10^3 molecule detection limit for RT-PCR.

Sample Type[a]	DNA Hybridization		rRNA Hybridization		FISH rRNA[b]		PCR/RT-PCR DNA/rRNA	
	Solid	Groundwater	Solid	Groundwater	Solid	Groundwater	Solid	Groundwater
Contaminated aquifer, engineered remediation	10^1	10^2	10^3	10^5	10^4	10^5	10^6	10^8
Contaminated aquifer	−/+	+	10^2	10^3	10^3	10^4	10^5	10^6
High-conductivity fracture, pristine rock	NA	−	NA	10^1	NA	10^3	NA	10^4
Pristine aquifer	−	−	+	+	10^1	10^2	10^3	10^3
Contaminated aquitard or vadose zone	−	NA	−	NA	+	NA	10^2	NA
Pristine aquitard or vadose zone	−	NA	−	NA	−	NA	10^1	NA

[a] Assume 10^9 cells g^{-1} in the contaminated aquifer with engineered bioremediation; 10^8 cells g^{-1} in the contaminated aquifer; 10^6 cells g^{-1}, pristine aquifer; 10^5 cells g^{-1}, contaminated aquitard or vadose zone; 10^4 cells g^{-1}, pristine aquitard or vadose zone. Assume groundwater populations of 10, 1, and 0.1% of the solid for contaminated aquifer with engineered bioremediation, contaminated aquifer, and pristine aquifer, respectively. Thus, ca. 10^8 cells mL^{-1} in the contaminated aquifer undergoing engineered bioremediation; 10^6 cells mL^{-1}, contaminated aquifer; 10^4 cells mL^{-1}, high K fracture in pristine rock; 10^3 cells mL^{-1}, pristine aquifer. We further assume that a sample concentration does not result in precipitates.

[b] Fluorescence *in situ* hybridization. Assume 1 g sand or coarse silt applied to slide, 100 fields of 100 μm×100 μm are examined, and little background fluorescence from sample. Assume groundwater concentrate is diluted or further concentrated to provide maximum number of cells applied to slide but not to exceed 1 × 10^5 cells applied to slide; 100 fields of 100 μm×100 μm are examined.

almost inescapable technique for the molecular detection of deep subsurface microorganisms *in situ*.

3.2 PCR Considerations at Low Target Concentrations

Some of the generalized pitfalls of PCR-based analyses are summarized by Wintzingerode et al., (1997); however the studies and findings summarized therein relate to relatively high-biomass or high-copy-number amplifications. The purpose of this section is to highlight some limitations of PCR specifically as they relate to low-biomass samples (i.e., $\leq 10^7$ cells in the volume being extracted) and low-copy-number amplifications. It is important to realize that PCR was developed for standard biotechnology and clinical applications, and was later applied to environmental samples. Therefore, it is not surprising that some of the limitations and pitfalls of PCR-based analyses to low-biomass environmental samples have been slow to be discovered and appreciated.

3.2.1 PCR Inhibition Enzyme inhibition due to coextracted sample contaminants is a common pitfall of PCR-based analyses of environmental samples (Al-Soud and Rådström, 1998; Kreader, 1996; Wilson, 1997). In addition to previously described phenomena, we frequently observe a profound copy-number effect when confronting subsurface environmental extracts and PCR inhibitors. That is, given an equivalent concentration of inhibitor, PCR reactions containing relatively high concentrations of template (i.e. ≥ 1 ng, or $\geq 10^6$ copies) are much more tolerant of contaminants (e.g. humic acids) than amplifications with low concentrations ($\leq 10^6$ copies) of target (unpublished). In practice, the most common strategies for overcoming enzyme inhibition include extensive nucleic acid purification methods or template dilution prior to PCR. Although neither strategy is limiting in a high-biomass context, additional purification steps must be balanced against (significant) losses of target nucleic acid, especially in low-biomass samples. Further, dilution of low-biomass nucleic acid extracts prior to the PCR may easily reduce the number of sampled genomes to a point that is below the detection limit of a given PCR assay (e.g., Chandler and Brockman, 1996; Chandler et al., 1997a, b). To estimate the relative effects of enzyme inhibitors on PCR amplification and differentiate between enzyme inhibition and lack of specific target in the sample, we are strong proponents of "spiked" controls (Chandler, 1998). A more dramatic consequence of template dilution is to push PCR kinetics into a region of unpredictable, random bias that confounds our ability to use PCR methods for quantitative and qualitative analyses, as described below.

3.2.2 PCR Bias In most of the literature, PCR bias refers to the preferential amplification of one allele or template over another coamplified template and has been observed in both environmental (Reysenbach et al., 1992; Wilson and Blitchington, 1996; Farrelly et al., 1995) and biomedical (Mutter and Boynton, 1995; Sarkar et al., 1990; Walsh et al., 1992) contexts. The sources of preferential amplification can include (but are not limited to) differential denaturation of templates (or alleles) due to % G+C content or secondary structure, differential primer annealing, differential amplicon lengths, degraded or impure template, and differential genome size or copy number (Mutter and Boynton, 1995; Walsh et al., 1992; Sarkar et al., 1990; Farrelly et al., 1995; Rychlik, 1995). It is even reported that simply freezing PCR buffer can result in PCR bias if certain precautions are not taken (Hu et al., 1992). In some cases, PCR bias resulting from systematic bias can be relieved through chemical additives, such as acetamide (Reysenbach et al., 1992), bovine serum albumin (BSA) or T4 gene 32 protein (Kreader, 1996). Recent studies suggest that bias in PCR product ratios is significantly affected by template rehybridization in or near the plateau phase of the PCR (the $C_0 t$ effect), a consequence of product accumulation that serves to accelerate the amplification of low-copy targets in later PCR cycles (Suzuki and Giovannoni, 1996; Suzuki et al., 1998; Mathieu-Daudé et al., 1996). Descriptions and examples of the $C_0 t$ effect, however, initiate the PCR with ≥ 10 ng of target template (10^{11} copies 100 bp in size) that is subsequently mixed into model communities of two or more target sequences.

How do the controlling factors of PCR bias change if the amplification starting point is ≤ 1 ng of total DNA (likely) containing ≥ 100 specific targets (e.g., 16S homologs), a situation typifying many nucleic acid extracts obtained from deep subsurface environments? This situation represents the opposite extreme where PCR bias is better described by molecular sampling error (Walsh et al., 1992) or PCR drift (Polz and Cavanaugh, 1998); random fluctuations in primer : template : polymerase complex formation result in inconsistent amplification of homologues, regardless of starting copy number. Under such a low-biomass or low-copy number situation, the likelihood that any *one* target will reach a plateau phase where product rehybridization competes for primer annealing/extension is highly unlikely, as noted by Suzuki and Giovannoni (1996), suggesting that skewed PCR product ratios are generated in the very early rounds of the PCR rather than as a consequence of the $C_0 t$ effect or other systematic amplification biases. We originally proposed a molecular sampling-error hypothesis to explain the discrepancy between clone libraries generated from identical DNA extracts prepared from low-biomass sediments differing only in the amount of target DNA introduced for PCR amplification (Chandler et al., 1997a). We further suggested that at a certain (albeit undefined) template concentration or copy number, molecular sampling error is superseded by "reproducible amplification biases" and the $C_0 t$ effect described above. This hypothesis is supported by a recent analysis of systematic amplification biases in multitemplate PCR (Polz and Cavanaugh, 1998).

Given our current understanding, we therefore cannot assume *a priori* that PCR product abundance is a quantitative reflection of *in situ* target abundance. Due to molecular sampling error at very low target concentrations, we also cannot assume that a PCR product distribution is an accurate reflection of *in situ* target diversity. These fundamental uncertainties surrounding multitemplate PCR have significant implications for the analysis of, and data interpretation from, low-biomass microbial communities by DGGE, SSCP, clone library screening, quantitative PCR, or any other technique that relies on a simultaneous preamplification of multiple targets (as in 16S rDNA profiling). From a pragmatic viewpoint, the effects of molecular sampling error at low template concentrations can be reduced by performing replicate PCRs on each nucleic acid extract and pooling those replicates prior to analysis. On the other hand, template dilution and molecular sampling error can be used to our advantage in both high- and low-biomass environments by increasing the diversity of PCR products that are recovered from any one nucleic acid extract. For a compromise solution to quantitative PCR techniques under low-biomass situations, a replicate limiting dilution approach based on external standards is recommended (Chandler, 1998). Practical solutions to PCR bias under high-biomass and high-copy-number amplifications are also described by Polz and Cavanaugh (1998).

3.2.3 *Chimeras*   A chimeric molecule is generated when one target fragment anneals to a homologous sequence and primes the next cycle of PCR amplification. The formation of chimeric PCR products is particularly troublesome for community-level 16S rDNA analyses (Wintzingerode et al., 1997), but is also relevant for the coamplification and detection of homologous genes. We discuss chimera formation here because we believe very low target concentrations in a PCR result in fundamentally different processes than in the presence of relatively high concentrations of target. That is, we assume that the underlying mechanisms of chimera formation are akin to hybridization kinetics, thermodynamics, and systematic errors associated with PCR bias, and are therefore related to the starting point (template concentration or copy number) of the PCR.

Wang and Wang (1996) reported a 30% frequency of chimera formation from highly related sequences after 30 cycles of amplification, using 100-pg 16S rDNA or 10-ng genomic DNA as template (1996, 1997). Decreasing cycle numbers or increasing elongation times reduce the rate of chimera formation, a result also observed in a study of HIV *tat* genes (Meyerhans et al., 1990). Consequently, chimera formation is usually attributed to elongation arrest, polymerase pausing, highly fragmented DNA, or other DNA damage (Pääbo et al., 1990; Wang and Wang, 1996; Rokkones et al., 1985; Meyerhans et al., 1990; Shuldiner et al., 1989). From these observations, one could argue that multitemplate amplifications (i.e., 16S rDNA profiling) from low-biomass samples, typically requiring 40 cycles to generate enough target for cloning or visualization on a gel, will be particularly prone to chimera formation. Such an assumption may not be factually true.

The frequency of chimera formation under low template concentrations ($\leq 10^5$ copies) has not been expressly investigated, but during an analysis of 242 16S rDNA clone sequences (500 bp from the $5'$ end) derived from a low-biomass sediment (Chandler et al., 1998), we only detected one chimeric molecule (0.4%) through secondary structure analysis (unpublished). While surely an underestimate of actual chimera formation due to the length of sequence investigated, this observation and consideration of PCR kinetics and thermodynamics can be used to formulate a working hypothesis of chimera formation during low-copy-number, multitemplate amplifications. We propose that chimera formation, like amplification bias described above, is ultimately an artifact of the copy-number effect and that low template concentrations will actually *discourage* the formation of chimeric products.

The formation of primer : template duplexes directly competes against target reannealing in every PCR cycle. On the one hand, primer : template formation is kinetically driven by the vast excess of primers over target sequences (0.2 μM primers in a standard PCR; 10^{13} copies). On the other hand, the rate-limiting step for solution-phase nucleic acid hybridization is the initial nucleation event, where short regions of double-stranded DNA anneal followed by strand "zippering" to form the complete duplex. Invariably, the kinetics of nucleation favor large DNA fragments over small DNA fragments because the thermodynamic stability of large duplex DNA can displace primer from the target strand and because the rate of PCR product (or target) reannealing increases as a function of length and concentration (Wetmur and Sninsky, 1995; Britten et al., 1974; Wetzmer and Davidson, 1968; Perry-O'Keefe et al., 1996). It has therefore been suggested that template reannealing (i.e., the C_0t effect) becomes a significant factor in multitemplate PCR when product concentrations approach 10 nM (Mathieu-Daudé et al., 1996), a situation common to a typical PCR assay. In those instances where chimera formation has been evaluated, the starting template concentration is generally ≥ 10 ng of genomic DNA template, or ≥ca. 10^8 copies of individual targets. For a 100-μL PCR and 10^8 copies of target, the starting concentration of target is therefore ca. 0.2 nM. Within six cycles at 100% amplification efficiency, then, these reactions enter a nucleic acid hybridization dynamic that favors the thermal and thermodynamic stability of template reannealing over primer : template complexation in each annealing phase of the reaction. If template reannealing (prematurely terminated strands, fragmented DNA, or full-length product) *is* the major cause of chimera formation, then it is no wonder that even after 10 cycles of PCR, 5% of examined clones can represent chimeric artifacts (Wang and Wang, 1996), and why relatively high template concentrations have been characterized as a "highly unfavorable condition" (Meyerhans et al., 1990) for multitemplate PCR amplification.

In a low-biomass community-profiling situation, a typical PCR reaction will contain subnanogram quantities of (sheared) genomic DNA representing tens to hundreds of homologous sequences. Given this starting point for the PCR (ca. 10^4 copies of 16S rDNA homologues), stringent temperature control, and limited time (30 sec to 1 min) for hybrid formation, the kinetics of primer : template complexation are strongly favored over template reannealing, mispriming from nontarget strands, or chimera formation. Further, sheared DNA or damaged template (Liesack et al., 1991) is theoretically irrelevant at this point because primer annealing, extension, and exponential amplification (versus linear amplification of truncated targets) will favor the formation of a complete sequence within a few cycles; this is the basic premise of PCR specificity and exponential amplification. In the same way that template dilution can minimize the C_0t effect and the resulting normalization of product abundance (Mathieu-Daudé et al., 1996), we therefore propose that template dilution can also minimize chimera formation during multitemplate PCR. Other practices we routinely employ for low-copy-number amplifications that reduce the formation of chimeric (and other) artifacts include "hot start" strategies, short primer annealing times (10–20 sec), stringent annealing temperatures (60–65 °C), biphasic PCR strategies, single-stranded binding proteins, and sufficient elongation times for complete primer extension during each cycle (Chandler et al., 1997a, b).

3.2.4 PCR Contamination

3.2.4 PCR Contamination Procedures to minimize or prevent contamination are published in many PCR manuals (e.g., Kwok, 1990). Adequate laboratory procedures, however, cannot mitigate the occurrence of "joyride" DNA and RNA in commercial enzymes (Schmidt et al., 1992; Kenzelmann and Muhlemann, 1997; Tanner et al., 1998). Numerous methods have been developed to counter DNA contamination in enzyme preparations (Meier et al., 1993; Frothington et al., 1992; Rochelle et al., 1992; Sarkar and Sommer, 1991, 1993; Thornton et al., 1992; Hilali et al., 1997), none of which are completely satisfactory under all amplification conditions. Perkin-Elmer responded to this challenge by offering a "low-DNA" *Taq* polymerase (LD-*Taq*), containing ≤ 10 *E. coli* genomes of contaminating DNA per amplification reaction. Nevertheless, *any* contaminating DNA (or RNA) can be problematic under low-biomass community-profiling conditions, because the contaminant may be mistaken for authentic, environmentally relevant organisms (Tanner et al., 1998).

The extent to which contaminating DNA impacts 16S rDNA clone libraries from environmental samples is difficult to assess because there are very few references in the literature specifically referring to 16S rDNA sequences recovered from both a "no sediment" control extraction (i.e., a processing blank) and from a no template PCR control. In the former, the "sample" is processed, cloned, amplified, and screened in parallel with the sample of interest. Utilizing PCR techniques optimized for low-copy-number amplifications (above) and LD-*Taq* polymerase, we obtained only two clones from blank extract and no template controls during a survey of deep subsurface microbial diversity (Chandler et al., 1998). One was identical to a clone retrieved from the sediment (99.4% similarity to *Burkholderia pickettii*), and the other had no counterpart in any database. This result contrasts with the work of Tanner et al., (1998) where obvious PCR product and numerous clones were obtained from no template reactions. This indicates that the incidence of DNA contamination is highly variable and that controls described above are the only way to distinguish contaminating 16S rDNA clones from authentic clones.

3.3 Sample Considerations

There is no difficulty in selecting a representative sample in a well-mixed laboratory culture or bioreactor. In contrast, a natural environment is a complex mosaic of many overlapping physical, hydrologic, geochemical, and microbiological properties and processes (and their interactions). Each property or process has a spatially hetero-geneous distribution and these distributions give rise to patterns at multiple scales or hierarchical levels. The most appropriate scale of observation for a particular scientific question or applied problem is the scale or hierarchical level that dominates the behavior of the system or problem of interest. For example, the most appropriate scales for studying global biogeochemical fluxes, bioremediation design and engineering, microbial alteration of physical properties, and the dynamics of microbial colonization would likely be the ecosystem, strata, laminae, and pore network, respectively.

Microbiological properties in soil and the subsurface vary by many orders of magnitude at multiple scales, often exhibit a log normal distribution, and contain "hotspots" (reviewed in Brockman and Murray, 1997a, b). This variability results from the overlapping spatially nonhomogenous distributions of properties that impact the microorganisms' chemical and physical environment and yields a continuum of zones that can vary from optimal to excluding. Different environments will vary in the presence/absence, relative volume, and spatial distribution of these different zones (Fig. 10.2). It is evident that changes in the size, number, and location of samples; use or nonuse of sample homogenization; fraction of sample that is assayed; and how common or rare a specific microorganism or metabolism is in a specific environment could easily result in different results and conclusions. For example, vadose zone environments with low moisture recharge require ≥ 100-g samples to integrate over a sufficient volume to minimize variation in the mean (i.e., minimize samples with no measurable activity) (Fig. 10.3). In contrast, 10-g samples provided a mean with the lowest variance for aerobically incubated samples from a site with recharge of 20 cm. There is a continuing need to better understand the spatial heterogeneity of specific microbial types in various subsurface environments.

Dynamic changes in water and nutrient flux and other chemical properties such as E_h and pH occur in many subsurface environments as the result of seasonal changes, storm events, and other natural and anthropogenic activity. Nonequilibrium processes often dominate a subsurface system in the presence of significant temporal variation or after a major change such as the introduction of a contaminant plume. This is particularly true of geochemical processes and attendant microbiological responses such as adaptation, genetic transfer, and growth and death of different microorganisms promoted by the new conditions. Thus, in environments undergoing significant change, another pitfall can be inadequate temporal resolution and the inability to correctly identify the important processes.

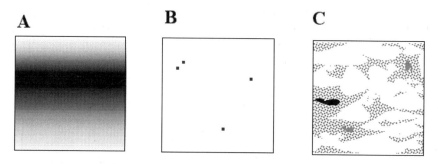

FIGURE 10.2 Examples of potential distribution patterns of a microbiological property. Increased darkness of shading represents a higher level of the property. (A) Gradient model; (B) rare microcolony model; (C) clumping model with two scales of excluding and nonexcluding conditions, plus a zone of optimal conditions.

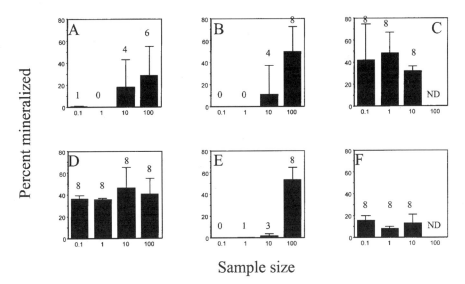

Sample size

FIGURE 10.3 Effect of sample size on determination of glucose/acetate/succinate mineralization in vadose zone environments with different amounts of average annual moisture recharge. Core material was not homogenized prior to sampling. Eight replicate 0.1-, 1-, 10-, and 100-g samples were assayed. Values above bars indicate the number of replicates in which activity was detected. Means and standard deviations are for the activity-positive samples. Panel (*a*): recharge of 15-μm, 60-day aerobic incubation. Panel (*d*): B horizon of surface soil (control), 7-day aerobic incubation. Panels (*b*) and (*e*): recharge of 1-mm, 60-day aerobic and anaerobic incubations, respectively. Panels (*c*) and (*f*): recharge of 20-cm, 3-day aerobic, and 26-day anaerobic incubation, respectively. ND=assay not done. Deoxygenated anaerobic incubations contained nitrate, amorphous Fe(III), and sulfate.

Both spatial and temporal variability is further complicated by the fact that supporting geologic, hydrologic, geochemical, and microbiological information is often collected at different locations and at different scales. Different locations refer to separate boreholes, samples meters apart in the same borehole, and samples from different geologic material in the same core. To a large degree, the analytical methods dictate the maximum size of an individual sample that is characterized. For example, a borehole pump test is used to determine hydraulic conductivity, a large core (several kilograms) is used for centrifugation and analysis of pore water from nonaquifer materials, and tens of grams of sediment are used for microbiological analysis.

Differences in location and scale would have no impact if subsurface properties were constant over large distances. However, we know that individual properties in the subsurface have continuity over certain distances and that different properties can be spatially correlated to one another over certain distances. Use of different scales of measurement and "distant" sampling locations for different measurements reduces the probability of understanding the system because much of the information content has been lost by decoupling the measurements. In addition, the spatial dependence of the data makes it difficult to apply classical statistical methods (e.g., analysis of variance) because classical methods assume independence of the data. Geostatistics is a specialized form of statistics that does not assume all samples are independent. Instead, geostatistics evaluates the degree of spatial dependence of a property, or the degree of spatial correlation between properties, as a function of distance between samples and builds probabilistic models of the spatial dependence or spatial correlation. Sampling plans for geostatistical analysis have special design considerations and require a large number of samples. However, the analysis can provide very important information for the design of improved sampling strategies, be used to predict values at unmeasured locations by kriging, co-kriging, or stochastic simulation, and provide new insight into ecological principles (Brockman and Murray, 1997a, b).

If the spatial structure of the properties being measured are not known (which is almost always the case in subsurface microbiological investigations), each sample should be analyzed for all properties of interest using (to the degree possible) similar subsample sizes. In some cases, technological limitations or sensitivity issues may require the use of larger subsamples for specific analyses. This approach does not preclude the use of larger or smaller samples for some measurements and collection of different sources of information from different sampling locations. Rather, it emphasizes the importance of *coupled information* at *similar scales* to optimize understanding of the system, and complementing this understanding with measurements that can be made at larger (or smaller) scales and/or additional locations. This strategy provides a stronger foundation from which to develop and test hypotheses, and to better extrapolate and interpret data between scales.

3.4 Extrapolating from Sequences to the Environment

The enormous variability of nucleic acid sequences (1.6×10^{60} possible variants for a 100-mer) give them unique identification power and information content, but this information also has an inherent weakness. It is important to realize that our knowledge of microbial diversity at the sequence level is limited: Probes and primers targeting specific genes require *a priori* knowledge of the target nucleic acid sequences. Moreover, from the standpoint of relating 16S or other sequences to specific metabolic processes (e.g., biogeochemical transformations), we are limited to those sequences that have been directly affiliated with the process in microorganisms which have been successfully cultivated and studied in the laboratory. Although there is a relatively large amount of knowledge on eubacterial 16S rRNA genes, for example, our knowledge base is still quite limited relative to the total estimated microbial diversity in soils and sediments. In contrast, there is much less knowledge of sequence divergence for genes involved in biogeochemical processes, contaminant biotransformation, and other cellular processes. The problems of limited sequence information and imperfect knowledge of probe/primer specificity have significant implications for the extrapolation of information from individual gene sequences to the structure, metabolic potentials, and metabolic processes of the microbial community.

The challenges of data interpretation arise from three sources of uncertainty. First, probes/primers are specific only in regard to current sequence information. Over the last decade, probes/primers have often been shown to be less specific than originally thought, a pattern that is likely to accelerate as a result of genomic sequencing efforts and other research. Therefore, although one must conduct research based on present-day database searches, use of probes in environmental contexts does not guarantee that *only* the target phylogenetic group or metabolic trait of interest is being detected. For this reason, collaborating chemical information or additional characterization of the hybridized band or amplicon using restriction enzymes or "redundant" probes or primers is (ideally) desirable.

Second, functionally equivalent proteins (carrying out the same process) may contain limited nucleic acid sequence identity. For example, different probes/primers are required to characterize the five (known) toluene degradation pathways and the two (known) variants of microbial nitrite reductase. As more sequence information becomes available, it is inevitable that additional probes and primer sets will be required to inclusively cover a metabolic trait or phylogenetic group of interest. Therefore, failure to detect a given sequence does not necessarily indicate the absence of the associated trait or group.

Third, geochemically equivalent endpoints can be mediated by phylogenetically divergent organisms and by different processes. For example, dissimilatory iron reduction (DIR), an important biogeochemical process in subsurface ecosystems, is performed by members of the genera *Geobacter, Desulfotomaculum, Thiobacillus, Shewanella, Bacillus, Pseudomonas, Pelobacter, Staphylococcus, Clostridium, Sulfolobus*, and others, including fungi (Ehrlich, 1996). Thus, it may not be possible to *ensure* accurate measurement of DIR potential or activity based solely on a limited suite of nucleic acid sequences. In addition, divergent processes can contribute to produce the same bulk geochemical measurement: microbial DIR, microbial iron oxidation, abiotic iron oxidation, and abiotic iron reduction. Small-scale pH and E_h distributions, the mineralogic form and small-scale distribution of Fe(II) and Fe(III), and the amount and small-scale distribution of reactants can all play a role in the relative contributions of these processes to a bulk geochemical measurement. Thus, systems with similar iron geochemistry may have very different DIR populations or activity, undergirding the importance of (accurate) culture-independent nucleic acid-based data. On the other hand, systems with similar hybridization or PCR data for DIR may have very different iron geochemistry. Together, these uncertainties highlight the requirement for joint chemical and nucleic acid information to understand biogeochemical processes and systems.

Iron reduction, polyaromatic and long-chain hydrocarbon degradation, and many other phenotypic traits and peripheral metabolisms are distributed in the phylogenetic tree such that 16S sequences cannot be associated with a particular metabolism or function. However, it should be noted that methanogenesis, obligate methanotrophy, sulfate reduction, and ammonia oxidation occur in closely related phylogenetic groups such that metabolic phenotype and phylogeny go hand in hand. For these metabolisms, probes and primers based on 16S rRNA sequences are fairly accurate predictors of phenotype and function. Therefore, data generated from these probes/primers are useful for addressing hypotheses and making inferences that link physiology, ecology, and geochemistry.

Aside from linking metabolic phenotype and phylogeny, a major goal of 16S characterization of clone libraries from community DNA or RNA is to illuminate heretofore unknown microbial diversity. Results of microbial diversity surveys in the subsurface show that many of the sequences represent undescribed close relatives of common soil microorganisms. Some sequences are very different from known microorganisms or sequences and are termed "novel." For example, a clade of organisms was identified from the subterranean groundwaters of the Oklo uranium deposit with no known relatives in the EMBL sequence database (Pedersen et al., 1996 b). A deeply rooted clade of eubacteria most closely related to Sargasso Sea clones and the phototroph *Chloroflexus* and a novel lineage of (presumably mesophilic) crenarchaea were discovered in a hydrologically remote deep subsurface paleosol very distant from any past or present hydrothermal activity (Chandler et al., 1998) (Fig. 10.4). Interestingly, the subsurface mesophilic crenarchaea were most closely related to sequences recovered from the thermophilic hot springs of Yellowstone National Park rather than to mesophilic crenarchaea sequences in near surface and open ocean environments (Buckley et al., 1998). A third example is the discovery of sequences representing six candidate novel eubacterial divisions at a shallow aquifer undergoing intrinsic bioremediation of hydrocarbons (Dojka et al., 1998).

Although such discoveries are important from the perspective of microbial diversity and a more comprehensive understanding of evolution, it should be emphasized that molecular systematics is a highly theoretical field and the microbial evolutionary tree is still extremely sparse relative to the huge amount of undiscovered diversity. Phylogenetic interpretations are guaranteed to evolve and change in the future with the huge influx of new 16S sequence information from diversity surveys and advancements in molecular systematics. Sequences that are currently "novel" and unique to the subsurface may be found to be common in surface soil and other environments as knowledge of microbial diversity increases. More important from the perspective of understanding subsurface processes, it is critical to conduct sampling for diversity surveys coincident with extensive chemical and physical characterization. This approach enables the possibility of establishing

→

FIGURE 10.4 Novel eubacterial and archaeal 16S rDNA clones recovered from a deep (188-m) subsurface paleosol in Washington state. (*a*) Phylogenetic affiliations of deep subsurface eubacterial clones inferred from maximum parsimony analysis of 1215 unambiguous nucleotide positions. This analysis is consistent with those reported elsewhere (Chandler et al., 1998; Buckley et al., 1998). GenBank accession numbers for subsurface eubacterial clones are AF005745-AF005750. (*b*) Phylogenetic affiliations of deep subsurface Archaea clones inferred from maximum likelihood analysis of 784 unambiguous nucleotides. This analysis is consistent with previous reports for similar sequences (Chandler et al., 1998; Giovannoni et al., 1996; Barns et al., 1996). Accession numbers for the deep subsurface archaeal clones are AF005752-AF005767. For both panels, sequences retrieved from GenBank are shown with their respective accession number. All other sequences were retrieved from the Ribosomal Database Project. Values at each bifurcation indicate the average bootstrap value deduced from 300 replications.

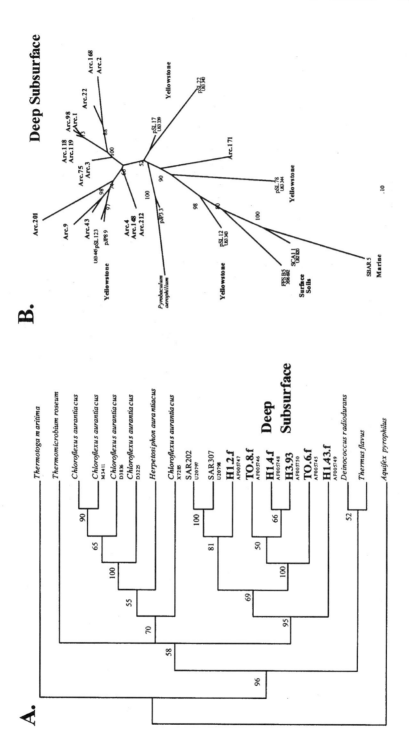

relationships or hypotheses between the presence of dominant sequences and key subsurface processes, as has been demonstrated for 16S rDNA (Dojka et al., 1998) and rRNA (Chandler et al., 1998).

The use of highly specific primers for DNA amplification can be problematic in environments with very low nutrient flux and/or clay content where living cells are "archived" (preserved but inactive) or extracellular DNA is protected from degradation (Aardema et al., 1983; Khanna and Stotzky, 1992; Ogram et al., 1994; Romanowski et al., 1991). In these conditions, a significant portion of the amplified product may be a result of past processes and, if interpreted to be the result of current conditions, would yield incorrect conclusions. On the other hand, sequences arising from archived cells and preserved DNA might provide insight into past conditions if the sequence(s) grouped tightly with a clade of microorganisms that have a common energy metabolism and that metabolism differs from the current geochemical conditions. For example, the presence of many methanogen sequences at a contaminated site with little methane and few methanotrophs could indicate that a large fraction of the readily bioavailable contaminant has been degraded.

4   POSSIBILITIES AND BIOGEOCHEMICAL IMPLICATIONS

Nucleic acid analysis is a superb tool for understanding basic biology, but nucleic acid analysis alone is of little value for understanding microbial ecology and biogeochemistry. Classic methods such as microbial enrichments and geochemical measurements will continue to be a primary source of information. A host of available technologies in the physical, chemical, and nucleic acid analysis arenas provide great promise for major advancements in the understanding of microbial processes as they occur in the environment. Some technologies can be applied in the field; other technologies are currently restricted to the laboratory but can be used to interrogate systems ranging from simple constructed batch systems to core material and full cores. The objective of this section is to discuss how these physical, chemical, and nucleic acid analysis technologies can be used in concert to more powerfully probe biogeochemical processes and microbial ecology in field and laboratory studies. By way of example, the focus will be on intact cores.

As noted in Section 3.3, a natural environment is a complex mosaic of a very large number of individual physical, hydrologic, geochemical, and microbiological properties and processes that are spatially heterogeneous, overlapping, and interactive. Although it is simple enough to make measurements, issues of optimal observation scale versus scale of characterization, spatial and temporal variability and resolution, and decoupled measurements reduce the ability to extract certain types of information from complex systems. A greater emphasis on spatially coupled measurements of important physical, hydrologic, geochemical, and microbiological properties at a similar scale, and where possible, using geostatistical methods would likely improve our understanding of specific mechanisms and relationships in complex systems. Another factor that impedes our understanding is that biological and chemical reactions are often slow in rate and nonlinear under many subsurface conditions. Approaches that address this problem include the use of introduced solutes and installation of sensors that minimally disrupt the system and provide long-term data, as well as more extensive manipulative experiments designed to accelerate an *in situ* process or promote a new process. Both of these general approaches—spatially coupled measurement properties at a similar scale and manipulation experiments in core in which *in situ* conditions have been preserved—are tractable with today's technologies.

Generally speaking, what is different today compared to one or more decades ago is a much greater ability to make more specific, sensitive, and rapid measurements across a much wider range of scales, particularly small scales. It is now possible to analyze core materials at the nanometer to millimeter scale (to varying degrees depending on the particular property or process being analyzed) and evaluate the spatial relationships between mineral type, metal species, characteristics of fluid transport, microorganisms of particular types, and reactivity of inorganic and organic contaminants. This enables previously intractable questions to be answered: "Where do the dominant processes occur?", "To what degree are the microorganisms capable of a transformation of interest co-located with certain mineral types, physical matrices, or contaminants?", and "Do injected chemicals have sufficient access to locations containing microorganisms capable of a transformation of interest, and over what time periods?"

Six types of measurements are potentially useful in studying the spatial and temporal distribution of solid- and/or liquid-phase analytes in intact cores (Fig. 10.5). Several of the specific technologies are capable of measuring both biological and nonbiological properties. The specific technologies shown are intended to be

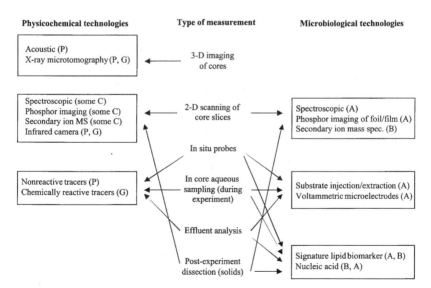

FIGURE 10.5 Type of measurement, specific technologies, and properties measured. A, microbial activity; B, microbial biomass; C, contaminants; G, geochemical and mineralogic properties; P, physical and hydrologic properties.

representative rather than inclusive. Three-dimensional imaging of cores and two-dimensional scanning of faces of core slices are particularly valuable because they are nondestructive and nonintrusive, thereby preserving the spatial integrity and heterogeneity of structural, mineralogic, geochemical, and biologic features. Data from three-dimensional imaging can be used to refine hypotheses and design experiments for core manipulation, select locations for *in situ* probes, and identify important post-experiment sampling locations. Two-dimensional scanning technologies offer the potential to make large numbers ($\geq 10^3$) of high-resolution (nanometer to millimeter) measurements across areas as large as tens of cm^3 in hours to days.

Spatially resolved maps of elements or molecules amenable to microbial transformation can be produced by hard and soft X-ray absorption and by vibrational (infrared) spectroscopy methods. These methods provide information on composition, concentration, and speciation of elements or molecules. Synchrotron light sources arc available at four national user facilities: Argonne National Laboratory, Brookhaven National Laboratory, Lawrence Berkeley National Laboratory, and Stanford Synchrotron Radiation Laboratory. In contrast to the vacuums needed for electron microprobes and electron microscopy, X-ray absorption can be performed in the presence of normal atmospheres and in aqueous samples and infrared can be performed in normal atmospheres in samples that do not contain free water. Radiolabeled substrates that form precipitates by microbial metabolism can be applied to fresh core faces, and two-dimensional maps of activity recorded by real-time imaging of the core face or contact foil (Krumholz et al., 1997; Ulrich et al., 1998). Secondary ion mass spectrometry can be used to map organic molecules and may also be useful for mapping membrane lipids indicative of microbial biomass (Todd et al., 1997).

Nonreactive tracers with different diffusion coefficients can be used to evaluate preferential flow and nonequilibrium mass transfer, and chemically reactive tracers can be used to evaluate specific mineral phases or surface-adsorbed reaction products (Jardine et al., 1998, 1999). Solid-state voltammetric microelectrodes can measure multiple redox pairs in microniches and the technology is, in principle, deployable in cores and the shallow subsurface (Luther et al., 1998, 1999). Metabolism-specific or general microbial substrates can be injected into cores or in the field, allowed to react for a given time period, extracted, and substrate and product analyzed to evaluate metabolic potential (Istok et al., 1997; Schroth et al., 1998).

Figure 10.6 illustrates one way in which technologies could be integrated and sequenced to enhance understanding of biogeochemical processes and relationships between environmental and microbiological properties. The figure and approach outlined below was generated during a workshop sponsored by the Natural and Accelerated Bioremediation Research program within the U.S. Department of Energy. It should be recognized that many variations are possible in number and arrangement of cores and core subsamples, and the features of the subsurface strata will determine the priority and specific combination of technologies applied. Multiple "replicate" cores are depicted for several reasons including the potential impact of X-rays and acoustic waves on microorganisms, inherent variability in adjacent cores, the need for a "control" core, and assessing the impact of tracer

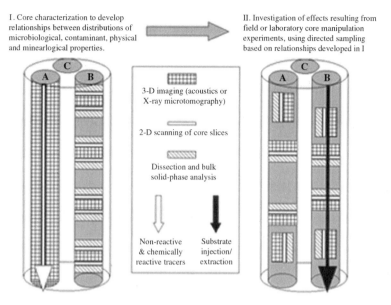

I. Core characterization to develop relationships between distributions of microbiological, contaminant, physical and minearlogical properties.

II. Investigation of effects resulting from field or laboratory core manipulation experiments, using directed sampling based on relationships developed in I

3-D imaging (acoustics or X-ray microtomography)

2-D scanning of core slices

Dissection and bulk solid-phase analysis

Non-reactive & chemically reactive tracers

Substrate injection/ extraction

FIGURE 10.6 Conceptual representation of approach for conducting studies in cores, including the need for pseudoreplication of cores.

studies on microbiology. Replicate cores could be from a single large-diameter core (as depicted in Fig. 10.4), vertically adjacent cores in a visually homogeneous unit, or cores at the same stratigraphic location in closely spaced boreholes.

The objective of core I.A. would be to develop relationships regarding the distribution of microbiological, physical, and mineralogical properties, and if present, contaminants. Physical and geochemical heterogeneity would be imaged by acoustics and X-ray microtomography (XMT). Nonreactive and chemically reactive tracers would then be used to evaluate transport behavior and hydrologic properties using breakthrough curves (effluent analysis) and, if deemed useful, in subsequent experiments by in-core aqueous sampling and/or probes installed in the core. Following tracer tests, the core would be dissected and analyzed to supplement tracer test data and calibrate the physical measurements to actual core physical properties. Core I.B. would be reserved for technologies devoted to two-dimensional scanning of core slices; bulk solid-phase analysis of mineralogy, contaminant speciation, signature lipid biomarkers, and nucleic acids; and XMT of subcores. Note that samples for these measurement approaches are shown as adjacent to one another to provide tight spatial linkage of the measurements. Stratifying the bulk solid-phase samples based on visual differences in texture, weathering, or mineralogy would provide a framework for determining the degree to which correlations found in the two-dimensional scanning data sets are lost or retained in the bulk solid-phase data sets. Subcores for XMT provide a means for extrapolating information in core I.B. to the full three-dimensional image of properties in core I.A. Core I.C. could be used as a replicate core for either I.A. or I.B.

The objective of core II would be to investigate changes in relationships and processes resulting from chemical manipulation of the core. Core II would need to be carefully archived to maintain the *in situ* chemical conditions. Sampling would be targeted, with sample location and orientation determined by the relationships discovered during analysis of core I, which may result in a fewer number of required samples. Core II.A. would be analyzed in the same manner as core I.B. and serve as a control for the manipulations performed in core II.B. and for relating core I to core II. If (in separate experiments) acoustic or XMT energy were found to have little to no impact on microbial viability and activity, one or both three-dimensional imaging methods could be used prior to other measurements to further evaluate similarities and differences between core I and II. Core II.B. would be used to measure microbial activity at the core scale and at specific locations in the core using one or more of the following approaches: effluent analysis, analysis of water sampled after injection of a substrate that is microbially transformed, aqueous sampling via ports in the core, and probes installed in the core. If substrate was added for a short period of time to determine initial activity, the same substrate or a different substrate(s) could be added over longer periods of time to evaluate the dynamic response in activity and redox conditions over time. After completion of the activity measurements, core II.B. would be analyzed in the same manner as cores I.B. and II.A. Core II.C. could be used as a replicate core for either II.A. or II.B.

The implications of these advanced technologies—particularly when used in tandem, at similar scales, and (as appropriate) through the careful use of select manipulations—has tremendous potential to increase the understanding of biogeo-chemical processes and relationships between environmental and microbiological properties. For example, they enable fine-scale spatial relationships between distri-butions of the oxidation state of a metal, the immediate chemical environment, pore structure, and biomass. Such knowledge is invaluable for predicting the utility and stability of toxic elements that can be stabilized by microbial transformations, and evaluation of *in situ* conditions that could be manipulated to optimize such transformations. Organic substrates that produce a spectroscopic signature or fluorescence upon microbial attack could be used to label specific microbial types *in situ* and provide additional information on microbial activity. Other microbiolo-gical technologies such as cell robotics workstations that enable physical removal of individual cells from surfaces using "laser tweezers," and confocal laser scanning microscopy used in conjunction with chemicals sensitive to metabolic processes, also have high value for studying biogeochemical mechanisms and relationships between microorganisms and their immediate environment.

5 TECHNOLOGY DEVELOPMENT FOR CULTURE-INDEPENDENT NUCLEIC ACID-BASED METHODS

Prior to 1995, a drawback of nucleic acid analysis in environmental samples was a lack of rapid and cost-effective analysis methods, particularly for sample processing.

Several ongoing technological developments are reducing this problem. For example, systems for automated purification of environmental samples including efficient removal of PCR inhibitors have been developed (e.g., Chandler et al., 1999, 2000). A miniaturized *Taq*-man PCR assay system capable of detecting as few as 5 cells in 9 min (40 cycles) has been demonstrated (Belgrader et al., 1999). The convergence of molecular biology and microelectronics has resulted in DNA microarrays (or DNA "chips") and readers (Chee et al., 1996; de Saizieu et al., 1998; Lockhart et al., 1996; Schena et al., 1996; Zhang et al., 1997), providing the technology for massively parallel hybridization analysis of microbial communities (Guschin et al., 1997) using genetic information in databases and from future 16S diversity surveys and sequenced bacterial genomes. Seamless sample-to-answer automated systems for analysis of environmental samples are on the horizon. In concert, these technologies will allow us to extract more information from a single sample than currently possible, while being more rapid and economical than existing techniques.

Although PCR-based analysis has enabled great strides in our understanding of microbial ecology, numerous limitations remain (see Sections 3.1 and 3.2). Consequently, for nucleic acid analysis to reach its full potential in deep subsurface ecosystems, radically new methods for the *direct* detection (i.e., without amplification) of nucleic acids in sediments are required. From where will these techniques arise? There are numerous technologies with the potential to realize direct detection of low-copy-number targets in environmental matrices. For example, electrokinetic nucleic acid concentration (Sosnowski et al., 1997) may provide the means for effective nucleic acid recovery from very dilute suspensions. Nucleic acid analogues (Egholm et al., 1993; Iyer et al., 1995; Peffer et al., 1993) with chemically modified backbones and electrostatic properties might overcome some of the humic acid inhibition, thermodynamic, and kinetic limitations of current nucleic acid hybridization and detection techniques. Extremely rapid analysis of individual cells in pure buffers using labeled oligonucleotide probes or antibodies and fluorescence-activated cell-sorting instruments is commonplace. Similar instrumentation with lasers, photon counters, and sophisticated electronics has been developed for the detection of a single fluorescently labeled gene in unamplified genomic DNA (Castro and Williams, 1997), an approach that could potentially be applied to environmental samples when linked to appropriate sample processing methodologies. Although sensitivities are currently poor compared to fluorescence methods, ever more powerful mass spectrometry techniques are being rapidly developed and applied specifically for the direct detection, sequencing, and identification of nucleic acids and whole microbial cells (Doktycz et al., 1995; Muddiman et al., 1996; Roskey et al., 1996; Krishnamurthy et al., 1996). Electrochemical sensors, relying on the redox potential of nucleotide bases in DNA, are able to detect nucleic acids with sensitivities that rival many current optical or fluorescent techniques (Fojita and Palecek, 1997; Jelen et al., 1997; Wang et al., 1996, 1997; Palecek and Fojita, 1994). Finally, detectors based on electron transport through DNA hybrids offer potential for overcoming the inhibitory or obscuring effects of humic acids and other soluble contaminants pervasive in terrestrial subsurface nucleic acid extracts.

6 SUMMARY

This chapter has focused on culture-independent nucleic acid techniques for the identification and characterization of *in situ* microbiota and the relationship between analytical techniques and the *original* environmental condition. Although molecular methods have distinct advantages over culture-based techniques, they also have numerous limitations. We therefore assert that studies of microbial ecology and biogeochemistry are better served by combining nucleic acid analysis with other approaches, including more standard techniques such as enrichment cultures and signature lipid biomarker analysis.

Throughout the chapter, we emphasized the necessity of collecting microbiological information coincident with extensive chemical and physical characterization. A number of issues related to sample representativeness and spatial and temporal resolution were examined. Linked measurements of multiple properties at similar scales and locations are important, regardless of whether the measurements are at the nanometer or field scale. Advanced technologies that provide information on environmental properties in bulk samples and by spatial analysis at very fine scales were presented. Attention was focused on how these technologies could be used in concert with microbiological techniques to more powerfully probe biogeochemical processes and microbial ecology in cores. Although such investigations are complex and technically challenging, they could provide critical and novel information and insight, with profound implications for a greater understanding of relationships between microorganisms and biogeochemical processes.

We emphasized that biomass can vary many orders of magnitude in subsurface environments, but many environments have biomass levels and/or other aspects of the sample that limit which molecular techniques are possible or applicable. In particular, limitations of PCR specifically as they relate to low-biomass samples and low-copy-number amplifications were discussed. Difficulties in interpretation of nucleic acid data and extrapolating information from individual gene sequences to the structure, metabolic potentials, and metabolic processes of the microbial community were discussed. Clearly, knowledge of gene sequences and metabolic processes will be greatly accelerated through whole genome sequence analysis and bioinformatics. With better sequence information comes better predictive and determinative power in a biogeochemical context, such that a broader cross section of microbial diversity and metabolic potential/activity can be appraised through direct nucleic acid analysis. Successful transfer of bioinformatics and DNA microarray technologies such as community profiling and expression (RNA) analysis to the environmental sciences, however, will ultimately require specific attention to the "pitfalls" of molecular analyses described here and in other reviews of environmental molecular biology.

Developments are well underway in automated processing of nucleic acids, rapid PCR, and DNA microarrays for analysis of environmental samples. We highlighted several technologies that might eventually contribute to the detection of low-copy-number sequences without nucleic acid amplification. These advancements enable highly multiplexed analyses that are also rapid and economical. Even with the explosion of sequence information and the power of analytical instrumentation being developed in other disciplines, however, molecular investigations of subsurface microbiota often continue to be handicapped by detection problems associated with low biomass and the inability to analyze large volumes. In the same way that significant advances in deep subsurface microbiology and biogeochemistry arose from the integration of microbiology, geology, hydrology, geochemistry, and engineering, the next generation of molecular tools for low-copy-number detection in environmental matrices will coalesce from nascent technologies in molecular biology, physics, chemistry, and electronics. Biology is in a unprecedented golden age driven by rapid technological advancement. Likewise, biogeochemistry and microbial ecology are entering an age of new possibilities driven by our increasing ability to understand microbial distributions of biomass and activities within the context of environmental properties, and across scales ranging from nanometers to tens of meters, in undisturbed and minimally disturbed samples.

Acknowledgments   This work was supported by the U.S. Department of Energy's (DOE) Natural and Accelerated Bioremediation Program (NABIR). Pacific Northwest National Laboratory is operated by Battelle Memorial Institute for the U.S. DOE under contract DE-AC06-76RLO 1830.

REFERENCES

Aardema BW, Lorenz MG, and Krumbein WE (1983) Protection of sediment-adsorbed transforming DNA against enzymatic inactivation. *Appl Environ Microbiol 46*:417–420.

Albrechtsen H-J and Winding A (1992) Microbial biomass and activity in subsurface sediments from Vejen, Denmark. *Microbial Ecol 23*:303–317.

Al-Soud WA and Rådströp (1998) Capacity of nine thermostable DNA polymerases to mediate DNA amplification in the presence of PCR-inhibiting samples. *Appl Environ Microbiol 64*:3748–3753.

Amy PS (1997) Microbial dormancy and survival in the subsurface. In Amy PS and Haldeman DL (eds): *The Microbiology of the Terrestrial Deep Subsurface*. New York: Lewis Publishers, pp 185–203.

Balkwill DL (1989) Numbers, diversity, and morphological characteristics of aerobic, chemoheterotrophic bacteria in deep subsurface sediments from a site in South Carolina. *Geomicrobiol J 7*:33–52.

Balkwill DL and Boone DR (1997) Identity and diversity of microorganisms cultured from subsurface environments. In Amy PS and Haldeman DL (eds): *The Microbiology of the Terrestrial Deep Subsurface*. New York: Lewis Publishers, pp 105–117.

Balkwill DL, Fredrickson JK, and Thomas JM (1989) Vertical and horizontal variations in the physiological diversity of the aerobic chemoheterotrophic bacterial microflora in deep southeast coastal plain subsurface sediments. *Appl Environ Microbiol 55*:1058–1065.

Balkwill DL and Ghiorse WC (1985) Characterization of subsurface bacteria associated with two shallow aquifers in Oklahoma. *Appl Environ Microbiol 50*:580–588.

Barns, SM, Fundyga, RE, Jeffries, MW, and Pace, NR (1944) Remarkable archaeal diversity detected in a Yellowstone National Park hot spring environment. *Proc. Natl. Acad. Sci. USA* *91*:1609–1613.

Belgrader PW, Benett W, Hadley D, Richards J, Stratton P, Mariella R Jr, and Milanovitch F (1999) PCR Detection of bacteria in seven minutes. *Science 284*:449–450.

Beloin RM, Sinclair JL, and Ghiorse WC (1988) Distribution and activity of microorganisms in subsurface sediments of a pristine study site in Oklahoma. *Microbial Ecol 16*:85–97.

Bock M, Kämpfer P, Bosecker K, and Dott W (1994) Isolation and characterization of heterotrophic, aerobic bacteria from oil storage caverns in northern Germany. *Appl Microbiol Biotechnol 42*:463–468.

Boivin-Jahns V, Ruimy R, Bianchi A, Daumas S, and Christen R (1996) Bacterial diversity in a deep-subsurface clay environment. *Appl Environ Microbiol 62*:3405–3412.

Bone TL and Balkwill DL (1988) Morphological and cultural comparison of microorganisms in surface soil and subsurface sediment at a pristine site in Oklahoma. *Microbial Ecol 16*:49–64.

Boone DR, Liu Y, Zhao ZJ, Balkwill DL, Drake GR, Stevens TO, and Aldrich HC (1995) *Bacillus infernus* sp. nov., an Fe(III)- and Mn(IV)-reducing anaerobe from the deep terrestrial subsurface. *Internat J Syst Bacteriol 45*:441–448.

Britten RJ, Graham DE, and Neufeld BR (1974) Analysis of repeating DNA sequences by reassociation. *Meth Enzymol 29*:363–418.

Brockman FJ (1995) Nucleic acid-based methods for monitoring the performance of *in situ* bioremediation. *Molec Ecol 4*:567–578.

Brockman FJ, Kieft TL, Fredrickson JK, Bjornstad BN, Li SW, Spangenburg W, and Long PE (1992) Microbiology of vadose zone paleosols in south-central Washington state. *Microbial Ecol 23*:279–301.

Brockman FJ, Li SW, Fredrickson JK, Ringelberg DB, Kieft TL, Spadoni CM, McKinley JP, and White DC (1998) Post-sampling changes in microbial community structure and activity in a subsurface paleosol. *Microbial Ecol 36*:152–164.

Brockman FJ and Murray CJ (1997a) Microbiological heterogeneity in the terrestrial subsurface and approaches for its description. In Amy PS and Haldeman DL (eds): *The Microbiology of the Terrestrial Deep Subsurface*. New York: Lewis Publishers, pp 75–102.

Brockman FJ and Murray CJ (1997b) Subsurface microbiological heterogeneity: Current knowledge, descriptive approaches, and applications. *FEMS Microbiol Rev 20*:231–247.

Brockman FJ, Murray CME, Bjornstad B, Balkwill D, Ringelberg D, Pfiffner S, and Griffiths R (1997) Microbial life in the unsaturated subsurface under conditions of extremely low recharge: An extreme environment. In *Instruments, Methods, and Missions for the Investigation of Extraterrestrial Microorganisms*. Hoover RB (ed), San Diego, CA, The International Society for Optical Engineering, pp 388–394.

Buckley DH, Graber JR, and Schmidt TM (1998) Phylogenetic analysis of nonthermophilic members of the kingdom *Crenarchaeota* and their diversity and abundance in soils. *Appl Environ Microbiol 64*:4333–4339.

Castro A and Williams JGK (1997) Single-molecule detection of specific nucleic acid sequences in unamplified genomic DNA. *Anal Chem 69*:3915–3920.

Chandler DP (1998) Redefining relativity: Quantitative PCR at low template concentrations for industrial and environmental microbiology. *J Industrial Microbiol 21*:128–140.

Chandler DP and Brockman FJ (1996) Estimating biodegradative gene numbers at a JP-5 contaminated site using PCR. *Appl Biochem Biotechnol 57/58*:971–982.

Chandler DP, Brockman FJ, and Fredrickson JK (1997a) Effect of PCR template concentration on the composition and distribution of total community 16S rDNA clone libraries. *Molec Ecol 6*:475–482.

Chandler DP, Brockman FJ, and Fredrickson JK (1998) Phylogenetic diversity of Archaea and Bacteria in a deep subsurface paleosol. *Microbial Ecol 36*:37–50.

Chandler DP, Holman DA, Brockman FJ, Grate JW, and Bruckner-Lea CJ (2000) Renewable microcolumns for solid-phase nucleic acid separations and analysis from environmental samples. *Trends Anal Chem 19*:314–321.

Chandler DP, Schreckhise RW, Smith JL, and Bolton H Jr (1997b) Electroelution to remove humic compounds from soil DNA and RNA extracts. *J Microbiol Meth 28*:11–19.

Chandler DP, Schuck BL, Brockman FJ, and Bruckner-Lea CJ (1999) Automated nucleic acid isolation and purification from soil extracts using renewable affinity microcolumns in a sequential injection system. *Talanta 49*:969–983.

Chee M, Yang R, Hubbell E, Berno A, Huang XC, Stern D, Winkler J, Lockhart DJ, Morris MS, and Fodor SPA (1996) Accessing genetic information with high-density DNA arrays. *Science 274*:610–614.

Colwell FS (1989) Microbiological comparison of surface soil and unsaturated subsurface soil from a semiarid high desert. *Appl Environ Microbiol 55*:2420–2423.

de Saizieu A, Certa U, Warrington J, Gray C, Keck W, and Mous J (1998) Bacterial transcript imaging by hybridization of total RNA to oligonucleotide arrays. *Nat Biotech 16*:45–48.

Dojka MA, Hugenholtz P, Haack SK, and Pace NR (1998) Microbial diversity in a hydrocarbon- and chlorinated-solvent-contaminated aquifer undergoing intrinsic bioremediation. *Appl Environ Microbiol 64*:3869–3877.

Doktycz MZ, Hurst GB, Habibi-Goudarzi S, McLuckey SA, Tang K, Chen CH, Uziol M, Jacobson KB, Woychik RP, and Buchanan MV (1995) Analysis of polymerase chain reaction-amplified DNA products by mass spectrometry using matrix-assisted laser desorption and electrospray: Current status. *Anal Biochem 230*:205–214.

Egholm M, Buchardt O, Christense L, Behrens C, Freier S, Driver DA, Berg RH, Kim SK, Norden B, and Nielsen PE (1993) PNA hybridizes to complementary oligonucleotides obeying the Watson-Crick hydrogen bonding rules. *Nature 365*:566–568.

Ehrlich HL (1996) *Geomicrobiology.* New York: Marcel Dekker.

Ekendahl S, Arlinger J, StÅhl F, and Pedersen K (1994) Characterization of attached bacterial populations in deep granitic groundwater from the Stripa research mine by 16S rRNA gene sequencing and scanning electron microscopy. *Microbiology 140*:1575–1583.

Farrelly V, Rainey FA, and Stackebrandt E (1995) Effect of genome size and *rrn* gene copy number on PCR amplification of 16S rRNA genes from a mixture of bacterial species. *Appl Environ Microbiol 61*:2798–2801.

Fojita M and Palecek E (1997) Supercoiled DNA-modified mercury electrode: A highly sensitive tool for the detection of DNA damage. *Anal Chim Acta 342*:1–12.

Francis AJ, Slayer JM, and Dodge CJ (1989) Denitrification in deep subsurface sediments. *Geomicrobiol J 7*:103–116.

Fredrickson JK, Garland TR, Hicks RJ, Thomas JM, Li SW, and McFadden KM (1989) Lithotrophic and heterotrophic bacteria in deep subsurface sediments and their relation to sediment properties. *Geomicrobiol J 7*:53–66.

Fredrickson JK, Li SW, Brockman FJ, Haldeman DL, Amy PS, and Balkwill DL (1995a) Time-dependent changes in viable numbers and activities of aerobic heterotrophic bacteria in subsurface samples. *J Microbiol Meth 21*:253–265.

Fredrickson JK, McKinley JP, Nierzwicki-Bauer SA, White DC, Ringelberg DB, Rawson SA, Li S-M, Brockman FJ, and Bjornstad BN (1995b) Microbial community structure and biogeochemistry of Miocene subsurface sediments: Implications for long-term microbial survival. *Molec Ecol 4*:619–626.

Frothington R, Blitchington RB, Lee DH, Greene RC, and Wilson KH (1992) UV absorption complicates PCR decontamination. *Bio Techniques 13*:208–210.

Fry NK, Fredrickson JK, Fishbain S, Wagner M, and Stahl DA (1997) Population structure of microbial communities associated with two deep, anaerobic, alkaline aquifers. *Appl Environ Microbiol 63*:1498–1504.

Fuller ME, Mu DY, and Scow KM (1995) Biodegradation of trichloroethylene and toluene by indigenous microbial populations in vadose sediments. *Microbial Ecol 29*:311–325.

Ghiorse WC and Balkwill DL (1983) Enumeration and morphological characterization of bacteria indigenous to subsurface environments. *Dev Industrial Microbiol 24*:213–223.

Ghiorse WC and Wilson JT (1988) Microbial ecology of the terrestrial subsurface. *Adv Appl Microbiol 33*:107–172.

Giovannoni, SJ, Rappé, MS, Vergin, KL, and Adair, NL (1996) 16S rRNA genes reveal stratified open ocean bacterioplankton populations related to the Green Non-Sulfur bacteria. *Proc. Natl. Acad. Sci. USA 93*:7979–7984.

Guschin DY, Mobarry BK, Proudnikov D, Stahl DA, Rittmann BE, and Mirzabekov AD (1997) Oligonucleotide microchips as genosensors for determinative and environmental studies in microbiology. *Appl Environ Microbiol 63*:2397–2402.

Haldeman DL and Amy PS (1993) Bacterial heterogeneity in deep subsurface tunnel at Rainier Mesa, Nevada test site. *Microbial Ecol 25*:183–194.

Haldeman DL, Amy PS, Ringelberg D, and White DC (1993) Characterization of the microbiology within a 21 m^3 section of rock from the deep subsurface. *Microbial Ecol 26*:145–159.

Haldeman DL, Amy PS, White DC, and Ringelberg DB (1994) Changes in bacteria recoverable from subsurface volcanic rock samples during storage at 4 °C. *Appl Environ Microbiol 60*:2697–2703.

Hazen TC, Jimeniz L, Lopez de VG, and Fliermans CB (1991) Comparison of bacteria from deep subsurface sediment and adjacent groundwater. *Microbial Ecol 22*:293–304.

Hicks RJ and Fredrickson JK (1989) Aerobic metabolic potential of microbial populations indigenous to deep subsurface environments. *Geomicrobiol J 7*:67–78.

Hilali F, Saulnier P, Chachaty E, and Andremont A (1997) Decontamination of polymerase chain reaction reagents for detection of low concentrations of 16S rRNA genes. *Molec Biotechnol 7*:207–216.

Hu C-Y, Allen M, and Gyllensten U (1992) Effect of freezing of the PCR buffer on the amplification specificity: Allelic exclusion and preferential amplification of contaminating molecules. *PCR Meth Appl 2*:182–183.

Hugenholtz P and Pace NR (1996) Identifying microbial diversity in the natural environment: a molecular phylogenetic approach. *TIBTECH 14*:190–197.

Istok JD, Humphrey MD, Schroth MH, Hyman MR, and O'Reilly KT (1997) Single-well, "push-pull" test method for *in situ* determination of microbial metabolic activities. *Ground Water 35*:619–631.

Iyer M, Norton JC, and Corey DR (1995) Accelerated hybridization of oligonucleotides to duplex DNA. *J Biol Chem 270*:14712–14717.

Jardine PM, O'Brien R, Wilson GV, and Gwo JP (1998) Experimental techniques for confirming and quantifying physical nonequilibrium processes in soils. In Selim HM and Ma L (eds): *Physical Nonequilibrium in Soils: Modeling and Application*. Chelsea MI: Ann Arbor Press, pp 243–271.

Jardine PM, Sanford WE, Gwo JP, Reedy OC, Hicks DS, Riggs RJ, and Bailey WB (1999) Quantifying diffusive mass transfer in fractured shale bedrock. *Water Resources Res 35*:2015–2030.

Jelen F, Tomschick M, and Palecek E (1997) Adsorptive stripping square-wave voltammetry of DNA. *J Electroanal Chem 423*:141–148.

Jones RE, Beeman RE, and Suflita JM (1989) Anaerobic metabolic processes in the deep terrestrial subsurface. *Geomicrobiol J 7*:117–130.

Kenzelmann M and Muhlemann K (1997) Pitfalls of PCR: Cross-reactivity with joyride *E. coli* nucleic acid. *Bio Techniques 23*:204–206.

Khanna M and Stotzky G (1992) Transformation of *Bacillus subtilus* by DNA bound on monmorillonite and effect of DNase on the transforming ability of bound DNA. *Appl Environ Microbiol 58*:1930–1939.

Kieft TL, Amy PS, Brockman FJ, Fredrickson JK, Bjornstad BN, and Rosacker LL (1993) Microbial abundance and activities in relation to water potential in the vadose zones of arid and semiarid sites. *Microbial Ecol 26*:59–78.

Kieft TL, Fredrickson JK McKinley JP, Bjornstad BN, Rawson SA, Phelps TJ, Brockman FJ, and Pfiffner SM (1995) Microbiological comparisons within and across contiguous lacustrine, paleosol, and fluvial subsurface sediments. *Appl Environ Microbiol 61*:749–757.

Kieft TL and Phelps TJ (1997) Life in the slow lane: Activities of microorganisms in the subsurface. In Amy PS and Haldeman DL (eds): *The Microbiology of the Terrestrial Deep Subsurface*. New York: Lewis Publishers, pp 137–163.

Kieft TL, Wilch E, O'Connor K, Ringelberg DB, and White DC (1997) Survival and phospholipid fatty acid profiles of surface and subsurface bacteria in natural sediment microcosms. *Appl Environ Microbiol 63*:1531–1542.

Kreader CA (1996) Relief of amplification inhibition in PCR with bovine serum albumin or T4 gene 32 protein. *Appl Environ Microbiol 62*:1102–1106.

Krishnamurthy T, Ross PL, and Rajamani U (1996) Detection of pathogenic and non-pathogenic bacteria by matrix-assisted laser desorption/ionization time-of-flight mass spectrometry. *Rap Comm Mass Spec 10*:883–888.

Krumholz LR, McKinley JP, Ulrich GA, and Suflita JM (1997) Confined subsurface microbial communities in cretaceous rock. *Nature 386*:64–66.

Liesack W, Weyland H, and Stackebrandt E (1991) Potential risks of gene amplification by PCR as determined by 16S rDNA analysis of a mixed-culture of strict barophilic bacteria. *Microbial Ecol 21*:191–198.

Lockhart DJ, Dong H, Byrne MC, Folletti MT, Gallo MV, Chee MS, Mittmann M, Wang C, Kobayashi M, Horton H, and Brown EL (1996) Expression monitoring by hybridization to high-density oligonucleotide arrays. *Nature Biotechnol 14*:1675–1680.

Luther GWI, Brendel PJ, Lewis BL, Sundby B, Lefranois L, Silverberg N, and Nuzzio DB (1998) Simultaneous measurement of O2, Mn Fe, I- and S(-II) in marine porewaters with a solid-state voltammetric micro-electrode. *Limnol Ocean 43*:325–333.

Luther GWI Reimers CE, Nuzzio DB, and Lovalvo D (1999) *in situ* Deployment of voltammetric, potentiometric and amperometric microelectrodes from a ROV to determine O2, Mn, Fe, S(-2) and pH in porewaters. *Environ Sci Technol 33*:4352–4356.

Madsen EL and Bollag JM (1989) Aerobic and anaerobic microbial activity in deep subsurface sediments from the Savannah River plant. *Geomicrobiol J 7*:93–102.

Mathieu-Daudé F, Welsh J, Vogt T, and McClelland M (1996) DNA rehybridization during PCR: The "C_0t effect" and its consequences. *Nucl Acids Res 11*:2080–2086.

Meier A, Persing DH, Finken M, and Bottger EC (1993) Elimination of contaminating DNA within polymerase chain reaction reagents: Implications for a general approach to detection of uncultured pathogens. *J Clin Micro 31*:646–652.

Meyerhans A, Vartanian J-P, and Wain-Hobson S (1990) DNA recombination during PCR. *Nucl Acids Res 18*:1687–1691.

Muddiman DC, Wunschel DS, Liu C, Pasa-Tolic L, Fox KF, Fox A, Anderson GA, and Smith RD (1996) Characterization of PCR products from Bacilli using electrospray ionization FTICR mass spectrometry. *Anal Chem 68*:3705–3712.

Mutter GL and Boynton KA (1995) PCR bias in amplification of androgen receptor alleles, a trinucleotide repeat marker used in clonality studies. *Nucl Acids Res 23*:1411–1418.

Ogram A, Sun W, Brockman FJ, and Fredrickson JK (1995) Isolation and characterization of RNA from low-biomass deep-subsurface sediments. *Appl Environ Microbiol 61*:763–768.

Ogram AV, Mathot MI, Harsh JB, Boyle J, and Pettigrew CAJ (1994) Effects of DNA polymer length on its adsorption to soils. *Appl Environ Microbiol 60*:393–396.

Ogram AV and Sayler GS (1988) The use of gene probes in the rapid analysis of natural microbial communities. *J Industrial Microbiol 3*:281–292.

Olson BH and Tasi Y-L (1992) Molecular approaches to environmental management. In Mitchell R (ed): *Environmental Microbiology.* New York: John Wiley & Sons, pp 239–263.

Pääbo S, Irwin DM, and Wilson AC (1990) DNA damage promotes jumping between templates during enzymatic amplification. *J Biol Chem 265*:4718–4721.

Palecek E and Fojita M (1994) Differential pulse voltammetric determination of RNA at the picomole level in the presence of DNA and nucleic acid components. *Anal Chem 66*:1566–1571.

Parkes RJ, Cragg BA, Bale SJ, Getliff JM, Goodman K, Rochelle PA, Fry JC, Weightman AJ, and Harvey SM (1994) Deep bacterial biosphere in Pacific Ocean sediments. *Nature 371*:410–413.

Pedersen K, Arlinger J, Ekendahl S, and Hallbeck L (1996a) 16S rRNA gene diversity of attached and unattached bacteria in boreholes along the access tunnel to the Åspö hard rock laboratory, Sweden. *FEMS Microbial Ecol 19*:249–262.

Pedersen K, Arlinger J, Hallbeck L, and Pettersson C (1996b) Diversity and distribution of subterranean bacteria in groundwater at Oklo in Gabon, Africa, as determined by 16S rRNA gene sequencing. *Molec Ecol 5*:427–436.

Peffer HJ, Hanvey JC, Bisi JE, Thomson SA, Hassman CF, Noble SA, and Babis LE (1993) Strand-invasion of duplex DNA by peptide nucleic acid oligomers. *Proc Natl Acad Sci USA 90*:10648–10652.

Perry-O'Keefe H, Yao X-W, Coull JM, Fuchs M, and Egholm M (1996) Peptide nucleic acid pre-gel hybridization: An alternative to Southern hybridization. *Proc Natl Acad Sci USA 93*:14670–14675.

Phelps TJ, Murphy EM, Pfiffner SM, and White DC (1994) Comparison between geochemical and biological estimates of subsurface microbial activities. *Microbial Ecol 28*:335–349.

Phelps TJ, Raione EG, White DC, and Fliermans CB (1989) Microbial activities in deep subsurface environments. *Geomicrobiol J 7*:79–92.

Polz MF and Cavanaugh CM (1998) Bias in template-to-product ratios in multitemplate PCR. *Appl Environ Microbiol 64*:3724–3730.

Power M, van der Meer JR, Tchelet R, Egli T, and Eggen R (1998) Molecular-based methods can contribute to assessments of toxicological risks and bioremediation strategies. *J Microbiol Meth 32*:107–119.

Reysenbach A-L, Giver LJ, Wickham GS, and Pace NR (1992) Differential amplification of rRNA genes by polymerase chain reaction. *Appl Environ Microbiol 58*:3417–3418.

Rochelle PA, Weightman AJ, and Fry JC (1992) DNase I treatment of Taq DNA polymerase for complete PCR decontamination. *Bio Techniques 13*:520.

Rokkones E, Alestrom P, Skjervold H, and Gautvik KM (1985) Development of a technique for microinjection of DNA into salmonid eggs. *Acta Physiol Scand 124*:417.

Romanowski G, Lorenz MG, and Wackernagel W (1991) Adsorption of plasmid DNA to mineral surfaces and protection against DNAse I. *Appl Environ Microbiol 57*:1057–1061.

Roskey MT, Juhasz P, Smirnov IP, Takach EJ, Martin SA, and Haff LA (1996) DNA sequencing by delayed extraction-matrix-assisted laser desorption/ionization time of flight mass spectrometry. *Proc Natl Acad Sci USA 93*:4724–4729.

Rychlik W (1995) Priming efficiency in PCR. *Bio Techniques 18*:84–90.

Sarkar G, Cassady J, Bottema CDK, and Sommer SS (1990) Characterization of polymerase chain reaction amplification of specific alleles. *Anal Biochem 186*:64–68.

Sarker G and Sommer SS (1991) Parameters affecting susceptibility of PCR contamination to UV inactivation. *Bio Techniques 10*:590–594.

Sarkar G and Sommer SS (1993) Removal of DNA contamination in polymerase chain reaction reagents by ultraviolet irradiation. *Meth Enzymol 218*:381–389.

Sayler GS and Layton AC (1990) Environmental application of nucleic acid hybridization. *Ann Rev Microbiol 44*:625–648.

Schena M, Shalon D, Heller R, Chai A, Brown PO, and Davis RW (1996) Parallel human genome analysis: Microarray-based expression monitoring of 1000 genes. *Proc Natl Acad Sci USA 93*:10614–10619.

Schmidt TM, Pace B, and Pace NR (1992) Detection of DNA contamination in Taq polymerase. *Bio Techniques 11*:176–177.

Schroth MH, Istok JD, Conner GT, Hyman MR, and O'Reilly KT (1998) Spatial variability in aerobic respiration and denitrification rates in a petroleum contaminated aquifer. *Ground Water 36*:924–937.

Shuldiner AR, Nirula A, and Roth J (1989) Hybrid DNA artifact from PCR of closely related target sequences. *Nucl Acids Res 17*:4409.

Sinclair JL and Ghiorse WC (1989) Distribution of aerobic bacteria, protozoa, algae, and fungi in deep subsurface sediments. *Geomicrobiol J 7*:15–32.

Sosnowski RG, Tu E, Butler WF, O'Connell JP, and Heller MJ (1997) Rapid determination of single base mismatch mutations in DNA hybrids by direct electric field control. *Proc Natl Acad Sci USA (1994)*:1119–1123.

Stahl DA (1995) Application of phylogenetically based hybridization probes to microbial ecology. *Molec Ecol 4*:535–542.

Stahl DA (1997) Molecular approaches for the measurement of density, diversity, and phylogeny. In Hurst CJ, Knudsen GR, McInerney MJ, Stetzenbach LD, and Walter MV (eds): *Manual of Environmental Microbiology*. Washington, DC: ASM Press, pp 102–114.

Stapleton RD, Ripp S, Jimenez L, Cheol-Koh S, Fleming JT, Gregory IR, and Sayler GS (1998) Nucleic acid analytical approaches in bioremediation: Site assessment and characterization. *J Microbiol Meth 32*:165–178.

Steffan RJ and Atlas RM (1991) Polymerase chain reaction: Applications in environmental microbiology. *Ann Rev Microbiol 45*:137–161.

Stevens TO and Holbert BS (1995) Variability and density dependence of bacteria in terrestrial subsurface samples: Implications for enumeration. *J Microbiol Meth 21*:283–292.

Suzuki M, Rappé MS, and Giovannoni SJ (1998) Kinetic bias in estimates of coastal picoplankton community structure obtained by measurement of small-subunit rRNA gene PCR amplicon length heterogeneity. *Appl Environ Microbiol 64*:4522–4529.

Suzuki MT and Giovannoni SJ (1996) Bias caused by template annealing in the amplification of mixtures of 16S rRNA genes by PCR. *Appl Environ Microbiol 62*:625–630.

Tanner MA, Goebel BM, Dojka MA, and Pace NR (1998) Specific ribosomal DNA sequences from diverse environmental settings correlate with experimental contaminants. *Appl Environ Microbiol 64*:3110–3113.

Thornton CG, Hartley JL, and Rashtchian A (1992) Utilizing uracil DNA-glycosylase to control carryover contamination in PCR-characterization of residual UDG activity following thermal cycling. *Bio Techniques 13*:180–184.

Todd PJ, McMahon JM, Short RT, and McCandlish CA (1997) Organic SIMS of biological tissue. *Anal Chem 69*:529A–535A.

Ulrich GA, Martino D, Burger K, Routh J, Grossman EL, Ammerman JW, and Suflita JM (1998) Sulfur cycling in the terrestrial subsurface: Commensal interactions, spatial scales and microbial heterogeneity. *Microbial Ecol 36*:141–151.

Walsh PS, Erlich HA, and Higuchi R (1992) Preferential PCR amplification of alleles: Mechanisms and solutions. *PCR Meth Appl 1*:241–250.

Wang GC and Wang Y (1996) The frequency of chimeric molecules as a consequence of PCR co-amplification of 16S rRNA genes from different bacterial species. *Microbiology 142*:1107–1114.

Wang GC-Y and Wang Y (1997) Frequency of formation of chimeric molecules as a consequence of PCR coamplification of 16S rRNA genes from mixed bacterial genomes. *Appl Environ Microbiol 63*:4645–4650.

Wang J, Cai X, Rivas G, Shiraishi H, Faris PAM, and Dontha N (1996) DNA electrochemical biosensor for the detection of short DNA sequences related to the human immunodeficiency virus. *Anal Chem 68*:2629–2634.

Wang J, Rivas G, Cai X, Dontha N, Shiraishi H, Luo D, and Valera FS (1997) Sequence-specific electrochemical biosensing of *M. tuberculosis* DNA. *Anal Chim Acta 337*:41–48.

Ward DM, Bateson MM, Weller R, and Ruff-Roberts AL (1992) Ribosomal RNA analysis of microorganisms as they occur in nature. *Adv Microbial Ecol 12*:219–286.

Ward N, Rainey FA, Goebel B, and Stackebrandt E (1995) Identifying and culturing the "unculturables": A challenge for microbiologists. In Allsopp D, Colwell RR, and Hawksworth DL (eds): *Microbial Diversity and Ecosystem Function*. Wallingford, UK: CAB International, pp 89–110.

Wetmur JG and Sninsky JJ (1995) Nucleic acid hybridization and unconventional bases. In Innis MA, Gelfand DH, and Sninsky JJ (eds): *PCR Strategies*. San Diego, CA: Academic Press, pp 69–83.

Wetzmer JG and Davidson N (1968) Kinetics of renaturation of DNA. *J Molec Biol 31*:349–370.

White DC and Ringelberg DB (1997) Utility of the signature lipid biomarker analysis in determining the *in situ* viable biomass, community structure, and nutritional/physiologic status of deep subsurface microbiota. In Amy PS and Haldeman DL (eds): *The Microbiology of the Terrestrial Deep Subsurface*. New York: CRC Lewis Publishers, pp 119–136.

Wilson IG (1997) Inhibition and facilitation of nucleic acid amplification. *Appl Environ Microbiol 63*:3741–3751.

Wilson KH and Blitchington RB (1996) Human colonic biota studied by ribosomal DNA sequence analysis. *Appl Environ Microbiol 62*:2273–2278.

Wintzingerode Fv, Göbel UB, and Stackebrandt E (1997) Determination of microbial diversity in environmental samples: Pitfalls of PCR-based rRNA analysis. *FEMS Microbiol Rev 21*:213–229.

Zhang L, Zhou W, Velculescu VE, Kern SE, Hruban RH, Hamilton SR, Vogelstein B, and Kinzler KW (1997) Gene expression profiles in normal and cancer cells. *Science 276*:1268–1272.

11

THE DEEP BIOSPHERE: LESSONS FOR PLANETARY EXPLORATION

CHRISTOPHER P. MCKAY

Space Science Division, NASA Ames Research Center, Moffett Field, California

1 INTRODUCTION

The search for past or present life on the other planets of our solar system is necessarily based on our understanding of life on Earth. The majority of life forms on the Earth live in the comfortable regions of the surface biosphere where liquid water, sunlight, and the essential chemical elements for life are abundant. However, such an environment is found nowhere else in the solar system. Thus, to find relevant analogs for life on other planets, we must turn to life in extreme and obscure environments on the Earth. In this vein, the subsurface environment provides many lessons relevant to planetary exploration, particularly with respect to energy sources other than sunlight and long-term dormancy.

In all environments on Earth, including the subsurface, the requirements for life can be itemized as (McKay, 1991): (1) a source of energy, usually sunlight; (2) carbon; (3) liquid water; and (4) some combination of other elements such as N, P, S, etc.

Life needs a source of energy, as do all organized open systems that maintain themselves separate from the environment. On Earth, sunlight is the primary source of energy for the surface biosphere and for most of the known subsurface biosphere.

Subsurface Microgeobiology and Biogeochemistry, Edited by James K. Fredrickson and Madilyn Fletcher.
ISBN 0-471-31577-X. Copyright 2001 by Wiley-Liss, Inc.

Photosynthesis can proceed at low light levels, 0.01% sunlight, corresponding to a distance from the sun of 100 times the distance of Earth. Subsurface organisms use solar energy indirectly by using organic material and oxygen carried from the surface that together form the basis for growth in most subsurface ecosystems. Even the deep-sea hydrothermal vents rely on oxygen produced by photosynthesis at the surface to oxidize the hydrogen sulfide emanating from the vents. However, there are organisms that can utilize chemical energy (e.g., reacting hydrogen plus carbon dioxide to form methane) as a basis for primary production. Deep below the Columbia River Basin basalt flows, one small ecosystem exists that appears to be based entirely on this process—truly independent of the surface biosphere and the yellow rays of the sun (Stevens and McKinley, 1994).

Life needs carbon. Carbon is the basic building block of biochemistry, and although it is not impossible to imagine biochemical schemes based on other elements (e.g., silicon), searching for carbon-based life appears to be the only practical option. In our own solar system, carbon is abundant. Earth and the other terrestrial planets are depleted in carbon compared to the rest of the solar system (McKay, 1991).

Liquid water is the quintessential requirement for life. It provides the medium in which biochemical reactions take place and contributes to the shape and stability of biomolecules. In addition, it transports nutrients to organisms and takes waste products away. While ubiquitous on Earth, it is not present on the surfaces of any other object in our solar system at the present time. Because the presence of liquid water is the most restrictive of the requirements for life, the search for life beyond the Earth is, operationally, a search for liquid water.

Certainly, life requires some other elements beyond water and carbon. The complete list of elements utilized by life on Earth is extensive and probably not all are essential. The minimal essential list may consist of no more than nitrogen, phosphorous, and sulfur. These and other elements are common throughout the solar system.

Two other limits to life are relevant in considering subsurface life. These are the high temperature limit and the limit on survival under dormant conditions. In the presence of sufficient pressure to maintain liquid water, the upper temperature limit for life appears to be set by the stability of biomolecules against thermal degradation. The current upper limit for demonstrated growth is 113°C (Blochl et al., 1997). Survival in a dormant state presents another limit for life. Thermal degradation and radiation damage set the upper limits for dormancy. The longest demonstrated dormancy period is 25 Myr for bacteria encased in amber (Cano and Borucki, 1994). Theoretical arguments suggest that dormancy may be possible for more than 100 Myr and in the case of radiation resistant strains, up to 900 Myr.

In our search for life in subsurface environments in other worlds, we must, therefore, seek liquid water at temperatures less then 113°C containing chemical energy sources suitable for anaerobic chemosynthesis. Periods of dormancy between habitable states should be 100 Myr or less. Our goal is defined; launch the spacecraft, prepare the drills, and let the search begin.

2 OTHER WORLDS

The surfaces of the other worlds of our solar system do not appear favorable to life. The surfaces of Mercury and Venus are much too hot to hold liquid water. The small bodies of the outer solar system including comets, asteroids, Europa, and the other jovian satellites all lack atmospheres. Titan, the largest satellite of Saturn, has an atmosphere of nitrogen and methane. The surface temperature, however, is $-180°C$: too cold for liquid water and an unlikely habitat for life. Europa and Mars are the best candidates for life. In both of these cases, the subsurface is the most promising region.

2.1 Europa

The recent results from the Galileo spacecraft provide considerable evidence for an ocean of water underneath the surface ice on Europa. These include:

1. The presence of large blocks, several kilometers in size, that may have once been mobile icebergs (Carr et al., 1998; Pappalardo et al., 1999)
2. The morphology of the largest impacts that seem inconsistent with impact into solid ice, but can be explained by impact into an ice layer $\sim$ 10 to 15 km thick over water (Moore et al., 1998; Pappalardo et al., 1999)
3. Features indicating flow of surface materials and softening of terrain, presumably due to subsurface warming (Greeley et al., 1998)
4. Electromagnetic induction indicating the presence of a conducting layer, presumably a salty liquid-water layer (Khurana et al., 1998)
5. The possibility of salts on the surface (McCord et al., 1998, 1999)
6. Indications that the surface layers are dominated by water or ice, based on surficial geological processes (Pappalardo et al., 1998, 1999; Sullivan et al., 1998; Geissler et al., 1998; Greenberg et al., 1998) and gravity data (Anderson et al., 1998)
7. The young (10 Myr old) age of the surface determined from crater statistics (Zahnle et al., 1998)

It is important to note that there is no direct evidence of liquid water on Europa today. Nonetheless, the considerable evidence consistent with the presence of a liquid-water layer leads us to consider the implications of such an ocean. If we interpret the Europa data in a way that supports the presence of an ocean, then these would be the characteristics of that ocean:

- The surface is geologically young.
- The ice is 10–15 km thick.
- There are 100 km of ocean under the ice (pressure at ocean bottom is 1500 atm, equivalent to 15 km deep on Earth).
- The ocean is saline, stratified, and charged with CO_2.

- There are hot plumes causing localized surface melting.

Could an ocean on Europa have life? There are two limiting conditions that must be satisfied for life to be present on Europa: (1) there must be a way for the origin of life on Europa or the transfer of life from another site and (2) there must be an energy source in the ocean to provide for primary productivity.

Some, but not all, models for the origins of life on Earth could apply to a subice ocean on Europa (Davis and McKay, 1996). These include a submarine origin of life at hydrothermal vents (Corliss et al., 1979) and the origin of life during impact melting of a frozen ocean on Earth (Bada et al., 1994). It is possible that life could be carried to Europa from Earth or Mars. However, given the large distance, the gravitational effect of Jupiter, the high jovian radiation level, and the thick ice on Europa, such transfer is not likely even over the age of the solar system.

There have also been speculations about biological energy sources for life in an ocean on Europa, including photosynthesis (Reynolds et al., 1983) and chemical energy (Jakosky and Shock, 1998).

At 5 AU, the level of sunlight on Europa's surface is adequate to support photosynthesis. Indeed, on Europa, as on the Earth, available sunlight exceeds geothermal heating by many orders of magnitude. However, the mean temperature of Europa is $-180°C$ and the maximum surface temperature $< -130°C$. Clearly, at these temperatures, there is no liquid water at the surface and no liquid water present to the depths of 50 m. Below this, the light levels are probably less than photosynthetic limits.

Because of the thick ice crust (well over 1 km) and the indication of possible hydrothermal activity, the possibility of chemosynthetic energy sources from vents on the bottom of Europa's ocean should be considered.

Life at deep-sea vents on the Earth is often described as being independent of the surface biosphere. However, the life forms observed in abundance at these vents (crabs, worms, etc.) are part of an ecosystem based on chemoautotrophic bacteria that consume hydrogen sulfide outgassing from the vent and oxidize this with oxygen dissolved in the ambient seawater. This oxygen derives ultimately from photosynthesis on the surface of the Earth. On Europa, hydrogen sulfide may be emanating from geothermal vents, but there is not likely to be a source of oxygen in the ocean water. Without oxygen, the energetics of hydrogen sulfide consumption are not favorable. However, recent work (Kinkle and Radu, personal communication) indicates that hydrogen sulfide can react with carbon dioxide in the presence of soluble ferrous iron, and this can form the basis for a chemoautotrophic ecosystem. Carbon dioxide is likely to be present in the ocean of Europa (Crawford and Stevenson, 1988).

2.2 Mars

Mars is the most favorable case for life and even on this planet the surface appears to be profoundly lacking liquid water. Early in its history, Mars may have had surface conditions suitable for life (McKay, 1986, 1997). But now, water exists in the atmosphere at an average concentration equivalent to a condensed layer of ice on the surface about 10 μm thick. Frost was observed on the ground at the Viking lander site. The polar caps also contain a relatively large volume of water ice. However, the atmospheric pressure on Mars is so low (0.6–1 kPa; compared to 100 kPa for Earth) that water is essentially never liquid (Ingersoll, 1970; Kahn, 1985). From the point of view of liquid water, the surface of Mars is drier than the driest desert on Earth.

There is reason to believe that the subsurface of Mars may contain more salubrious environments than the surface. The huge volcanos on Mars are direct evidence that Mars was geologically active in the past. Relatively young fluvial features on the surface, such as shown in Fig. 11.1, indicate possible hydrothermal activity. Gulick (1998) modeled how volcanic systems could generate liquid-water flows even under the present climate conditions (Fig. 11.2). The youngest of the martian meteorites indicates a surface lava flow on Mars less than 200 Myr ago (McSween, 1994). The presence of ground ice on Mars and recent volcanism together lend credence to the possibility of subsurface hydrothermal systems that could support life (Boston et al., 1992). Furthermore, direct evidence for hydro-

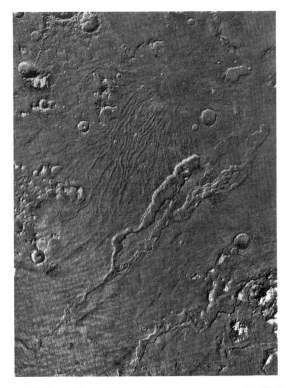

FIGURE 11.1 Dao Vallis on Mars, 40 km wide near its source on the flanks of the volcano Hadriaca Patera (visible at the top of the photo). It has been suggested that this feature could have been caused by volcanic melting of ground ice. Sustained hydrothermal activity may have been present at this site. (NASA Viking photograph.)

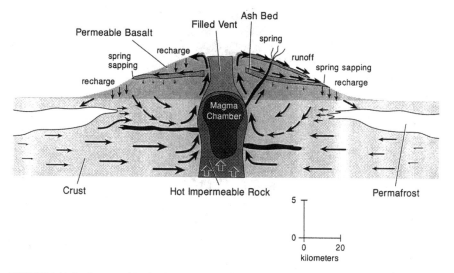

FIGURE 11.2   Proposed hydrothermal system capable of generating subsurface and surface liquid-water flows on the present Mars (from Gulick, 1998). Water reaction with martian basalt would be a source of hydrogen, which together with atmospheric carbon dioxide, could support methanogen-based ecosystems. (Used with permission of the American Geophysical Union.)

thermal fluids comes from isotopic studies of the martian meteorites (Karlsson et al., 1992; Leshin et al., 1994). Finally, Clifford and colleagues (e.g. Clifford, 1993; Clifford and Parker, 2000) suggested the presence of a global-scale subsurface liquid-water system on Mars.

Boston et al., (1992) speculated on the type of life that might be found in subsurface hydrothermal systems on Mars. They suggested that the most likely basis for subsurface primary production would be hydrothermally supplied hydrogen reacting with atmospheric carbon dioxide. They also considered other possible chemical reactions. The discovery on Earth of a subsurface system based on hydrogen and carbon dioxide (Stevens and McKinley, 1995) greatly supports their suggestion. It also provides a specific analog for study. The source of hydrogen for the subsurface methanogens in the Columbia River Basalts (Stevens and McKinley, 1995) is the oxidation by water of ferrous iron in the surrounding basaltic rocks. This system represents the first and, to date, the only example of a microbial ecosystem completely independent of sunlight, organic material, or oxygen from the Earth's surface. Kral et al., (1998) have shown that chemotrophic methanogens consume hydrogen down to levels as low as 4 Pa. Such systems could be possible on Mars, where basaltic rocks and carbon dioxide are expected to be plentiful.

However, there are serious problems with postulating subsurface life on Mars. Although there may well be active subsurface volcanism, it is not probable that hydrothermal ecosystems could maintain biological continuity over the history of Mars. There is no direct evidence for the globally connected subsurface hydrothermal system on Mars, as suggested by Clifford (1993). Hence, a new hydro-thermal site coming into existence would be isolated from life in previous, distant sites. Natural radiation from uranium, thorium, and potassium would limit the expected life span of dormant cells on Mars to much less than a billion years. Any microbes remaining at a site would likely be dead when a new hydrothermal system erupted. For subsurface life on Mars to be probable requires either that the planet have some sort of globally connected groundwater system or that hydrothermal systems be continuous over the long lifetime of the planet. Although these are not impossible, they are not supported by the admittedly meager, geological evidence we do have about subsurface Mars. These caveats notwithstanding, deep drilling for life associated with hydrothermal sites remains a key goal for future exploration. Subsurface analogs on Earth are important in guiding the search for life in these sites.

Subsurface permafrost regions of Mars provide an alternative possibility for a life search. Below the annual temperature wave, the surface of Mars is frozen to depths of one to several kilometers (Fig. 11.3). Poleward about 40°, ground ice would be stable in this frozen zone while more equatorial regions would be losing ice to sublimation. It has been shown that bacteria are culturable after more than 3 million years frozen in Siberian permafrost (Gilichinsky et al., 1991; Rivkina et al., 1998). Samples have been obtained aseptically without drilling fluids from drill holes (up to 50-m depth) in this permafrost. They contain bacteria at levels of up to $\sim 10^8$ culturable units per gram soil (Vorobyova et al., 1997). The current temperature of the permafrost is $-10°C$. One key question is the state of the bacteria at this low temperature. Rivkina et al. (2000) experimentally addressed this question by incubating permafrost samples at temperatures from $+5°$ to $-20°C$. To detect microbial activity, ^{14}C-labeled acetate was physically mixed into the permafrost without fully melting it. Subsequently, counts were determined from lipids extracted from replicates and compared to a baseline established after initial preparation. The initial mixing was accomplished with the samples held at subzero temperatures. It is important to realize that this mixing refreshes the environment surrounding any microorganisms in the sample. Two features of these results are of interest. First, after an initial lag period, there is a rapid increase in incorporation of labeled acetate into lipids. However, this rapid uptake is not sustained and the counts level off. Rivkina et al. (2000) attribute this leveling off to the exhaustion of local resources surrounding each microorganism in the sample. This scenario explains why the final

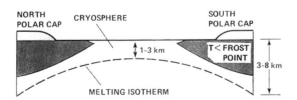

FIGURE 11.3 Distribution of permafrost and stable ground ice on Mars (adapted from Fanale, 1976).

asymptotic value varies exponentially with temperature: The thin film of unfrozen water that exists in soil–ice mixtures down to $-40°C$ decreases exponentially with decreasing temperature below zero.

These results indicate that the limitation to microbial activity over long periods of time at subfreezing temperatures is the transport of nutrients to and removal of waste material from the microorganisms, and not the intrinsic reductions in the rate of metabolism due to low temperature. The bacteria found in permafrost after millions of years have spent this time in a metabolic state that is well below their potential at that temperature, that is, they are starving.

3 LIMITS TO DORMANCY IN THE SUBSURFACE

Even when metabolism stops, viability can remain. Dormancy seems to be associated with profound dehydration (as in amber), starvation, and deep freeze. Two effects limit the long-term viability of a dormant microorganism: thermal decay kT and ionizing radiation eV.

In principle, the second law of thermodynamics can be used to directly estimate the minimum free energy that an organism needs to maintain viability. Developing a thermodynamic model for a bacterial cell is not simple, but it is interesting nonetheless to outline how such a task might proceed. A statistical model of a cell would begin with the dry components of a living cell: proteins 55%, RNA 20.5%, DNA 3.1%, lipid 9.1%, lipopolysaccharide 3.4%, peptidogycan 2.5%, glycogen 2.5%, and other soluble metabolites 2.9% and a description of their chemical state. Clearly, most of the possible arrangements of these biomolecules do not constitute a living system. We can quantify this by considering a multidimensional phase space that describes all possible states of the atoms and molecules that compose a bacteria. Only a tiny region of this phase space corresponds to a viable cell—either growing or dormant. These viable states are complex, ordered states of low entropy. Even if no external forces acted on this system, random processes would cause a drift in phase space that would eventually take the system outside the viable zone. This is an inevitable result of the second law of thermodynamics. Minimal survival can be viewed as the process by which an organism collects just enough free energy from its environment to overcome this entropic slide away from viable conditions.

Using only the methods of statistical physics, it is possible to compute the rate at which entropy increases in a bacteria at any finite temperature. Clearly, this represents the physical minimum free energy that, on average, must be available to the cell to remain viable. Biological constraints may require that the actual free energy to survive be higher than this physical lower limit. The entropy S of the system is related to the Gibbs free energy dG and energy flow dQ by the first law of thermodynamic as follows:

$$dG = dQ + TdS \tag{1}$$

If we assume there is no external energy flow, then the rate of free energy that must be generated to compensate for entropy increase is

$$\frac{dG}{dt} = T\frac{dS}{dt} \tag{2}$$

There is very strong temperature dependence on the minimal free-energy flow, due to both the coefficient in the above equation and the strong, usually exponential, dependence of dS/dt on temperature.

To develop an approach to quantitative modeling, we begin by considering a viable cell at some specific and viable point in phase space. To determine the rate of entropic drift away from this unique point, we can list all possible reactions or changes in state away from this initial condition. Such reactions would include racemization of an amino acid, ionization of a molecule, denaturing of a protein, breaking of a DNA or RNA strand, dimerization of thymine in DNA, breaking of a covalent bond, rearrangement of hydrogen bonding pairs, etc. Many of these state changes may not take the organisms away from the viable zone. They do indicate the rate of entropy drift away from any specific point in the phase space and, hence, do provide a way to compute the rate of drift. Each possible reaction is associated with a probability that it will occur. We take these probabilities as being given by the Boltzmann statistical weighing factors only. The probability P that a reaction from state a to state b will occur is given as

$$P_{a\to b} = \frac{g_b}{g_a}e^{-\Delta E/kT} \tag{3}$$

where ΔE is the energy change from a to b, k Boltzmann's constant, T absolute temperature, and g_a and g_b the statistical weighting factors for states a and b, respectively. The g factors count the number of possible modes in a state, all of which have the same energy level. If two states have the same energy ($\Delta E = 0$) but one state has more modes (larger g), then the probability of being in that state is higher.

The temperature dependence of the thermal degradation can be illustrated for the case of amino acid racemization. Bada and McDonald (1995) showed that amino acid racemization varies exponentially with temperature, dropping by a factor of ~ 7 for every 10°C reduction in temperature. In addition, they found that liquid water accelerates decay by a factor of over 100. Dry materials have greatly reduced racemization rates, and this presumably applies to deeply frozen material.

Kanavarioti and Mancinelli (1990) followed this logic to argue that, at the low temperatures expected on Mars ($<-70°C$), biomolecules would have lifetimes against thermal decay that exceed billions of years.

Our current model of a bacteria cell is only conceptual; we cannot determine a numerical value for dS/dt as a function of temperature. To do so would require a catalog of all possible molecules in the cell, and an activation energy and decay rate (as per Eq. 3) assigned to each bond in each molecule. Using this model, we would then be able to compute the rate at which an organism experiences entropy-increased decay. In order to determine how this limits an organism's survival, it would be necessary to understand how much thermal decay, or increase in entropy, an organism could sustain and still remain viable. An estimate of this might be obtained from the level of radiation required to render a organism unable to survive, rather than just being unable to reproduce.

A second factor that limits long-term survival, even at extremely low temperatures, is natural radiation. Natural radiation is caused by gamma particles emitted by naturally occurring, long-lived radioisotopes. These are thorium, uranium, and potassium. In Siberian permafrost, the measured level of natural radiation deep (25 m) below the surface is ~ 0.2 rad year^{-1}. About half of this comes from uranium and thorium, with the rest from potassium 40.

Crustal materials on Mars would have similar levels of uranium and thorium, and possibly lower levels of potassium, based on theoretical considerations of the source material that formed both planets, and on direct measurements of the martian meteorites (Laul et al., 1986).

Radiation doses of 2 Mrad are sufficient to kill most organisms in soil. Approximately 18 Mrad will kill even the most radiation-resistant strains, such as *Deinococcus radiodurans* (Minton, 1994; Battista, 1997). Dehydration resistance is known to enhance radiation survival in bacteria (Mattimore and Battista, 1996). Interestingly, there is experimental evidence that a considerable part of the damage done by radiation is associated with free-radical formation due to liquid water (Swarts et al., 1992). Therefore, in the dehydrated state, damage is reduced. Recent data suggest that freezing at low temperatures ($-70°C$) also enhances survival by a factor of ~ 5, presumably also because of the absence of liquid water (J. Daly, personal communication).

Taking 2 Mrad as the typical lethal dose for soil bacteria, at a rate of 0.2 rad year^{-1}, a lethal dose of radiation is accumulated in about 10 Myr. Survival in frozen or dehydrated conditions could be 10 times longer, 100 Myr. Thus, bacterial survival in 3.5-Myr-old permafrost is not a test of the ultimate survival against radiation and is not even close to the limits of *D. radiodurans*. Even if the permafrost bacteria are not able to metabolize enough to repair radiation-induced damage, they could still survive the 3.5 million years of accumulated damage. There might be some elimination of less resistant strains, but there would be no sterilization of the soil.

4 CONCLUSIONS

Given the inhospitable surface conditions on the other planets, the search for life elsewhere in our solar system is likely to be in subsurface environments. Mars and Europa provide the most interesting targets.

Key questions that would help guide this search can be learned from studies of the subsurface environments on Earth. These questions include:

1. Are there other microbial ecosystems on Earth that are completely isolated from the surface sunlight, organic material, and oxygen?
2. What are the possible energy sources for anaerobic primary production in isolated microbial ecosystems on Earth?
3. What is the upper limit for temperature survival of individual organisms and of entire microbial ecosystems?
4. What is the longest dormancy period demonstrable for Earth microorganisms?
5. What is the maximum possible dormancy under possible extraterrestrial conditions including dehydration, freezing, and starvation?

Answering these questions via detailed studies of ecosystems here on Earth would provide a strong scientific base for studies of life under the surfaces of other planets. In addition, it would address questions related to the ability of life to spread from planet to planet, carried within rocks and dust particles.

REFERENCES

Anderson JD, Schubert G, Jacobsen RA, Lau EL, Moore WB, and Sjogren WL (1998) Europa's differentiated internal structure: Inferences from four Galileo encounters. *Science 281*:2019–2020.

Bada JL, Bigham C, and Miller SL (1994) Impact melting of frozen oceans on the early Earth: Implications for the origin of life. *Proc Natl Acad Sci 91*:1248–1250.

Bada JL and McDonald GD (1995) Amino acid racemization on Mars: Implications for the preservation of biomolecules from an extinct martian biota. *Icarus 114*:139–143.

Battista JR (1997) Against all odds: The survival strategies of *Deinococcus radiodurans*. *Ann Rev Microbiol 51*:203–224.

Blochl E, Rachel R, Burggraf S, Hafenbradl D, Jannasch HW, and Stetter KO (1997) *Pyrolobus fumarii*, gen. and sp. nov., represents a novel group of archaea, extending the upper temperature limit for life at 113 C. *Extremophiles 1*:14–21.

Boston PJ, Ivanov MV, and McKay CP (1992) On the possibility of chemosynthetic ecosystems in subsurface habitats on Mars. *Icarus 95*:300–308.

Cano RJ and Borucki MK (1995) Revival and identification of bacterial spores in 25- to 40-million-year-old Dominican amber. *Science 268*:1060–1064.

Carr MH, Belton MJS, Chapman CR, Davies ME, Geissler P, Greenberg R, McEwen AS, Tufts BR, Greeley R, Sullivan R, Head JW, Pappalardo RT, Klaasen KP, Johnson TW, Kaufman J, Senske D, Moore J, Neukum G, Schubert G, Burns JA, Thomas P, and Veverka J (1998) Evidence for a subsurface ocean on Europa. *Nature 391*:363–365.

Clifford SM (1993) A model for the hydrologic and climatic behavior of water on Mars. *J Geophys Res 98*:10,973–11,016.

Clifford SM and Parker TJ (2000) The evolution of the martian hydrosphere: Implications for the fate of a primordial ocean and the current state of the northern plains. *Icarus* (in press).

Corliss JB, Dymond J, Gordon LI, Edmond JM, von Herzen RP, Ballard RD, Green K, Williams D, Bainbridge A, Crane K, and van Andel TH (1979) Submarine thermal springs on the Galapagos Rift. *Science 203*:1073–1083.

Crawford GD and Stevenson DJ (1988) Gas-driven water volcanism and resurfacing of Europa. *Icarus 73*:66–79.

Davis WL and McKay CP (1996) Origins of life: A comparison of theories and application to Mars. *Orig Life Evol Biosph 26*:61–73.

Geissler PE, Greenberg R, Hoppa G, Helfenstein P, McEwen A, Pappalardo R, Tufts R, Ockert-Bell M, Sullivan R, Greeley R, Belton MJS, Denk T, Clark B, Burns J, Veverka J, and the Galileo Imaging Team (1998) Evidence for non-synchronous rotation of Europa. *Nature 391*:368–369.

Gilichinsky DA, Vorobyova EA, Erokhina LG, Fyordorov-Dayvdov DG, and Chaikovskaya NR (1992) Long-term preservation of microbial ecosystems in permafrost. *Adv Space Res 12* (4):255–263.

Greeley R, Sullivan R, Klemaszewski J, Homan K, Head JW, Pappalardo RT, Veverka J, Clark BE, Johnson TV, Klaasen KP, Benton M, Moore J, Asphaug E, Carr MH, Neukum G, Denk T, Chapman CR, Pilcher CB, Geissler PE, Greenberg R, and Tufts R (1998) Europa: Initial Galileo geological observations. *Icarus 135*:4–24.

Greenberg R, Geissler P, Hoppa G, Randall Tufts B, Durda DD, Pappalardo R, Head JW, Greeley R, Sullivan R, and Carr MH (1998) Tectonic processes on Europa: Tidal stresses, mechanical response, and visible features. *Icarus 135*:64–78.

Gulick VC (1998) Magmatic intrusions and a hydrothermal origin for fluvial valleys on Mars. *J Geophys Res 103*:19,365–19, 387.

Ingersoll AP (1970) Mars: Occurrence of liquid water. *Science 168*:972–973.

Jakosky BM and Shock EL (1998) The biological potential of Mars, the early Earth, and Europa. *J Geophys Res 103*:19, 359–19, 364.

Kahn R (1985) The evolution of CO_2 on Mars. *Icarus 62*:175–190.

Kanavarioti A and Mancinelli RL (1990) Could organic matter have been preserved on Mars for 3.5 billion years? *Icarus 84*:196–202.

Karlsson HR, Clayton RN, Gibson EK Jr, and Mayeda TK (1992) Water in SNC meteorites: Evidence for a martian hydrosphere. *Science 255*:1409–1411.

Khurana KK, Kivelson MG, Stevenson DJ, Schubert G, Russell CT, Walker RJ, and Polanskey C (1998) Induced magnetic fields as evidence for subsurface oceans in Europa and Callisto. *Nature 395*:777–780.

Kral TA, Brink KM, Miller SL, and McKay CP (1998) Hydrogen consumption by methanogens on the early Earth. *Orig Life Evol Biosphere 28*:311–319.

Laul JC, Smith MR, Wänke H, Jagoutz E, Dreibus G, Palme H, Spettel B, Burchele A, Lipschultz ME, and Verkouteren RM (1986) Chemical systematics of the Shergotty meteorite and the composition of its parent body (Mars). *Geochim Cosmochim Acta 50*:909–926.

Leshin LL, Hutcheon ID, Epstein S, and Stolper EM (1992) Water on Mars: Clues from deuterium/hydrogen and water contents of hydrous phases in SNC meteorites. *Science 265*:86–90.

Mattimore V and Battista JR (1996) Radioresistance of *Deinococcus radiodurans*: Functions necessary to survive ionizing radiation are also necessary to survive prolonged desiccation. *J Bacteriol 178*:633–637.

McCord TB, Hansen GB, Fanale FP, Carslon RW, Matson DL, Johnson TV, Smythe WD, Crowley JK, Martin PD, Ocampo A, Hibbitts CA, Granahan JC, and the NIMS Team (1998)

Salts on Europa's surface detected by Galileo's near infrared mapping spectrometer. *Science* *280*:1242–1245.

McCord TB, Hansen GB, Matson DL, Johnson TV, Crowley JK, Fanale FP, Carlson RW, Smythe WD, Martin PD, Hibbitts CA, Granahan JC, and Ocampo A (1999) Hydrated salt minerals on Europa's surface from the Galileo near-infrared mapping spectrometer (NIMS) investigation. *J Geophys Res 104*:11827–11852.

McKay CP (1986) Exobiology and future Mars missions: The search for Mars' earliest biosphere. *Adv Space Res 6*:269–285.

McKay CP (1991) Urey Prize lecture: Planetary evolution and the origin of life. *Icarus 91*:93–100.

McKay CP (1997) The search for life on Mars. *Orig Life Evol Biosphere 27*:263–289.

McSween HY (1994) What have we learned about Mars from SNC meteorites. *Meteoritics 29*:757–779.

Minton KW (1994) DNA repair in the extremely radioresistant bacterium *Deinococcus radiodurans*. *Mol Microbiol 13*:9–15.

Moore JM, Asphaug E, Sullivan RJ, Klemaszewski JE, Bender KC, Greeley R, Geissler PE, McEwen AS, Turtle EP, Phillips CB, Tufts BR, Head JW III, Pappalardo RT, Jones KB, Chapman CR, Belton MJS, Kirk RL, and Morrison D (1998) Large impact features on Europa: Results of the Galileo nominal mission. *Icarus 135*:127–145.

Pappalardo RT, Belton MJS, Breneman HH, Carr MH, Chapman CR, Collins GC, Denk T, Fagents S, Geissler PE, Giese B, Greeley R, Greenberg R, Head JW, Helfenstein P, Hoppa G, Kadel SD, Klaasen KP, Klemaszewski JE, Magee K, McEwen AS, Moore JM, Moore WB, Neukum G, Phillips CB, Prockter LM, Schubert G, Senske DA, Sullivan RJ, Tufts BR, Turtle EP, Wagner R, and Williams KK (1999) Does Europa have a subsurface ocean? Evaluation of the geological evidence. *J Geophys Res 104*:24,015–24,055.

Pappalardo RT, Head JW, Greeley R, Sullivan RJ, Pilcher C, Schubert G, Moore WB, Carr MH, Moore JM, Belton MJS, and Goldsby DL (1998) Geological evidence for solid-state convection in Europa's ice shell. *Nature 391*:365–367.

Reynolds RT, Squyres SW, Colburn DS, and McKay CP (1983) On the habitability of Europa. *Icarus 56*:246–254.

Rivkina E, Gilichinsky D, Wagener S, Tiedje J, and McGrath J (1998) Biogeochemical activity of anaerobic microorganisms from buried permafrost sediments. *Geomicrobiology 15*:187–193.

Rivkina EM, Friedmann EI, McKay CP, and Gilichinsky DA (2000) Metabolic activity of permafrost bacteria below the freezing point. *Appl Environ Microbiol 66*:3230–3233.

Stevens TO and McKinley JP (1995) Lithoautotrophic microbial ecosystems in deep basalt aquifers. *Science 270*:450–454.

Sullivan R, Greeley R, Homan K, Klemaszewski J, Belton MJS, Carr MH, Chapman CR, Tufts R, Head JW III, Pappalardo R, Moore J, Thomas P, and the Galileo Imaging Team (1998) Episodic plate separation and fracture infill on the surface of Europa. *Nature 391*:371–373.

Vorobyova E, Soina V, Gorlenko M, Minkovskaya N, Zalinova N, Mamukelashvili A, Gilichinsky D, Rivkina E, and Vishnivetskaya T (1997) The deep cold biosphere: Facts and hypothesis. *FEMS Microbiol Rev 20*:277–290.

Zahnle K, Dones L, and Levison HF (1998) Cratering rates on the Galilean satellites. *Icarus 136*:202–222.

INDEX